The geology of the Irish Sea

BRITISH GEOLOGICAL SURVEY

United Kingdom Offshore Regional Report

The geology of the Irish Sea

D I Jackson, A A Jackson, D Evans, R T R Wingfield, R P Barnes
and M J Arthur

with contributions by
M F Howells, R A Hughes and M G Petterson

LONDON HMSO 1995

Production of this report was funded by the Department of Energy (now incorporated into the Department of Trade and Industry) and the Natural Environment Research Council

The coastline used on many maps and diagrams in this book is based on Ordnance Survey mapping

Bibliographic reference

JACKSON, D I, JACKSON, A A, EVANS, D, WINGFIELD, R T R, BARNES, R P, and ARTHUR, M J. 1995. *United Kingdom offshore regional report: the geology of the Irish Sea.* (London: HMSO for the British Geological Survey.)

First published 1995

Dd 301831 C20 01/96

ISBN 0 11 884507 1

Contents

FIGURES

Foreword

This report is one of the series of United Kingdom Offshore Regional Reports, which integrates the results of surveys carried out by the British Geological Survey and commercial organisations with other published data. The series complements the 1:250 000-scale map series covering the United Kingdom Continental Shelf (see inside back cover). The survey work, map compilation, and report production carried out by BGS has been funded largely by the former Department of Energy, now part of the Department of Trade and Industry.

The area of this report is that part of the Irish Sea lying between the North Channel and St George's Channel. It is thought to contain rocks from the geological systems, ranging from Precambrian schists and gneisses to Cretaceous chalk and Paleogene basalts; many of these rocks subcrop beneath a locally thick cover of Quaternary sediments. Offshore, Carboniferous and Permo-Triassic strata dominate, and have considerable economic interest. The Carboniferous rocks contain coal; offshore exploitation of these reserves has long been established as extensions of onshore coalfields, and advances in technology, including coal gasification, may renew commercial interest offshore. The Carboniferous rocks are also the source for hydrocarbons, both oil and gas, that have been successfully exploited in Permo-Triassic reservoirs, notably at the Morecambe Gasfield. Halite has been worked in the coastal perimeter, and the Irish Sea may contain the largest resource of Triassic halite in the British Isles. The Quaternary cover provides a large resource of aggregate, some of which has been extracted from nearshore areas.

The report is also important for the insights it provides into the recent and present sedimentary processes at work in the Irish Sea. Given that the adjacent land areas include a number of major conurbations, all of which discharge effluent to this semi-enclosed sea, it is vital that we understand the processes of sediment movement. This study will contribute to that understanding.

Peter J Cook, DSc
Director
British Geological Survey

November 1994

ACKNOWLEDGEMENTS

Chapters in this report have been produced by the following authors:

Introduction	— A A Jackson and D I Jackson
Crustal structure	— D Evans and M J Arthur
Pre-Carboniferous	— R P Barnes and A A Jackson, with contributions from D I Jackson, M F Howells, R A Hughes and M G Petterson
Structure, Carboniferous, Permian and Triassic	— D I Jackson and A A Jackson
Jurassic and Cretaceous	— D I Jackson, D Evans, and A A Jackson
Paleogene and Neogene	— D I Jackson, D Evans and A A Jackson
Quaternary	— R T R Wingfield
Economic geology	— D I Jackson, R T R Wingfield, A A Jackson, and D Evans.

The offshore Regional Report Series is co-ordinated by the Marine Geology Group, and edited by D Evans and A G Stevenson.

In addition to the work of the authors, the report has drawn extensively upon the knowledge and expertise of other BGS staff, not only within the marine sphere, but also from the Land Survey and specialists within the fields of sedimentology, biostratigraphy, cartography and publication.

In particular, the authors and editors are grateful to the following for constructive comment: N Aitkenhead, D Beamish, T D J Cameron, D J Evans, R W Gallois, G S Kimbell, A C Morton, I B Paterson, J J Pattison, I F Smith, P Stone, G Warrington and A A Wilson.

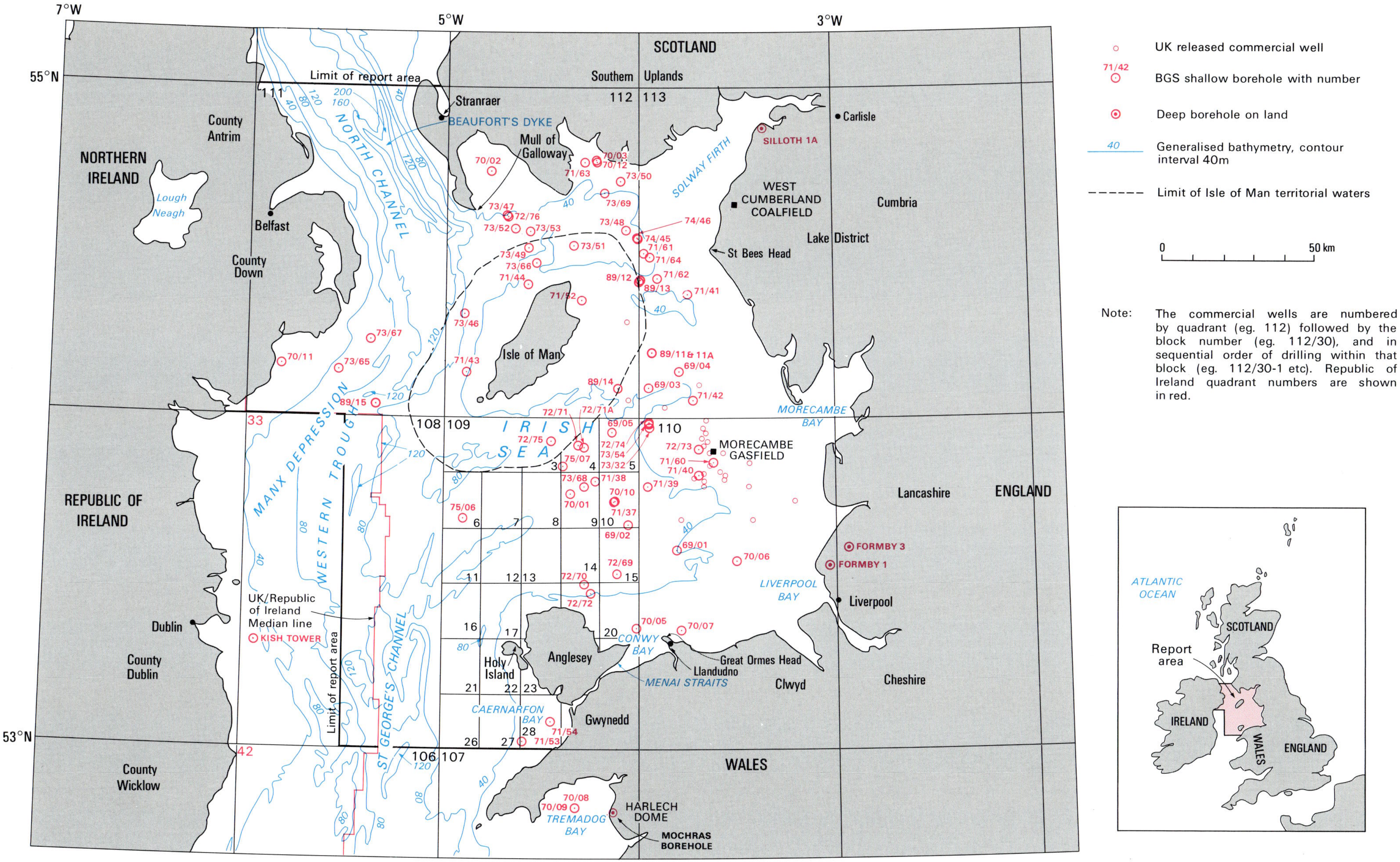

Figure 1 Location map of the report area showing licence blocks, released commercial wells, BGS shallow boreholes, and generalised bathymetry.

1 Introduction

The Irish Sea lies at the heart of the British Isles, and is surrounded by Ireland, southern Scotland, north-west England and North Wales. The sea has restricted outlets to the Atlantic Ocean through the North Channel to the north, and via the St George's Channel to the south. The offshore area covered by this report is bounded to the north and south respectively by latitudes 55°N and 53°N, within which it extends to the coasts of Wales, England, Scotland and Northern Ireland, and surrounds the Isle of Man (Figure 1). To the south of 54°N, the western limit is 5°30'W. The entire offshore area was termed the North Irish Sea by Bott (1964) and Bacon and McQuillin (1972), but the Admiralty usage of Irish Sea is followed in this account. The offshore area to the north is described in a corresponding report by Fyfe et al. (1993), and the area to the south by Tappin et al. (1994).

Bathymetrically, the area is largely formed of platforms with generally less than 60 m water depth, within which there are very localised enclosed deeps down to 137 m (Figure 1). An extensive deep-water zone is formed by a sinuous, north-to-south trough, 30 to 70 km wide, that runs from off Islay in the north, through the North Channel, the Manx Depression and the St George's Channel, to the Celtic Deep beyond the southern limit of the report area (see Figure 67). In this trough, here named the Celtic Trough, water depths are generally 60 to 120 m but, as on the platforms, deeper waters lie in separate, elongate, enclosed deeps. The portion of the Celtic Trough that lies in the report area is termed the Western Trough. The most notable deeps are located within the bathymetrically complex Beaufort's Dyke in the North Channel, which is locally 318 m in depth (Figure 1 and see Figure 66).

This report is concerned mainly with the offshore geology, but necessarily includes reference to adjacent onshore rocks. Summaries of the land geology are to be found in Holland (1981a) for Ireland, Craig (1991) for Scotland, Moseley (1978) for the Lake District, and Wood (1969) and Owen (1974) for Wales. The geology of the Isle of Man is included in the BGS Regional Geology guide for Northern England (Taylor et al., 1971); other guides for adjacent districts include Edwards and Trotter (1978), Smith and George (1961), Greig (1971) and Wilson (1972). The coverage of published BGS maps is shown inside the back cover of this report.

GEOLOGICAL SUMMARY

Large sections of the Irish Sea coastline lie close to, and run subparallel to, important basin-margin faults or other structural lines. The solid geology outcrop pattern of the Irish Sea broadly divides along a line from the Mull of Galloway, through the Isle of Man, to Anglesey (Figure 2). A number of Permo-Triassic and younger basins are present in the east (Figure 3 and see Figure 16); the largest of these basins is the East Irish Sea Basin, to the north of which lies the Solway Firth Basin. In the western portion, Precambrian to Carboniferous strata crop out predominantly, although the smaller Permo-Triassic North Channel, Peel and Caernarfon Bay basins occur respectively to the north and south of the older rocks. In the south-west of the report area, the eastern portion of the Kish Bank Basin contains sediments of Carboniferous to Paleogene age. The level of knowledge of these geological entities is, however, varied; outside a small area centred on the Morecambe Gasfield (Figure 1), detailed geology and stratigraphy are poorly known, and are likely to be substantially modified as more data become available.

Precambrian and Lower Palaeozoic rocks subcrop on the shelf around Anglesey, and Lower Palaeozoic rocks occur off the coasts of Ireland, southern Scotland and the Isle of Man. Within the Lower Paleozoic rocks lies the Iapetus suture, or convergence zone, that marks the join between two ancient continents that were brought together during the Caledonian orogeny in Silurian to Early Devonian times. To the north of the convergence zone is part of the old Laurentian craton in Scotland and the north of Ireland, whereas the ancient Avalonian craton lies to the south of it in England and Wales. A thick, unconformable Carboniferous sequence crops out extensively in the western part of the area, and is found at depth beneath the Permian and Triassic strata in the East Irish Sea and Solway Firth basins (Figure 3). Largely marine, Dinantian limestones pass up into Namurian marine to paralic mudstones and fluviodeltaic sandstones, and thence into Westphalian, paralic to floodplain, coal measures.

In the east of the area, Permo-Triassic rocks crop out beneath Quaternary sediments from the Solway Firth Basin in the north to the coast of North Wales in the south (Figure 2). In the western half, they floor the North Channel and Caernarfon Bay basins, and probably the Peel and Central Irish Sea basins. Thus they cover about 50 per cent of the offshore area. The Lower Permian arenaceous succession appears to infill a series of isolated topographic depressions many of which lie on the perimeter of the Irish Sea; the sequence is thickest near Formby on the Lancashire coast and offshore of North Wales. The Upper Permian sediments, which show a more uniform thickness and more even distribution including the centre of the East Irish Sea Basin, are largely marginal-marine and evaporitic deposits. In the Triassic sequence, the consistently thick sandstone of the Sherwood Sandstone Group is overlain by up to 3000 m of Mercia Mudstone Group, one of the thickest Upper Triassic successions recorded in north-west Europe (Jackson et al., 1987). Thick halites are developed in the northern and central parts of the East Irish Sea Basin.

An outlier of Lower Jurassic rocks occurs in the Keys Basin within the East Irish Sea Basin, and another is present in the Kish Bank Basin (Figure 2 and see Figure 16). Other Jurassic outliers are postulated but not proven by drilling. Cretaceous rocks may overlie the Lias in some outliers, but their presence in the report area is speculative except for the limited extension from the onshore Northern Ireland succession. Despite the widespread occurrence of Paleogene intrusive and extrusive igneous rocks in Northern Ireland, there is offshore evidence only of limited early Tertiary igneous activity (Al Shaikh, 1969; Kirton and Donato, 1985). Tertiary sediments are restricted to the Central Irish Sea and Kish Bank basins.

The very variable water depths in the western part of the Irish Sea (Figure 1) appear to be largely controlled by Quaternary processes. The Quaternary succession is thickest,

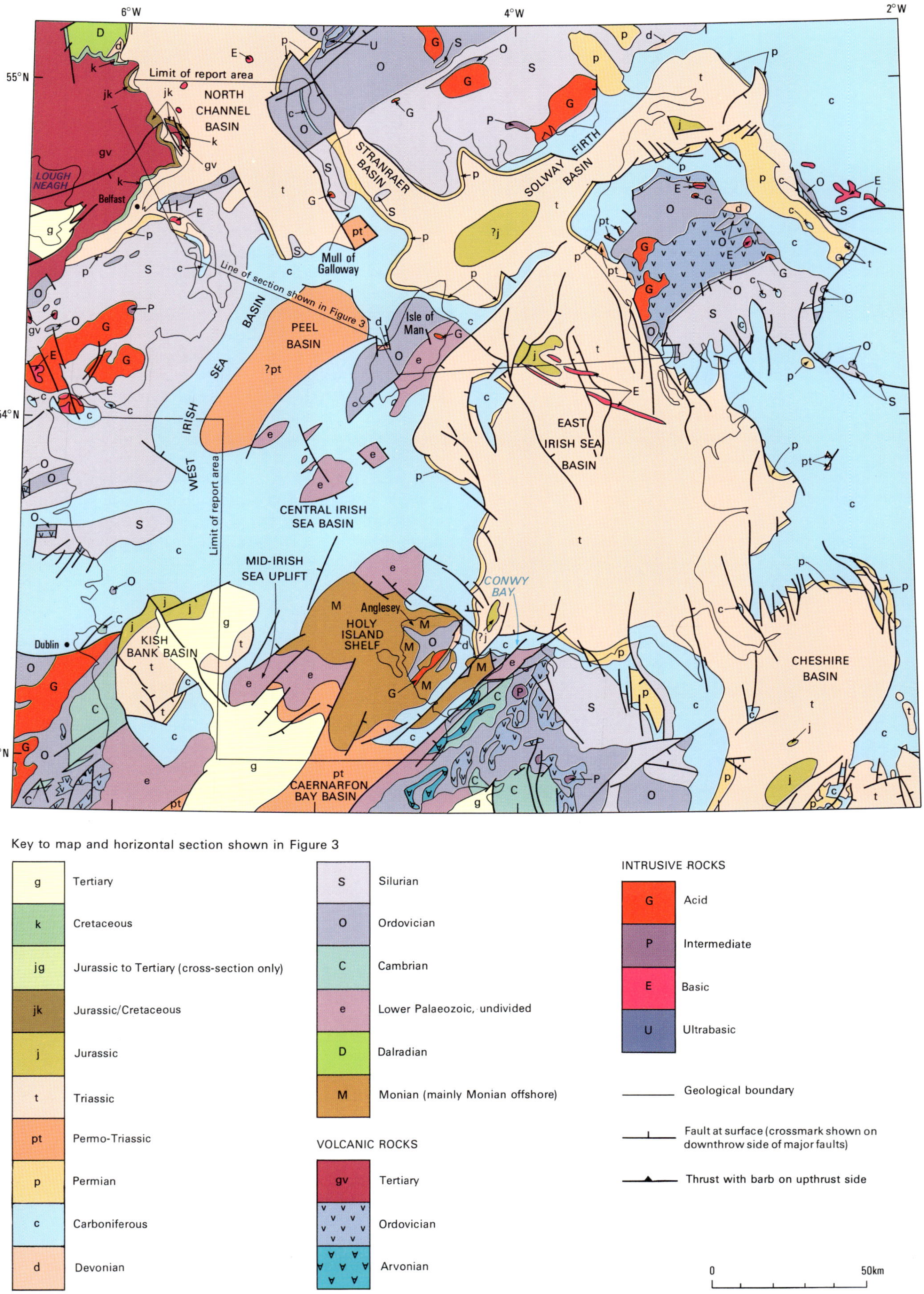

Figure 2 Generalised Pre-Quaternary geology of the Irish Sea. Based on 1:1 000 000 Geology of the United Kingdom, Ireland and the adjacent Continental Shelf (South Sheet). In places this map differs from diagrams illustrated in this report.

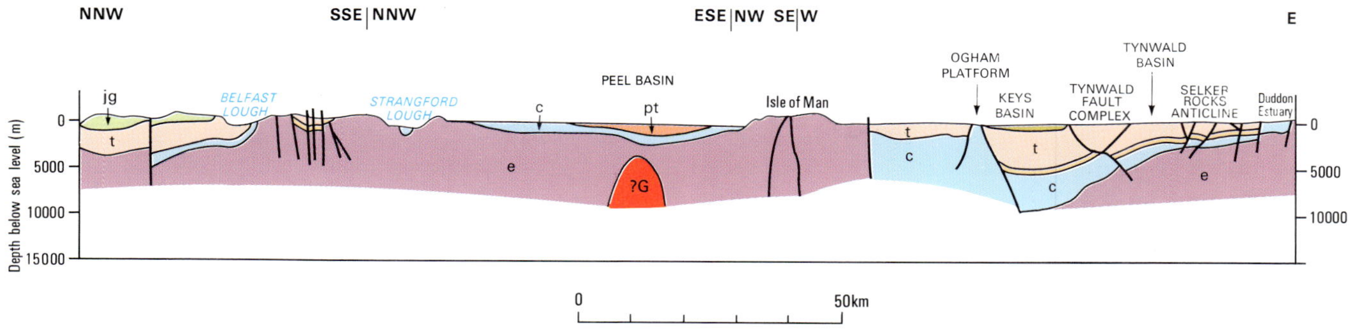

Figure 3 Generalised section across the Irish Sea. For location and key see Figure 2.

up to 300 m, in the Western Trough (see Figure 65), whereas the platforms generally have substantially thinner deposits. The Quaternary sequences offshore consist of three major cycles (Wingfield, 1989), probably ranging back to the mid-Pleistocene, although deposits of the last glacial episode and postglacial times are the most widespread. At the sea bed lie variably thick sediments, many of which are being actively reworked in the modern environment.

HISTORY OF RESEARCH

In the nineteenth and early twentieth centuries, ideas on the geology of the Irish Sea were based on extrapolation from onshore geology. Boreholes drilled on the Isle of Man between 1891 and 1907 led to the conclusion that salt-bearing Triassic marls lie beneath the Irish Sea (Boyd-Dawkins, 1894; Gregory, 1920); Lamplugh (1903) postulated that the Irish Sea had probably been a trough during the Permian, and that New Red Sandstone (Permo-Triassic) surrounds the Isle of Man.

Greenly (1919) produced the first offshore geological sketch map of the Irish Sea in an enlightened and deductive essay based on the variation and contrast in glacial-erratic content in Devensian drift between the northern and eastern coasts of Anglesey. He envisaged Anglesey as a Tertiary anticline, and predicted with remarkable accuracy the western limit of the East Irish Sea Basin. His proposal of a Cretaceous outlier resting on thin Jurassic in the East Irish Sea Basin has not been proved to date, but the limits of the Jurassic in the Keys Basin outlier bear an uncanny resemblance to his predictions, albeit displaced to the north-north-west.

Apart from some early geophysical observations (Browne and Cooper, 1950), research into the offshore geology of the area began in the late 1950s when the Geological Survey of Great Britain (subsequently the Institute of Geological Sciences and now the British Geological Survey — BGS) carried out an aeromagnetic survey that indicated the distribution of sedimentary basins, igneous rocks and basement. The first seismic survey appears to have been a local sparker survey carried out by the National Coal Board (British Coal) in 1960, tracing the offshore structure of the West Cumberland Coalfield (Figure 1; Clarke et al., 1961).

An early refraction-seismic experiment indicated a thinner crust beneath the eastern Irish Sea than that underlying the Southern Uplands (Agger and Carpenter, 1964; Bott, 1965); gravity surveys also demonstrated that the crust thins towards the eastern Irish Sea (Bott, 1964; 1968). Five major negative Bouguer gravity anomalies superimposed on the regional gravity high were ascribed to the Stranraer, Solway Firth, Peel, East Irish Sea and Kish Bank basins (see Figure 18). These were interpreted as being filled with Carboniferous and Permo-Triassic strata (Mansfield and Kennett, 1963; Bott, 1964; Bott and Young, 1971).

Immediately south of the report area, the Llanbedr (Mochras Farm) borehole, commonly termed the Mochras borehole (Figure 1), was drilled at the coast of Tremadog Bay in the late 1960s. This was a turning point in the history of research into the geology of the Irish Sea and other offshore areas. After much preliminary geophysical work (Powell, 1956; Griffiths et al., 1961), the borehole proved that within 3 km of outcropping Cambrian rocks of the Harlech Dome, a very thick, downfaulted sequence of concealed Mesozoic rocks (greater than 1338 m) underlies 609 m of Pleistocene and Tertiary strata (Wood and Woodland, 1968).

Subsequent research into the offshore solid geology falls into three main categories: studies carried out by universities, regional survey work carried out by BGS, and commercial exploration for hydrocarbons. Much progress has also been made in the investigation of Quaternary sediments, and the deep crust has been imaged by BIRPS (British Institutions' Reflection Profiling Syndicate) deep-seismic reflection surveys (Klemperer and Hobbs, 1992).

Working from the University of Wales at Aberystwyth, Al-Shaikh (1970) identified in Caernarfon Bay, two north-easterly trending low gravity anomalies interpreted as basins filled with Carboniferous and Permo-Triassic deposits; these are now known as the Caernarfon Bay and Central Irish Sea basins (see Figure 16). Later, Dobson et al. (1973), also from Aberystwyth, concluded that the highs between these basins, termed the Holy Island Shelf and the Mid-Irish Sea Uplift, consist of Precambrian and Lower Palaeozoic rocks. Dobson and Whittington (1979) published a map of the Kish Bank Basin, as a half-graben hinged on the south-eastern flank, and filled with deposits of Late Carboniferous, Permo-Triassic, Early Jurassic and Paleogene age.

The BGS programme, funded largely by the then Department of Energy (now the Department of Trade and Industry), carried out the first systematic survey of the Irish Sea (Wright et al., 1971). The geophysical methods used included airgun, sparker, pinger, sidescan sonar, magnetometer and gravity meter. Sampling was carried out using vibrocorer, gravity corer and Shipek grab. A shallow-drilling programme ran from 1969 to 1975, and intermittently thereafter (Parkin and Crosby, 1982). Summary results of micropalaeontological analyses of the BGS shallow boreholes proving solid rock are included in Wilkinson and Halliwell (1979), and the survey results are incorporated into the BGS 1:250 000 map series (see inside back cover). McQuillin et al. (1969) reported on a deep-seismic reflection survey which proved a major syncline containing Duckmantian (Westphalian B) sediments separating the Manx Slates in the north from Mona Complex and Ordovician rocks near Anglesey. Bacon and McQuillin (1972) described the results of seismic-refraction lines, and interpreted the low gravity anomaly at the Peel Basin as being due to a granite and associated minor intrusions. Later reviews of BGS work include those by Wright (1975) and Fletcher and Ransome (1978).

Caston (1975; 1976), at the Institute of Oceanographic Sciences (IOS), interpreted a detailed Ministry of Defence sidescan sonar survey in the North Channel; she identified Tertiary igneous dykes intruding Permo-Triassic rocks, and discussed the nature of the sea-bed sediments and their movement. Also in the North Channel, BGS used a submersible to examine the sea bed and to sample dolerite from an igneous knoll (Eden et al., 1973; McLean and Deegan, 1978).

Surveys by the oil industry in the Irish Sea have provided much information on the offshore geology. The first phase of exploration ran from about 1969 to 1983, and included the discovery of the Morecambe Gasfield in 1974. A significant advance in knowledge of the geology of the East Irish Sea Basin was made by Colter and Barr (1975) and Colter (1978), who presented the first results of deep commercial drilling by Hydrocarbons Great Britain (British Gas). The infill deposits of the East Irish Sea Basin were found to comprise 2400 m of Triassic rocks above more than 500 m of Permian sediments resting on Carboniferous rocks. These authors demonstrated the close geological similarities between the Permo-Triassic rocks of the East Irish Sea Basin and the onshore succession in Lancashire and Cheshire. They also recognised that the Permian basins, containing marl/evaporite and sandstone, mimic the Zechstein and Rotliegend basins of the southern North Sea.

Colter (1978) reported on the discovery of the Morecambe Gasfield, the second largest gasfield on the UK Continental Shelf. He described the half-graben structures of the East Irish Sea Basin, and demonstrated that the Permo-Triassic sequence had been buried to a depth of over 4000 m, and that 2000 m of sediment had subsequently been removed following uplift and erosion (see also Lewis et al., 1992). Ebbern (1981) provided further details on the geology of the gasfield and the structure of the faulted anticlinal trap. Bushell (1986) and Stuart and Cowan (1991) presented sedimentological interpretations of the reservoir sandstones, and histories of the hydrocarbon generation. BGS has provided a further synthesis of the geology of the East Irish Sea Basin (Jackson et al., 1987; Jackson and Mulholland, 1993), and has published a Special Edition 1:250 000 Solid Geology map of the East Irish Sea Basin.

Jenner (1981) gave a description of the structure, stratigraphy and hydrocarbon prospectivity of the Kish Bank Basin. Seismic interpretation indicated a 3000 m-thick Permian to Triassic succession with a similar seismic signature to that observed in the East Irish Sea Basin. McArdle and Keary (1986) provided additional details of the Westphalian sequence.

Quaternary research has chiefly been carried out by universities and geological surveys. As early as the nineteenth century, it had been inferred from onshore evidence that ice from Scotland had crossed the Irish Sea to emplace the 'Irish Sea Drift' deposits of the Isle of Man, eastern Ireland and Wales (Eyles and McCabe, 1989). The value of the early offshore surveys in the 1960s was limited both by the shallow penetration of grabs, dredges and corers, and by relatively unsophisticated geophysical instruments. Offshore research, mainly at BGS, IOS, and the University of Wales at Aberystwyth and Bangor (Menai Bridge), tended to concentrate on aspects of the sea-bed deposits and on sea-bed facies maps (Belderson, 1964; Pantin, 1977, 1978). Commercial nearshore boreholes for engineering projects did however provide information on the Quaternary deposits (Naylor, 1964; Knight, 1977), as did the boreholes of the BGS offshore programme after 1969 (Parkin and Crosby, 1982). Based both on these boreholes and improved shallow-seismic profiles, research by workers at BGS, Aberystwyth, and the Geological Survey of Ireland, have increasingly allowed comparison of the offshore sequences with those on land.

2 Crustal structure

Using a variety of geophysical techniques, it is possible to identify not only the base of the crust at the Mohorovičić Discontinuity (Moho), but also to discern major divisions within the crust. Early work largely employed seismic-refraction techniques, but more recently very deep seismic-reflection profiles, particularly the WINCH (Western Isles-North Channel) lines, have been collected by BIRPS (British Institutions' Reflection Profiling Syndicate). The reflection profiles image reflectors below the Moho (Brewer et al., 1983; Hall et al., 1984). In addition to seismic data, useful information on the nature of the crust beneath the Irish Sea has also been provided by gravity, magnetic and conductivity studies.

The lithosphere in this region largely attained its present configuration during the later stages of the Caledonian orogeny, as the Iapetus Ocean closed during Early Palaeozoic times. Sedimentary and volcanic rocks deposited on either margin of the ocean were tectonically imbricated and deformed as the North American (Laurentian) and palaeo-European (Avalonian) continental plates collided (e.g. Chadwick and Holliday, 1991). Identification of the resultant Iapetus suture and convergence zone (Soper et al., 1987), which separates rocks originally belonging to each of these two ancient continents, has been the focus of much recent investigation.

The general features of the crust in the Irish Sea, as revealed from the relatively poor-quality BIRPS profiles, are shown on Figure 4. Strata within the uppermost, post-Devonian, sedimentary basins produce good reflections, whereas the underlying Lower Palaeozoic rocks show few reflectors. The middle crust shows a variety of gently dipping events, and the lower crustal zone includes many, good, flat reflections in the 5 km above the Moho.

This pattern can be compared with that revealed by LISPB (Lithospheric Seismic Profile Britain) refraction work in the Southern Uplands area, which also shows three broad crustal divisions based on ill-defined velocity determinations (Bamford et al., 1978). Upper crust of 5.8 to 6.2 km/s velocity with its base at about 10 km depth lies above middle crust (crystalline basement) of 6.3 to 6.5 km/s velocity, whose base at some 20 km depth corresponds approximately to the top of the reflective lower crust on BIRPS profiles. The lower crust of 7 km/s velocity rests upon the Moho, which lies at around 30 km depth.

THE MOHO

The depth to the Moho in the report area generally ranges from 27 to 30 km. Refraction studies show that under the Southern Uplands–Solway Firth region, the Moho occurs at 30 km depth (Bamford et al., 1978), and it is reported to rise from 32 km depth under central Ireland to only 28 km some 50 km east of Ireland (Jacob et al., 1985). South of the report area in the south Irish Sea, the base of the lower crust lies at 30 km, above an 8.1 km/s-velocity mantle (Blundell and Parks, 1969). The northern Irish Sea is an area of mainly positive Bouguer anomalies (see Figure 18), with values increasing towards it from the interiors of both Ireland and Great Britain as a result of the crustal thinning (Bott, 1968).

The reflection character of the Moho in Britain is variable; it is a good reflector in the Caledonian Foreland of the north-west Scottish Highlands, but in the Irish Sea, where the crustal geology of the Caledonian orogen is more complex, it can be discontinuous and difficult to identify (Brewer et al., 1983). Nevertheless, the data show that the

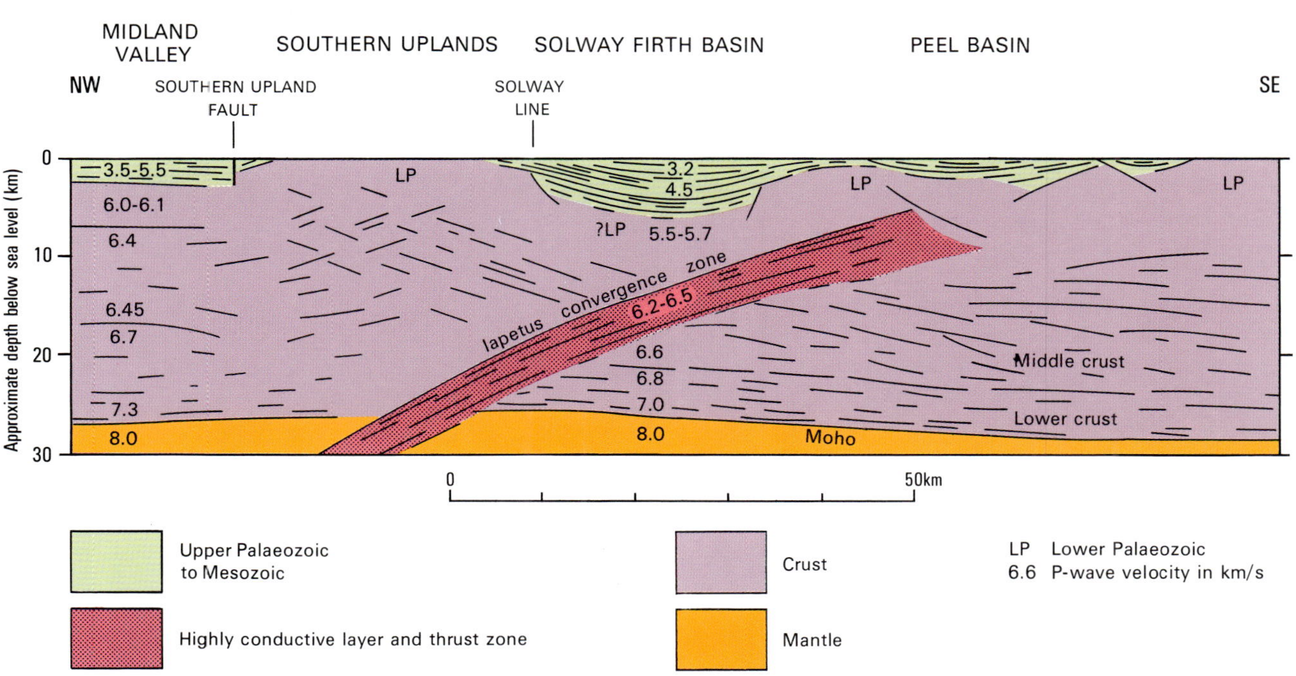

Figure 4 Speculative crustal section across the north-east Irish Sea. After Beamish and Smythe (1986).

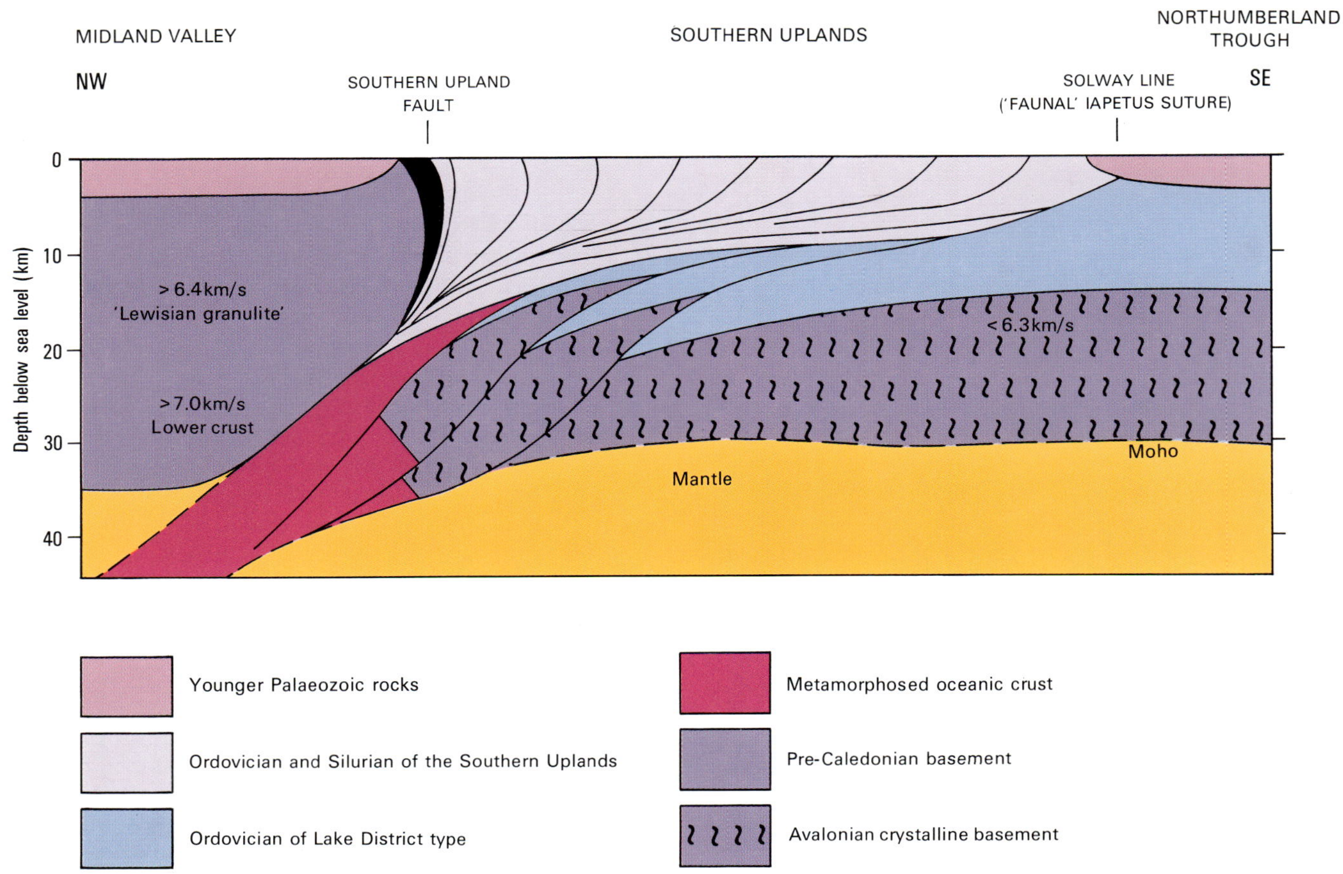

Figure 5 Simplified model for the crustal structure of southern Scotland. After Leggett et al. (1983).

Moho shallows by several kilometres under the zone of post-Devonian sedimentary basin development (Matthews, 1986), including beneath the Solway Firth Basin where it lies at a depth of some 26 km (Figure 4), suggesting north-west to south-east crustal extension during basin formation (Beamish and Smythe, 1986). In this way, the low-density sediments in such basins may be isostatically compensated by the regional rise of dense mantle material (Meissner et al., 1986).

THE CRUST

There is a distinct band of lower-crustal reflectors in a 5 km-thick zone above the Moho (Figure 4) in the Irish Sea (Beamish and Smythe, 1986). Modelling has led to the concept that these lower-crustal reflections are produced by constructive interference between reflections from alternating high- and low-velocity laminations only a few metres or decimetres thick, and a few hectometres or kilometres wide (Blundell and Raynaud, 1986; Matthews, 1986; Matthews and Cheadle, 1986; Wenzel et al., 1987).

Interpreters have generally agreed that there is a real difference in lower-crustal reflectivity on either side of the Iapetus convergence zone (Beamish and Smythe, 1986). To the north, the lower crust of the former Laurentian continent has a velocity of some 7.3 km/s (Figure 4), and the boundary between the crust and mantle shows an apparently gradual transition from lower-crustal to upper-mantle velocities. To the south there is a sharper velocity change at the base of the Avalonian crust. This difference between the lower crust of the two domains is consistent with the existence of a suture at depth beneath the Southern Uplands. The lithospheric velocity

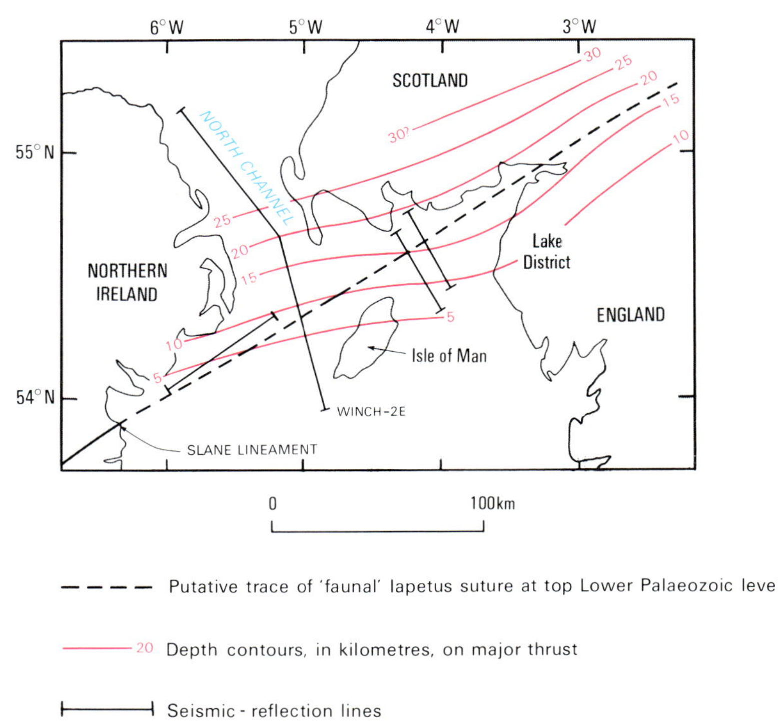

Figure 6 Depth to the major shear zone or thrust, seen on deep-seismic profiles, that is associated with the Iapetus suture. After Beamish and Smythe (1986) and Todd et al. (1991).

structure of the northernmost part of the paratectonic Caledonides is typical of aseismic continental platforms, whereas that under central and southern England is comparable with that of shield areas (Clark and Stuart, 1981).

The middle crust north of the Southern Uplands has a velocity greater than 6.4 km/s, akin to that of Lewisian granulites found in the north of Scotland, whereas to the south of the Southern Uplands it has a velocity less than 6.3 km/s, comparable with that of the Pentevrian basement of Brittany (Figure 5). These observations do not however demonstrate extensive structural continuity on either side of the Solway line. Geologically based plate-tectonic models suggest that the crusts north and south of the Southern Uplands belong to different terranes which amalgamated with the closure of the Iapetus Ocean (Bluck, 1985). Palaeomagnetism also suggests that the terranes were once widely separated (Piper, 1978). The terranes are defined largely by their faunal and structural traits, as well as by the geochemical character of late Silurian to Early Devonian granites on either side of the suture (Thirlwall, 1988).

Iapetus suture

The near-surface lineament termed the Solway line (Phillips et al., 1976; McKerrow and Soper, 1989) has been proposed as the 'faunal' Iapetus suture because it separates the Lower Palaeozoic succession of the Southern Uplands from that of the Lake District (Chadwick and Holliday, 1991; Todd et al., 1991). The surface trace of the Solway line is concealed beneath the Upper Palaeozoic and Mesozoic rocks that crop out to the south-east of the Southern Uplands. This 'faunal' suture is presumed to dip north-westwards in the model of Leggett et al. (1983), which predates the collection of BIRPS data (Figure 5), and is based on land mapping studies and interpretations of modern and ancient subduction zones,

Deep-reflection lines show a weak north-westerly dipping feature interpreted as a shear zone or thrust zone below the Solway Firth and Peel basins (Figures 4 and 6); this feature has also been identified as the Iapetus suture (Hall et al., 1984; Beamish and Smythe, 1986; Matthews and Cheadle, 1986), and is the 'geophysical' suture of Todd et al. (1991). The same, or a similar, crustal shear zone can be identified beneath northern England and the central North Sea (Chadwick and Holliday, 1991), as well as west of Ireland (Klemperer, 1989). If this inferred thrust zone is extrapolated to the surface, its trace lies some distance south of the southern margin of the Northumberland Trough, within, or to the south of, the Isle of Man and the Lake District (Beamish and Smythe, 1986; Figure 6). Chadwick and Holliday (1991) argued that it is most likely to crop out in the Lake District, perhaps as the Causey Pike and associated thrusts.

Conductivity (the reciprocal of resistivity) data can be derived from studying the electric and/or magnetic fields produced by electrical (telluric) currents flowing through rocks. Banks (1986) defined a high-conductivity anomaly which is interpreted as a high-conductance sheet or slab dipping north-north-westwards under the Southern Uplands from a depth of about 10 km near the centre of the Northumberland Trough. This anomaly (Figure 4) has been associated with the seismically identified thrust zone offshore (Beamish, 1986; Beamish and Smythe, 1986). The apparent dip of the thrust zone beneath the Irish Sea is 15° to 20° to the north-north-west on seismic-reflection data, and beneath northern England/southern Scotland it is calculated from conductivity data to be 15° to 26° in the same direction (Beamish and Smythe, 1986). A Northumberland Trough geomagnetic anomaly is interpreted by Banks (1986) as being largely due to the edge effects of the same, dipping, conductive body.

Chadwick and Holliday (1991) presented a model for the late Caledonian continental collision that explains the loca-

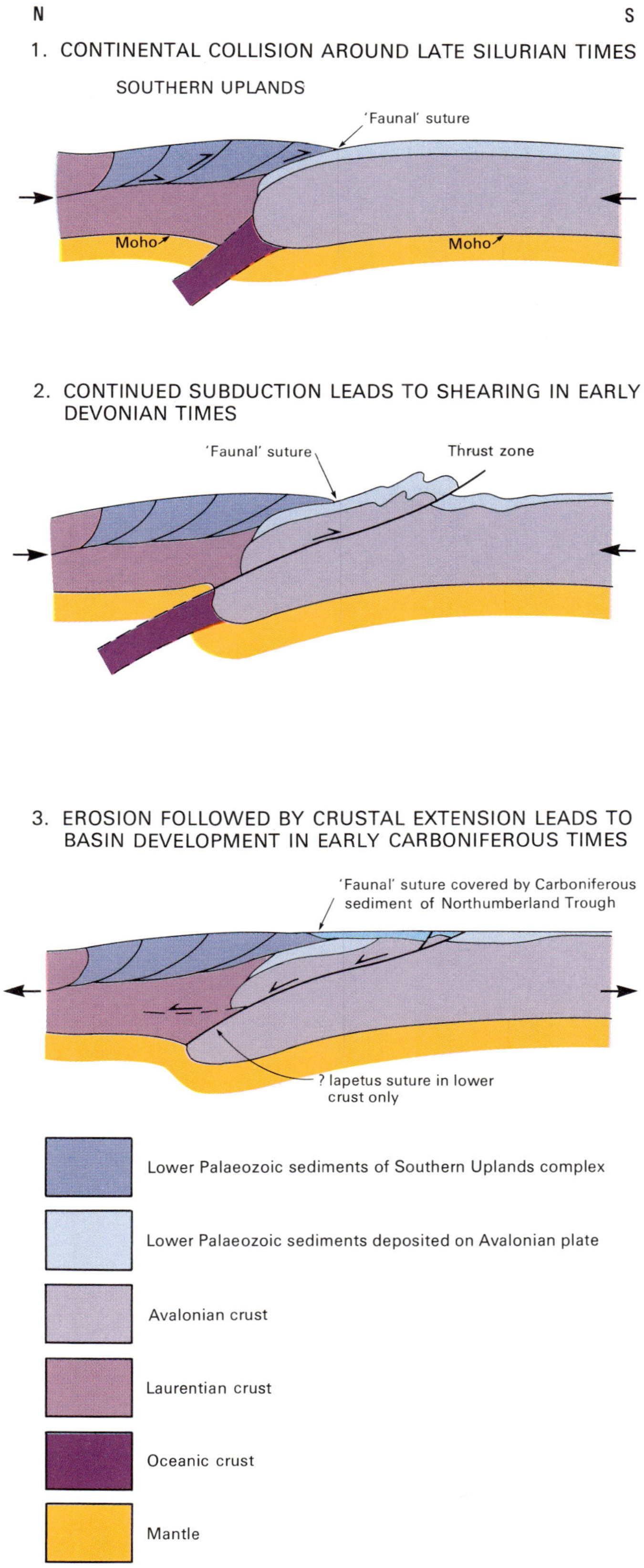

Figure 7 Schematic illustration of Caledonian continental collision, and the formation of the Iapetus convergence zone. After Chadwick and Holliday (1991).

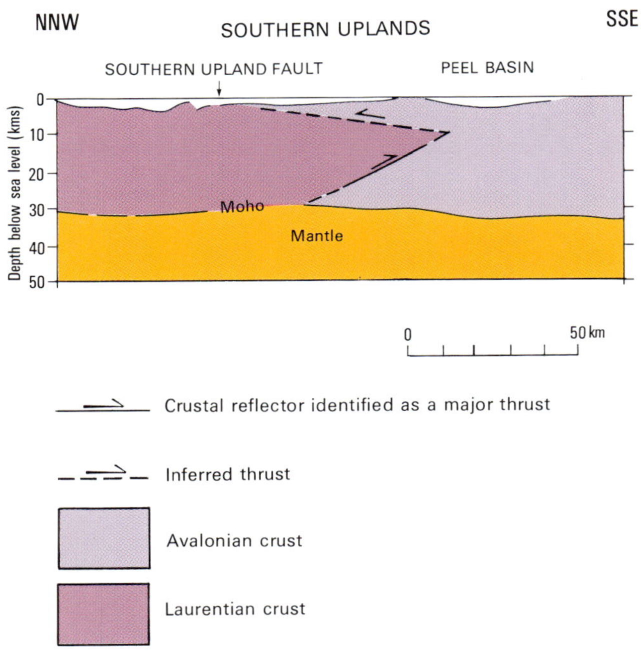

Figure 8 Thin-skinned crustal interpretation of the WINCH-2 deep-seismic profile. After Hall et al. (1984).

tional discrepancy between the trace of the 'faunal' suture and the thrust zone (Figure 7). In their model, the Avalonian crust was subducted beneath Laurentian crust, but the thrust zone developed within Avalonian crust, and not at the crustal boundary which lies to the north. The thrust zone may therefore represent the Iapetus suture only at lower crustal levels, where juxtaposed velocity differences have been identified (Beamish and Smythe, 1986). The faunal suture is now hidden by Upper Palaeozoic and Mesozoic sediments, which were deposited in the Solway Firth Basin and Northumberland Trough, and which developed due to subsidence on steeply dipping faults in the hanging wall above the thrust.

Southern Upland Fault

The Southern Upland Fault is a major fault that at the surface separates the Lower Palaeozoic rocks of the Southern Uplands from the largely Upper Palaeozoic rocks of the Midland Valley to the north. The fault was not seen at depth on BIRPS seismic-reflection lines in the North Channel (Hall et el., 1984). This is consistent with it having a vertical dip, as expected for a wrench fault, or alternatively it may be either an insignificant structure at depth, or the acoustic impedance contrast across it may be undetectable (Brewer et al., 1983; Hall et al., 1984). Davidson et al. (1984) found no evidence for large vertical displacement of basement across the fault, although a change in magnetic anomalies has been identified, particularly in Ireland (Max et al., 1983). The area to the north contains low- to moderate-frequency magnetic anomalies which have a small amplitude and are mainly positive; the positive and negative zones within it probably indicate different lithotectonic blocks that were produced by wrenching along the margin of the craton that lay to the north of the Iapetus Ocean. To the south of the fault are low-frequency, low-amplitude, mainly positive magnetic anomalies; the crust here was probably shortened and thickened by oblique collision during the Caledonian orogeny.

Commercial seismic-reflection lines parallel to, and east of, the WINCH profile, show weak, southward-dipping upper-crustal reflectors which originate from beneath the Southern Uplands (Figure 4; Beamish and Smythe, 1986). These southerly dipping reflectors provide support for a suggestion that the Southern Upland Fault could dip gently southwards under the Southern Uplands to meet the seismically identified northward-dipping thrust zone, such that the Southern Uplands accretionary prism would be only a few kilometres thick (Figure 8; Hall et al., 1984; Hall, 1986a; b). Part of the southern terrane in this thin-skinned/flake-tectonic model would thus be obducted on to the northern terrane, as the wedge-like northern terrane was driven into the southern terrane.

3 Pre-Carboniferous

Lower Palaeozoic and older rocks occur around much of the Irish Sea in Ireland, southern Scotland, the Lake District, North Wales, and on the Isle of Man (Figure 9). The offshore extent of these rocks has been delineated mainly from sparker records, on which they have a structureless internal character that distinguishes them from bedded younger strata (Wright et al., 1971). Their existence is confirmed by sampling in only very few places; information on these rocks therefore comes almost entirely from onshore.

PRECAMBRIAN

The largest occurrence of Precambrian rocks in southern Britain is on and around both Anglesey and the Llŷn Peninsula in north-west Wales (Figure 9). The Monian rocks of Anglesey (Gibbons, 1983; 1989) are Vendian (latest Precambrian) in age, and have a pronounced north-easterly structural grain. They comprise gneisses, schists and igneous rocks with calc-alkaline affinities, overlain by flysch and a variety of extrusive igneous rocks. Components of the complex have been dated at about 613 Ma (Tucker and Pharaoh, 1991). The igneous rocks are mostly of ensialic volcanic-arc type, although some, as within the Gwna Group, show affinities with ocean-floor basalts.

The contacts between markedly contrasting elements are tectonic shears that have been interpreted to represent terrane boundaries; at least four north-easterly trending suspect terranes have been identified. From north-west to south-east these are: the Monian Supergroup, the Central Anglesey Gneiss and Coedana Granite (Coedana Complex), the south-east Anglesey blueschists, and the Sarn Complex (Anderton et al., 1992).

Rocks from both the Monian Supergroup and the Coedana Complex have been correlated with the basement rocks of Rosslare in south-east Ireland, suggesting that these rocks extend offshore between the two areas. Present evidence is indicative of a late Precambrian to early Cambrian age for the transcurrent faulting which generated the shear zones that separate the terranes (Anderton et al., 1992).

Offshore, a sample of greenish grey psammitic rock from Holyhead Bay off Anglesey has been assigned to the Monian on the basis of lithological similarity (Wright et al., 1971). The Holy Island Shelf (Dobson et al., 1973) has been shown as mainly Monian on Figures 2 and 9 (after the BGS Anglesey Solid Geology sheet), where it is less extensive than indicated by Wright et al. (1971). Monian schists have also been sampled close to the Anglesey coast at Amlwch (BGS Anglesey Solid Geology sheet).

LOWER PALAEOZOIC

The Lower Palaeozoic succession which fringes the Irish Sea is largely composed of marine mudstones and turbiditic sandstones, with intercalated volcanic rocks of basic to acidic composition. The succession has been preserved at a low regional metamorphic grade (e.g. Smith et al., 1991), despite local structural complexity. The strata were deposited near the margins of the Precambrian to Early Palaeozoic Iapetus Ocean, which extended from Scandinavia through the British Isles to Newfoundland (Soper and Hutton, 1984). Sedimentary rocks of Ordovician to Silurian age, deposited on the northern margin of that ocean, now form most of the Southern Uplands terrane, which developed in response to northward-directed subduction (Leggett et al., 1979; Stone et al., 1987). Similar rocks comprise the Longford–Down Massif of Ireland. Sedimentary rocks of Ordovician age now exposed in north-west England, the Isle of Man and Wales, accumulated in ensialic basins close to the southern margin of the Iapetus Ocean (Anderton et al., 1979; McKerrow, 1988). Within these basins, volcanic activity was periodically intense, particularly during Ordovician times. Continued subduction closed the Iapetus Ocean by late Silurian to Early Devonian times.

Deformation related to the closure of the Iapetus Ocean continued progressively later in the south than in the north. It ranged from late Ordovician to Wenlock (Silurian) times in the Southern Uplands; in north-west England it continued until Ordovician to Emsian (Devonian) times, and in North Wales until Eifelian (Devonian) times (Barnes et al., 1989). Over a long period of time, various intrusive igneous rocks were emplaced contemporaneously with the extrusive sequences, culminating in the intrusion of granitic plutons during the Early to Mid-Devonian.

Southern Uplands and Northern Ireland

A dominantly turbiditic sandstone facies prevails in the Southern Uplands, where it ranges in age from Llandeilo to late Wenlock. The outcrop pattern is controlled by north-easterly trending, strike-parallel, subvertical faults which define structural blocks (Figure 10) in which the beds generally young towards the north-west and are near vertical. These faults developed as thrusts penecontemporaneously with the deposition of the turbidites as they prograded over a condensed pelagic mudstone sequence which formed the main *décollement* horizon (Leggett et al., 1979). This mudstone sequence, the Moffat Shale Group, is consequently preserved as a discontinuous outcrop at the southern margin of many older blocks (Figure 10), and the volcanic substrate on which it was deposited is preserved only locally as tectonic slivers within the imbricated mudstones (Leggett et al., 1979). The Moffat Shale Group itself ranges from late Llandeilo to mid-Llandovery in age, and at its fullest development is about 60 m thick. It is composed of grey or black mudstones and silty mudstones with thin layers of bentonitic clay.

The Leadhills Group and Gala Group greywackes which overlie the Moffat Shale Group (Figure 10) range in age from Llandeilo to late Llandovery, and become progressively younger towards the south-east in successive structural blocks. The Gala Group passes both upwards and laterally into the Hawick Group (late Llandovery to early Wenlock), a quartzofeldspathic greywacke succession with interbedded siltstone. Higher in the Hawick Group, interbedded red mudstone bands are a characteristic feature of late Llandovery and earliest Wenlock deposits, above which a distinctive,

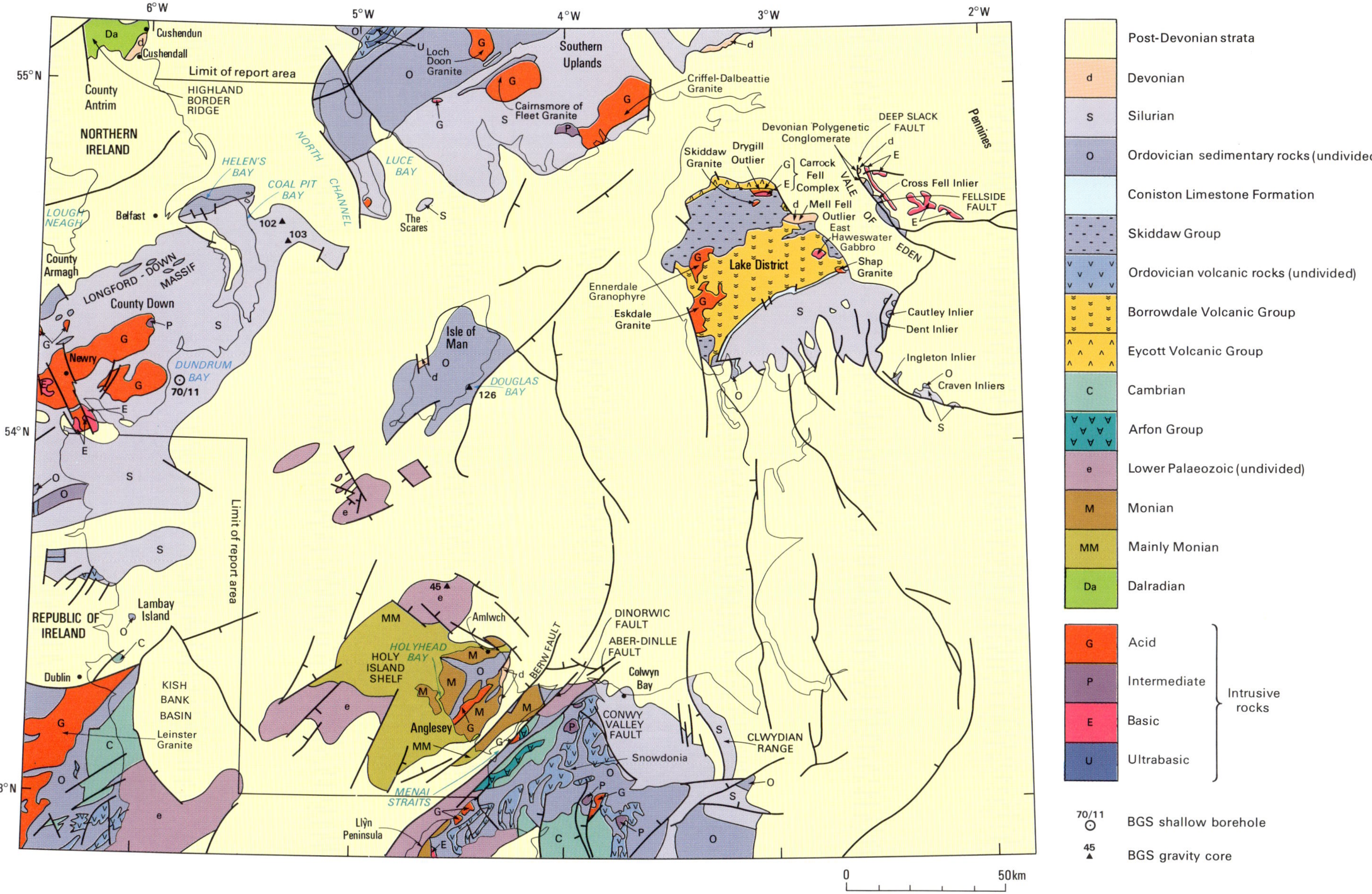

Figure 9 Distribution of Precambrian and Lower Palaeozoic rocks in the Irish Sea and adjacent areas.

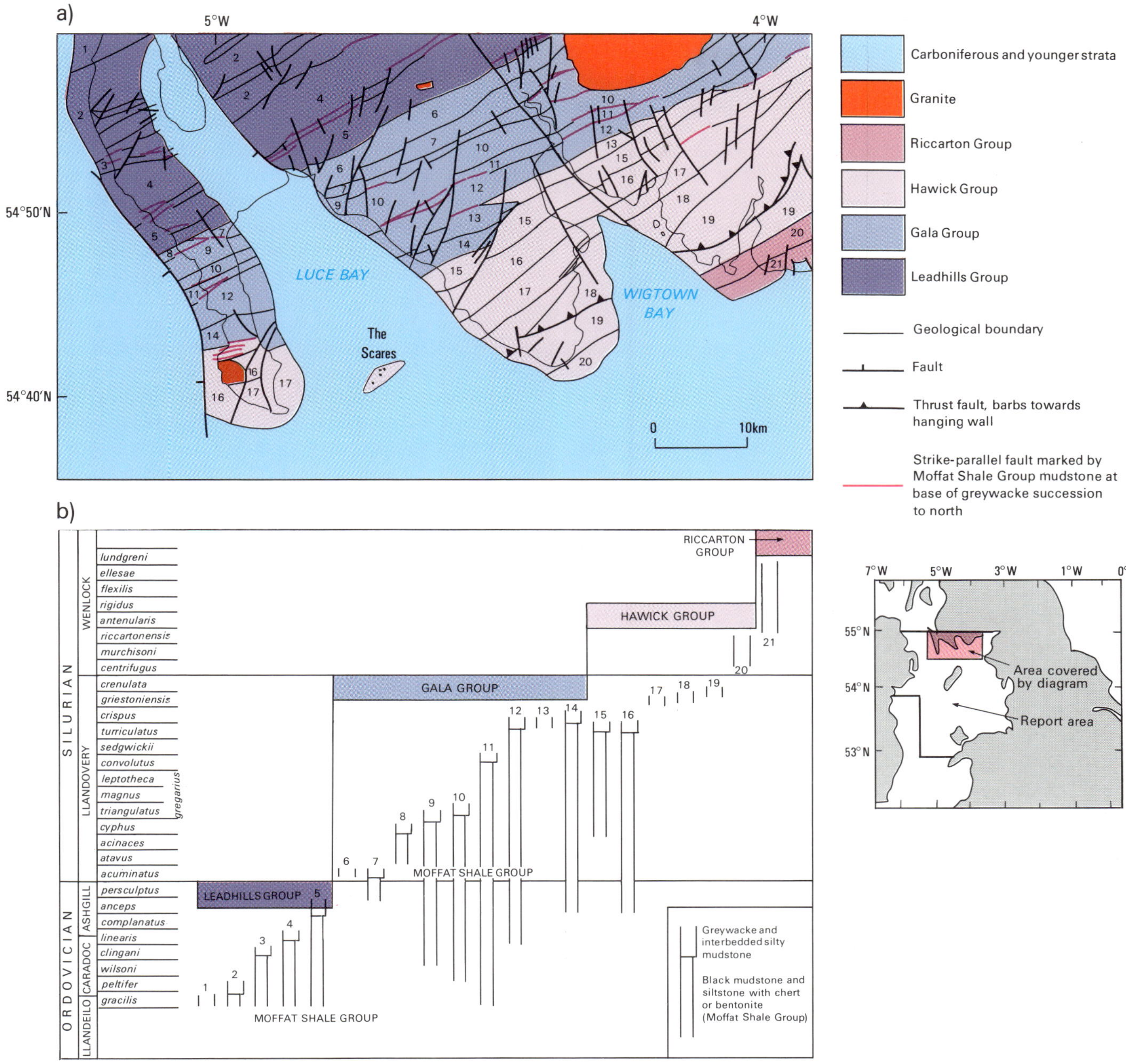

Figure 10 a) Structure of part of the Southern Uplands, south-west Scotland. b) Biostratigraphical relationships in south-west Scotland. Numbers refer to the fault-bounded blocks shown on the map above.

hemipelagic, laminated, carbonaceous siltstone is found. One of the major strike-parallel faults separates the relatively deformed Hawick Group from the less deformed and lithologically more variable turbidite assemblage of the Riccarton Group, which contains a significant proportion of laminated hemipelagite interbedded with the greywacke. This association of late Llandovery red mudstone succeeded by Wenlock carbonaceous siltstone is also identified in the Lake District, and provides evidence of the first sedimentary links between the opposing sides of the Iapetus Ocean (Barnes et al., 1989).

Deformation of the Southern Uplands sedimentary sequence was diachronous, and contemporaneous with the deposition of younger sediments (Barnes, 1989). The earliest intrusive rocks are dyke swarms which, though widespread, are mostly concentrated in the Hawick Group outcrop; lamprophyres are most common, with subordinate intermediate and felsic lithologies (Barnes et al., 1986). The last phase of intrusion included the Newry, Cairnsmore of Fleet, and Criffel/Dalbeattie granites. These tend to form higher ground and may have isostatically uplifted the blocks which they intruded; each is associated with a low gravity anomaly (see Figure 18).

In Northern Ireland, the Longford–Down Massif is the strike continuation of the Southern Uplands (e.g. Barnes et al., 1987), and both the facies and tectonic setting of the Lower Palaeozoic rocks are similar to those of the Southern Uplands Massif. Turbiditic greywackes dominate the sequence above an attenuated black shale sequence equivalent to the Moffat Shale Group. The rocks become progressively younger towards the south, but as in the Southern Uplands, are steeply dipping, commonly overturned, and predominantly young northwards within individual fault blocks. The oldest rocks, of Llanvirn age, occur only locally in County Armagh; elsewhere the strata range in age from Llandeilo to early Wenlock, with Silurian rocks by far the most widespread (Figure 9; Smith et al., 1991). The introduction of greywacke occurred progressively later towards the south; the greywackes contain clasts comparable with those of the

Precambrian volcanic rocks of Anglesey and the Welsh Borders (Manning et al., 1970).

Offshore, a relatively narrow subcrop of Lower Palaeozoic rocks occurs around south-west Scotland (Figure 9), with a slightly wider subcrop off Northern Ireland extending into the North Channel. In Luce Bay, the offshore reef of The Scares, formed of greywackes of Llandovery age (Wright et al., 1971), is seen on seismic records as a fault-bounded footwall inlier to the Luce Bay Basin. In Dundrum Bay, BGS borehole 70/11 recovered 6.5 m of sandstone and cleaved mudstone, presumed from regional considerations to be Silurian in age (Wright et al., 1971). Two gravity cores in the North Channel proved dark grey and purplish brown siltstones, one of which yielded a well-preserved acritarch assemblage which includes *Percultisphaera stiphrospinata* (Lister), and clearly indicates a Ludlow age (BGS Biostratigraphy Group). This age is anomalous for the Southern Uplands terrane, as the youngest beds proved onshore are late Wenlock. Similar lithologies of the same age are, however, exposed in inliers within the Midland Valley of Scotland to the north of the Southern Upland Fault (Selden and White, 1983), where they mark the transition to terrestrial Old Red Sandstone redbeds.

North-west England and the Isle of Man

A three-fold subdivision of Lower Palaeozoic rocks is applicable throughout the Lake District and adjacent areas (Figure 11) and, although these rocks do not crop out extensively offshore, they may underlie much of the central part of the report area. Early Ordovician and locally older marine sedimentation produced the Skiddaw, Manx and Ingleton groups. Subaerial to submarine volcanism followed in Llandeilo to Caradoc times to create the Eycott and Borrowdale volcanic groups. Further marine sedimentation during the late Ordovician and Silurian formed the Windermere Group.

Skiddaw, Manx and Ingleton groups

These groups crop out in geographically discrete areas, and range in age from Tremadoc to Llanvirn. Mudstones dominate both the Manx Group in the Isle of Man, and the Skiddaw Group which is exposed in the Lake District and adjacent inliers. Coarser grained lithologies of the Ingleton Group are exposed in inliers south-east of the Lake District (Figure 9).

The Skiddaw Group is up to 5000 m thick and consists of sparsely fossiliferous, turbiditic and hemipelagic greywackes, siltstones and mudstones. Two distinct lithostratigraphical sequences (Cooper and Molyneux, 1990) are separated by an east-north-easterly trending fault which controlled sedimentation patterns during the Ordovician. The northern sequence was deformed by major slumping during Llanvirn times (Webb and Cooper, 1988), and is overlain unconformably by the Eycott Volcanic Group. The southern sequence also shows evidence of major soft-sediment deformation, and is unconformably overlain by the Borrowdale Volcanic Group. The tectonic structure of the Skiddaw Group is dominated by upright major folds and a pervasive slaty cleavage modified by several subsequent deformational events which formed thrusts, minor folds, and a number of crenulation cleavages.

The Manx Group is largely equivalent to the Skiddaw Group, although the two cannot yet be correlated in detail. The succession is again dominated by mudstones, but includes turbiditic siltstones and sandstones and a thin sequence of andesitic volcanic rocks. The stratigraphy of Simpson (1963) may require major revision, for subsequent micropalaeontological work, which defined ages ranging from Tremadoc to Arenig/Llanvirn (Molyneux, 1979), found that Simpson's 'oldest' formation yielded the youngest faunas, whereas his 'youngest' formation yielded probable Tremadoc acritarchs. Numerous minor intrusions, ranging in composition from basic to acidic, suggest a similar intrusive history to that of the Southern Uplands.

Offshore, a BGS gravity core east of the Isle of Man proved 0.43 m of cleaved mudstone of the Manx Group, and a weathered, cleaved mudstone was collected by vibrocorer off the south coast of the island (Wright et al., 1971). To the south-west, some of the anticlinal Lower Palaeozoic inliers (Figure 9; BGS Anglesey Solid Geology sheet) are seen on poor seismic records to be horst blocks, possibly with a thin Dinantian cover locally.

The Ingleton Group is composed of conglomerates, sandstones and siltstones of Tremadoc to Arenig age, and is at least 600 m thick at outcrop (Arthurton et al., 1988).

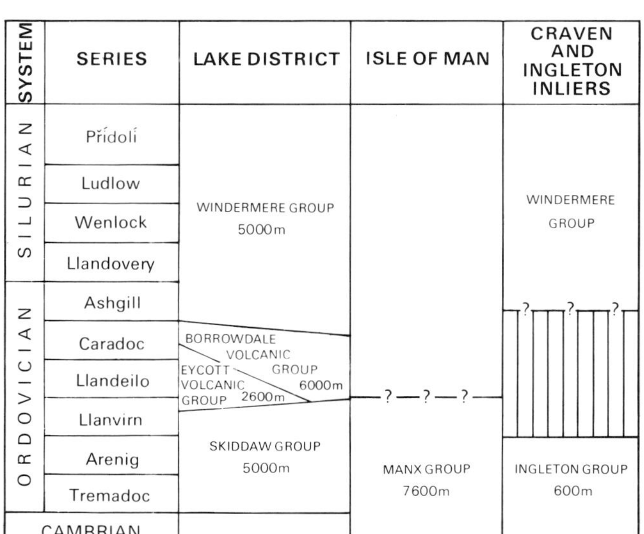

Figure 11 Simplified lithostratigraphy of the Lower Palaeozoic rocks in northern England and the Isle of Man. See text for details.

Eycott and Borrowdale Volcanic groups

In the north of the Lake District inlier, the Skiddaw Group is overlain unconformably by the Eycott Volcanic Group, which is composed of about 2600 m of volcanic rocks of possibly Llandeilo to early Caradoc age. The Eycott Volcanic Group also crops out within a small, fault-bounded block within the Cross Fell inlier (Arthurton and Wadge, 1981).

The lower part of the sequence consists of andesitic lavas and sills with thin beds of coarse-grained tuffs, volcaniclastic breccias and sandstones. Acidic volcaniclastic rocks dominate the upper part of the sequence. Although much of the volcanic succession is considered to be subaerial, siltstone and tuffaceous sandstone beds near the base have yielded a marine microflora indicating that the sequence is not older than late Llanvirn (Millward and Molyneux, 1992). A small, faulted inlier of middle Caradoc mudstone, the Drygill Shale, is considered to overlie and postdate the volcanic rocks, and thus gives a minimum age for them.

The Borrowdale Volcanic Group occupies the central part of the Lake District inlier, and small fault-bounded blocks also occur within the Cross Fell inlier. In the Lake District, it rests unconformably upon the Skiddaw Group, and is unconformably overlain by the Ashgill to Přídolí Windermere

Group. The upper part has yielded acritarchs of Caradoc age (Molyneux, 1988), and a Sm-Nd whole-rock late Llandeilo (Snelling, 1987) age of 457 ± 4 Ma (Thirlwall and Fitton, 1983).

The Borrowdale Volcanic Group was deposited in a dominantly subaerial environment (Millward et al., 1978; Branney, 1988). The lower part, up to 2700 m thick, is composed largely of lavas and associated sills with interbedded ignimbrites and volcanogenic sedimentary units. These volcanic rocks are largely andesitic, but range in composition from basaltic to dacitic, and were probably deposited from low-profile, shield-like volcanoes. The upper part of the succession is dominated by volcaniclastic sedimentary rocks interbedded with intermediate to silicic ignimbrites and associated pyroclastic deposits. Rapid, large, lateral thickness variations and nonsequences are a result of contemporaneous faulting and caldera subsidence (Branney, 1988).

Upper Ordovician and Silurian

Marine sedimentation resumed after the volcanic activity which produced the Eycott and Borrowdale volcanic groups, to form the Windermere Group. The oldest postvolcanic rocks, of mid-Caradoc age, occur in the north, where the Drygill Shale overlies the Eycott Volcanic Group. Along the southern margin of the main Borrowdale Volcanic Group and in the Craven inliers, the oldest postvolcanic rocks are of Ashgill age, and are succeeded by a more-or-less complete Silurian sequence. The Upper Ordovician succession is a shallow-water sequence of clastic and carbonate strata, that in Llandovery times was followed, in the Lake District, by deeper water, graptolitic mudstones. Subsequently, a substantial thickness of north-westerly derived siliciclastic turbidites was laid down during late Wenlock and Ludlow times (Kneller, 1991). In the Lake District, the succession ends with red siltstone of Přídolí age (Shaw, 1971).

The Lake District batholith

The major intrusions exposed in the Lake District (Figure 9) are the surface expressions of a much more extensive batholith which underlies much of the central Lake District, and extends eastwards under the northern Pennines (Bott, 1978; Lee, 1986). Two large intrusions, dominantly granite and granodiorite, are exposed in Eskdale and Ennerdale; the pretectonic Eskdale pluton has been dated by Rundle (1981) as about 430 Ma, and is of Llandovery age (Snelling, 1987). A host of minor intrusions, some associated with the plutonic centres, are principally acidic in composition, although basic and intermediate dykes are locally abundant. Small basic intrusions are exposed at Carrock Fell and East Haweswater. The granites at Shap and Skiddaw postdate the main cleavage, and are dated (Wadge et al., 1978) at about 390 Ma (Early Devonian — Snelling, 1987). The igneous rocks of the Lake District batholith are thought to terminate westwards against the bounding fault of the Lake District inlier, and not to extend offshore (Lee, 1989).

Republic of Ireland

Lower Palaeozoic rocks crop out in south-east Ireland (Brück et al. 1979; Holland, 1981 c; d), around the flanks of the Leinster Granite, where they unconformably overlie the pre-Caledonian, metamorphic basement of the Rosslare Complex (Phillips, 1981). The oldest rocks, the lower to middle Cambrian Bray Group, are dominated by sandstone turbidites that include thick quartzite units. The overlying Ribband Group is composed of a more-distal turbidite facies with interbedded volcanic rocks; its age ranges from late Cambrian to early Ordovician. The facies is similar to those of the Manx and Skiddaw groups, and have led to suggestions of a basin extending from south-east Ireland to Cumbria (e.g. Holland, 1981b).

Above the Ribband Group lies a widespread mid-Ordovician unconformity. This structural break is absent to the west and north-west of the Leinster Granite, where the Ribband Group is overlain by mainly Caradoc volcanic rocks which are in part laterally equivalent to, and pass up into, distal turbidites ranging up to Wenlock in age. Tectonic deformation of the sequence occurred during the late Silurian or early Devonian, when sinistral transcurrent faulting and north-westerly directed thrusting (Max et al., 1990) was accompanied by the emplacement of the syntectonic Leinster Granite.

The Lower Palaeozoic outcrop is thought to extend offshore (Figure 9), and it is possible that parts of the sequence extend beneath the Carboniferous and younger cover over much of the central part of the report area.

North Wales

The north-easterly structural trend in the Precambrian rocks is repeated within the Cambrian and Ordovician strata of North Wales. The lowermost Cambrian strata within the Arfon Group (Reedman et al., 1984) crop out within the Menai Straits fault system (Kokelaar et al., 1984), and overlie more than 2000 m of acidic ash-flow tuff which was ponded in a narrow graben, bounded on the north-western side by the Dinorwic Fault. The Cambrian sedimentary sequence reflects repeated control by contemporaneous faults (Webb, 1983; Kokelaar et al., 1984), and comprises markedly contrasting lithologies, including the slates which have been so economically important. Within Snowdonia, the basal Arenig sandstone lies with slight angular unconformity on Tremadoc strata, but to the north-west towards the Menai Straits, it rests upon the tuffs of the Arfon Group, and on Anglesey it rests on Precambrian, Monian rocks. The magnitude of this unconformity reflects the extent of uplift and erosion in the vicinity of the Menai Straits fault system.

The complex patterns of marine sedimentation which were established during Cambrian and early Ordovician times were broken by the initiation of volcanic activity during the early Caradoc (Howells et al., 1991). The siting of this activity was controlled by deep-seated fractures and the establishment of an extensional graben (Campbell et al., 1988). The volcanicity was predominantly acidic and basic, with relatively small influxes of magma of intermediate composition. The environment was mainly shallow marine to outer shelf, although two intervals of restricted emergence have been recognised. Acidic ash-flow tuffs form a major component of the sequences; the large volumes periodically erupted were emplaced over a wide area, and now form the basis of a detailed stratigraphy (Howells et al., 1991 and references therein).

Within the Snowdon Graben, the volcanic sequence is overlain by black, graptolitic, Caradoc to Ashgill mudstone with no indication of shallow-water reworking. Periodic influxes of turbiditic sandstone at the top of this sequence indicate the development of the broad, basinal environment that became established during Silurian times. Therein were deposited the strata which now crop out between the Conwy Valley Fault and the Clwydian Range (Warren et al., 1984). Turbiditic mudstones dominate, with intercalated pelagic and hemipelagic layers, and sporadic coarse-grained sandstones, as in the Denbigh Grits Group. Phases of instability, recorded by repeated slumped and disturbed beds, were probably initiated by the reactivation of faults which defined the local Silurian basin.

Off the North Wales coast, Lower Palaeozoic rocks are believed to occur close to the shore on either side of the Menai Straits and off Colwyn Bay (Figure 9); they have not been sampled. More extensive outcrops are mapped to the north and west of Anglesey (Figures 2 and 9), although sparse Ordovician chitinozoa, recovered from grey mudstone on the northern margin of the Holy Island Shelf by Wright et al. (1971), may have been reworked into Carboniferous strata. Cleaved sandstone, mudstone and siltstone recovered less than 5 km to the north of Amlwch have been assigned to the Lower Palaeozoic (BGS Anglesey Solid Geology sheet).

DEVONIAN

By the end of the Caledonian orogeny, the new landmass created by the closure of the Iapetus Ocean had been uplifted and subjected to subaerial erosion (Holland, 1981d). In most of Britain, Devonian rocks are thus part of a continental red-bed facies, termed the Old Red Sandstone (ORS). A warm, semiarid climate prevailed, with a markedly seasonal rainfall which supported major rivers. The region lay 15° to 25° south of the equator (Faller and Briden, 1978) on the margin of the major continent, with open sea to the south (Figure 12).

Marine Devonian deposits are now found in southern Ireland and south-west England, and can be traced across northern Europe from western France to Germany. Extensive outcrops of the continental ORS facies occur in the Munster Basin in Ireland and the Anglo-Welsh Basin, as well as farther north where sediments and lavas accumulated in the intracratonic, fault-bounded graben of the Midland Valley Basin in Scotland, and the Tyrone/Fermanagh Basin in Northern Ireland (Wilson, 1972; Cameron and Stephenson, 1985). This end-Caledonian rifting was associated with late-stage, acidic, andesitic and basaltic volcanism, so that lavas, tuffs and volcaniclastic sediments form a conspicuous part of the sequences in these northern basins.

The report area lay between these large marine and continental basins; only isolated outliers of Devonian rocks are now preserved on the surrounding land on Anglesey, the Isle of Man, in the eastern Lake District, southern Scotland, and just to the north of the report area at Cushendun and Cushendall in County Antrim (Figures 9 and 13). Devonian strata have not been recovered offshore, but seismic evidence and extrapolation of onshore geology indicate that they are likely to occur locally, especially in the western part of the area.

The Devonian rocks found around the report area are mainly Lower Devonian (Figure 13). In the Anglo-Welsh Basin, fluvial sediments accumulated on broad alluvial plains; thick pedogenic limestones indicate that stable conditions prevailed for prolonged periods. Pedogenic limestones form a significantly smaller proportion of the sequences to the north, where there is an increase in the proportion of cobble and boulder conglomerates deposited largely on alluvial fans (Allen and Crowley, 1983). The paucity of diagnostic fossils makes dating of the Old Red Sandstone facies difficult; it is predominantly Devonian, but ranges in age from Silurian to Early Carboniferous (Paterson and Hall, 1986; Marshall, 1991), and Upper Devonian rocks are present around the report area.

In County Antrim (Figures 9 and 13), the redbed conglomerate and sandstone sequence at Cushendun and Cushendall probably includes both Lower and Upper Devonian

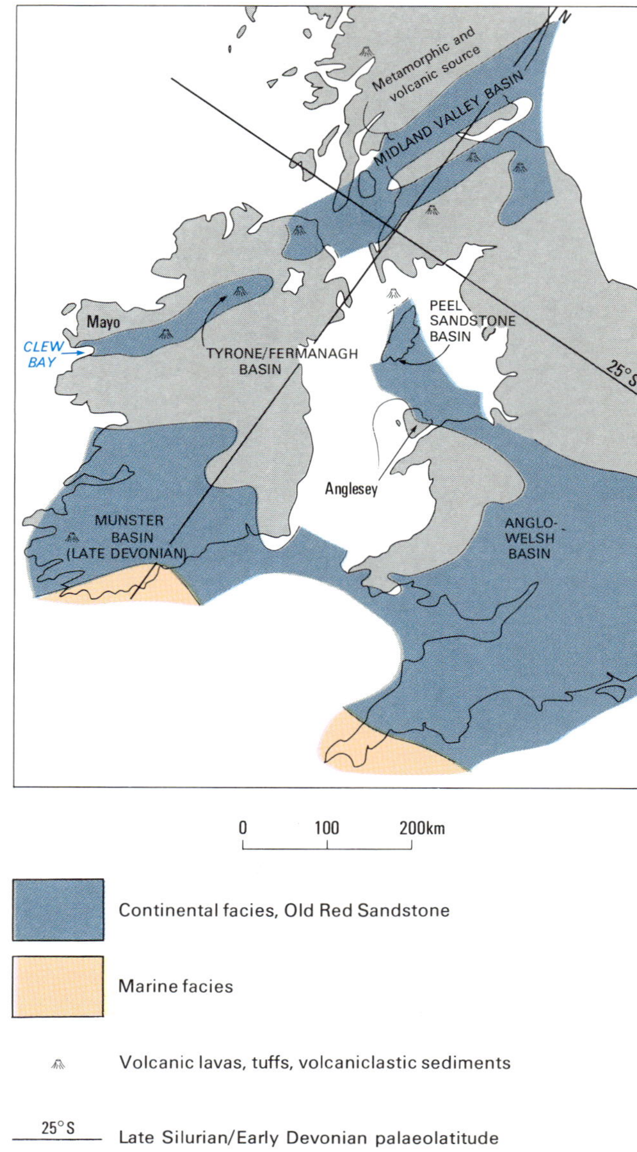

Figure 12 Generalised Devonian palaeogeography of the British Isles. Palaeocurrents were directed towards the south-west in Early Devonian times, and towards the north-east during the Late Devonian. Based on Simon and Bluck (1982) and Allen and Crowley (1983).

deposits (Simon, 1984a: b). The conglomerates near the top of the succession contain an increasing proportion of andesite debris (Wilson, 1972). The deposits are represented by the Cross Slieve Group (1370 m), overlain unconformably by the 400 m-thick Red Arch Formation (Simon, 1984a; b). This outcrop is possibly continuous with Devonian rocks thought to underlie the North Channel Basin (McLean and Deegan, 1978), although their extent is unknown.

The possibly Early Devonian (Capewell, 1955) Mell Fell Conglomerate of the Lake District (Wadge, 1978) consists mostly of coarse-grained, greywacke-cobble conglomerate passing upwards into coarse-grained sandstone in a single upward-fining cycle. Clasts are up to 1 m across, and sediment fining indicates a derivation from the south and west. Cornstones are absent, and the deposits resemble those of Cushendun.

North-west of Cross Fell on the eastern side of the Vale of Eden (Figure 9), the so-called Polygenetic Conglomerate consists of up to 35 m of steeply dipping, cobble conglomerate deposited in three partly coalescing alluvial fans (Arthurton

SERIES / STAGE		ANGLESEY	ISLE OF MAN	MELL FELL (LAKE DISTRICT)	COUNTY ANTRIM
UPPER DEVONIAN					Red Arch Fm.
MIDDLE DEVONIAN					
LOWER DEV.	Emsian	— ? — — ? —	PEEL SANDSTONE GROUP	Mell Fell Conglomerate	CROSS SLIEVE GROUP
	Siegenian	Traeth Lligwy Beds Porth y Mor Beds	PEEL SANDSTONE GROUP	Mell Fell Conglomerate	CROSS SLIEVE GROUP
	Gedinnian	Traeth Bach Beds Bodafon Beds		Mell Fell Conglomerate	CROSS SLIEVE GROUP

Figure 13 Lithostratigraphy of Devonian rocks around the report area. See text for details.

and Wadge, 1981). Clasts are locally derived from the west, and are up to 1 m across and coated in desert varnish, indicating an arid climate.

On the Isle of Man, the age of the Peel Sandstone Group is unconfirmed, but a consensus considers that these redbeds are of Old Red Sandstone facies, and probably of late Early Devonian (Siegenian to Emsian) age (Crowley, 1985). This is despite a provisional determination by Neves (in Ford, 1971) of lowermost Carboniferous spores, which may imply a correlation with the Upper Old Red Sandstone of the Midland Valley of Scotland, which ranges into the Carboniferous (Paterson and Hall, 1986). The Peel Sandstone Group occurs in a north–south-trending graben (Boyd-Dawkins, 1902) that extends offshore into the west Irish Sea. It is estimated that the Peel Sandstone Group (Crowley, 1985) is about 1000 m thick, possibly increasing to 1500 or 2000 m inland, but neither the top nor the base is seen. The sequence consists of upward-fining cyclothems of breccio-conglomerate with small, angular pebbles and a number of interbedded pedogenic limestones up to 2.4 m thick. The beds were deposited rapidly on a series of distal, coalescing, alluvial fans from rivers flowing towards the Isle of Man from a source possibly only 5 to 10 km to the north-west. The fauna within conglomeratic clasts of Silurian limestone shows an inverted stratigraphy which demonstrates progressive unroofing of a limestone shelf succession not presently found in the Southern Uplands.

On Anglesey, only the Lower Old Red Sandstone is represented (Figure 13); this ranges in age from late Silurian to Early Devonian (Allen and Crowley, 1983). Four lithostratigraphical subdivisions are present (Allen, 1965). The Bodafon Beds (45 m) at the base are pebble-conglomerates and sandstones interpreted as piedmont alluvial-fan deposits infilling irregularities of the Early Palaeozoic floor. The succeeding Traeth Bach Beds (130 m) are red, concretionary siltstones and pedogenic limestones interpreted as playa deposits. The disconformable Porth y Mor Beds (345 m) make up most of the succession; they progressively overlap the older beds towards the south-west, and are themselves overlapped by Dinantian strata. The Porth y Mor Beds consist of repeated cycles of pebbly conglomerates, sandstones and siltstones with mudstones and cornstones, and are interpreted as the deposits of a major river flowing to the south-east. The youngest unit, the Traeth Lligwy Beds (24 m), is formed of alternating siltstones and sandstones interpreted as lacustrine deposits, which are rare in the Old Red Sandstone of southern Britain.

The entire Anglesey succession accumulated on small alluvial fans along the west side of a broad, north-westerly trending valley bordered by the Anglesey uplands which were the source of exotic clasts. The heavy-mineral suite of zircon, rutile and tourmaline lacks garnet, and is thus unlike that of either the Peel Sandstone Group or the Welsh Border succession (Allen, 1965; Crowley, 1985). The disconformity at the base of the Porth y Mor Beds, and the overstep of Carboniferous strata, both increase in magnitude towards the south-west, indicating that the Lower Devonian succession thickens towards the Irish Sea (Allen, 1965). Thus it is possible that a substantial Early Devonian basin, or series of smaller basins, is concealed beneath the offshore Carboniferous.

4 Post-Caledonian structure

The Irish Sea was the site of major deposition not only from Cambrian to Devonian times, but also during the Carboniferous, Permian, and Triassic, when each episode of basin development corresponded to a time of subsidence, sedimentation and deformation (Figure 14; Jackson and Mulholland, 1993). There is limited evidence for other episodes of deposition, although the products are commonly no longer preserved.

During the Early Palaeozoic, deposition occurred along the margins of the Iapetus Ocean, which lay between the continents of Laurentia to the north, and Eastern Avalonia to the south (Soper et al., 1987). The Southern Uplands and Longford–Down massifs lay on the northern margin of this ocean, whereas the Lake District and Wales were on its southern margin. This episode of basin development was terminated by the closure of the Iapetus Ocean during the Caledonian orogeny; the suture or convergence zone which marks the junction of these ancient plate margins is believed to lie beneath the younger rocks of the report area (Figures 6 and 7; Chadwick and Holliday, 1991; Todd et al., 1991). The final cleavage event of Caledonian deformation is Emsian (Early Devonian) in age (Soper et al., 1987), and the general trend of the folds is north-easterly. This orogeny created the tectonic framework upon which the later Palaeozoic and early Mesozoic basins developed.

Uplift of the newly created continent of Laurasia occurred in the final stages of the Caledonian orogeny, and the resulting stress relaxation was accompanied by the emplacement of late- and postorogenic granites and granodiorites (Murphy, 1987). A continental depositional environment was established, and postorogenic molasse of Devonian age is preserved in the report area as the Old Red Sandstone. The contemporaneous coastline was situated to the south, passing through south-west England and southern Ireland (Figure 12).

During the Early Carboniferous, a tensional stress regime was established: this gave rise to a series of grabens and half-grabens which controlled sedimentation patterns throughout most of the Carboniferous Period (Gawthorpe et al., 1989). Intensive crustal thickening during the Late Carboniferous Variscan orogeny resulted in major nappes in central Europe (Giese et al., 1983). The British Isles were peripheral to the major Variscan events; the most northerly thrust front passes through southern Ireland and South Wales (Figure 15), and the effects of this orogeny diminish northwards. It reached a climax during the Stephanian in the Irish Sea, where the effects are slight compared with those of the Caledonian orogeny (Freshney and Taylor, 1980). Folds are upright and open, and cleavage is developed only locally. The trend of the Variscan structures is more dependent on the tectonic grain of the underlying basement, and on Dinantian lithology, than on the direct stresses that operated during this orogeny (Bott, 1967; Soper et al., 1987). In the southern part of the Pennine High in Derbyshire (Figure 15), there is typically an interference pattern between north–south-trending, fairly well-defined folds overlying the basins, and less-definite axial trends varying from easterly to north-easterly over the rigid blocks (Aitkenhead et al., 1985). A similar pattern may occur offshore in the report area.

From Early Permian to Tertiary times (Figure 14), there was a regime of lithospheric extension which broadly related to the opening of the Atlantic to the west. Episodes of active faulting in southern Britain, for example in the Early Permian and again in the Early Jurassic (Chadwick 1985), marked times of accelerated crustal subsidence and extension. Chadwick (1985) presented a schematic model for the development of the Permo-Triassic basins of the British Isles: initial crustal extension resulted in faulting and an elevation of the lithospheric isotherms. Gradual decay of this thermal anomaly resulted in a thermal-subsidence phase characterised by a general absence of normal faulting. Sediments deposited during this phase commonly overlap the earlier fault-bounded basin to produce a 'steer's head' profile in cross-section.

This model may be applied to the Irish Sea, where thick Lower Permian sediments and, locally, extrusive igneous rocks are confined to narrow grabens and half-grabens in Northern Ireland, south-west Scotland and the southern part of the East Irish Sea Basin. These depocentres are arranged peripherally to the centre of the East Irish Sea Basin, which may have coincided with the site of a major lithospheric thermal anomaly in earliest Permian times. In the Late Permian, decay of the thermal anomaly led to subsidence, and sediments of Late Permian age overstep those of the Early Permian; the unconformity at the base of the Upper Permian is most easily observed at basin margins. Overlap of the Upper Permian by the Early Triassic Sherwood Sandstone Group and overlap of Sherwood Sandstone Group by Mercia Mudstone Group sediments could be attributed to crustal flexuring during the thermal-subsidence phase. Rapid regional subsidence continued throughout the Triassic and into the Early Jurassic, punctuated by limited uplift of the Wales–Brabant Massif, and some renewed faulting.

Terrestrial conditions prevailed throughout most of the Triassic, but in Late Triassic/Early Jurassic times, a rise in sea level caused rapid flooding of the peneplaned landscape. Lias Group sediments are preserved in the Keys Basin and possibly in the Solway Firth and Berw basins, and crop out onshore in Northern Ireland and in the Carlisle and Cheshire basins (Figures 2 and 16).

Evidence from the Irish Sea and surrounding areas, especially Northern Ireland, indicates that at least seven periods of post-Triassic movement have occurred (Figure 14; Jackson et al., 1987). These can be loosely identified as end-Early Jurassic (Toarcian), at the Jurassic–Cretaceous boundary (the late-Cimmerian unconformity of the North Sea), end-Early Cretaceous, end-Cretaceous ('Laramide'), post-Paleocene and pre-mid-Oligocene, Miocene ('Alpine'), and finally at the Pliocene–Pleistocene boundary. Of these, the Jurassic/Cretaceous, end-Cretaceous (Laramide), and mid- to late Tertiary Alpine events are believed to be the most important. It is probable that the faulting was dominantly of normal extensional type, and primarily along north-westerly and north–south alignments similar to those active during the

Figure 14 Summary of major stratigraphical units and geological events in and around the report area. Timescale after Snelling (1985). Only selected stage names are shown for some systems.

Era	Period		Epoch		Age	Age (Ma)	Lithostratigraphical unit	Major tectonic events	Other events
CENOZOIC	QUAT.		HOLOCENE			0.01			
			PLEISTOCENE			1.67		Uplift	Glacials/Interglacials (Ice Age) Erosion
	TERTIARY	NEOGENE	PLIOCENE			5.3			
			MIOCENE			23		Alpine orogeny	
		PALEOGENE	OLIGOCENE	L		27	Lough Neagh Clays on 20m of weathered basalt	? Uplift on St Tudwal's/Galloway Ridge	Weathering and erosion
				E		34			
			EOCENE	L		39			
				M		45		? Uplift on St Tudwal's/Galloway Ridge	**54Ma** opening of NE Atlantic/ Reyjkanes Ridge
				E		53			**56-51Ma** Mourne granites and late north-west trending dykes
			PALEOCENE	L		59	**61-58Ma** Antrim Lava Group		**58-56Ma** Slieve Gullion igneous complex and dykes north-east trending dykes
				E					**65-61Ma** Fleetwood Dyke Group
MESOZOIC	CRETACEOUS		LATE		Maastrichtian	65	Ulster White Limestone Formation (chalk)	? Inversion, north-south axis (Laramide)	Erosion
					Campanian	72			
					Santonian	83			Transgression
					Coniacian	86	Hibernian Greensand Formation		
					Turonian	88			
					Cenomanian	91			
			EARLY		Albian	96		Earth movements	Eustatic sea-level rise
					Aptian	107			
					Barremian	114			
					Hauterivian	116			
					Valanginian	120			
					Ryazanian	128		Earth movements	Cimmerian unconformity
	JURASSIC		LATE		Portlandian	135			
					Kimmeridgian	139			
					Oxfordian	144			
			MIDDLE		Callovian	152			
					Bathonian	159			
					Bajocian	170			
					Aalenian	176			
			EARLY		Toarcian	180		Earth movements minor	
					Pliensbachian	188			
					Sinemurian	195	Lias Group		
					Hettangian	201			
	TRIASSIC		LATE		Rhaetian	205	Penarth Group		Widespread transgression, eustatic sea-level rise
					Norian	210			
					Carnian	220	Mercia Mudstone Group		
			MIDDLE		Ladinian	230		Basin subsidence	
					Anisian	235			
			EARLY		Scythian	240	Sherwood Sandstone Group {Ormskirk S. Fm. / St Bees S. Fm.}		Röt transgression Hardegsen Disconformity
PALAEOZOIC	PERMIAN		LATE		Tatarian	250			
					Kazanian	255	BS1, BS2, BS3, BS4, depositional cycles		
			EARLY		Kungurian	260			Bakevellia transgression
					Artinskian		Upper Breccia division		
					Sakmarian		Sandstone division		
					Asselian	300	Lower Breccia division	Thermal doming, rifting in marginal basins	Widespread volcanism, weathering and erosion
	CARBONIFEROUS	SILESIAN	STEPH.			305		Tilting and erosion (Variscan orogeny)	
			WESTPH.		Westphalian D		Syndepositional redbeds		
					Bolsovian				Symon Unconformity
					Duckmantian		Coal Measures		
					Langsettian	315			
			NAMURIAN		Yeadonian				
					Marsdenian				
					Kinderscoutian		Yoredale, Millstone Grit and Basinal facies		
					Alportian				Intra-Namurian unconformity
					Chokierian				
					Arnsbergian				
					Pendleian	325		Basin sag	
		DINAN.					Bowland Shale Group Worston Shale Group Chatburn Limestone Group Basement Beds	Rifting in response to regional extension	Basaltic lavas Scarlett Volcanic Formation
	DEVON.		LATE			355	Red Arch Formation		Cockermouth Lavas, Birrenswark Lavas
			MIDDLE			375			
			EARLY		Emsian	390	Mell Fell Conglomerate, Peel Sandstone Group, Cross Slieve Group	Caledonian orogeny	**390** Emplacement of Shap and Skiddaw granites
	SILUR.						Windermere Group and equivalents	Caledonian orogeny	**420** Emplacement of Eskdale and Ennerdale plutons
	ORDOV.					438	Borrowdale Volcanic Group		
							Eycott Volcanic Group		
						510	Skiddaw Group and equivalents		
	CAMB.						? ? Arfon Group		
	PRE-CAMBRIAN					570	Monian Supergroup	Cadomian orogeny	

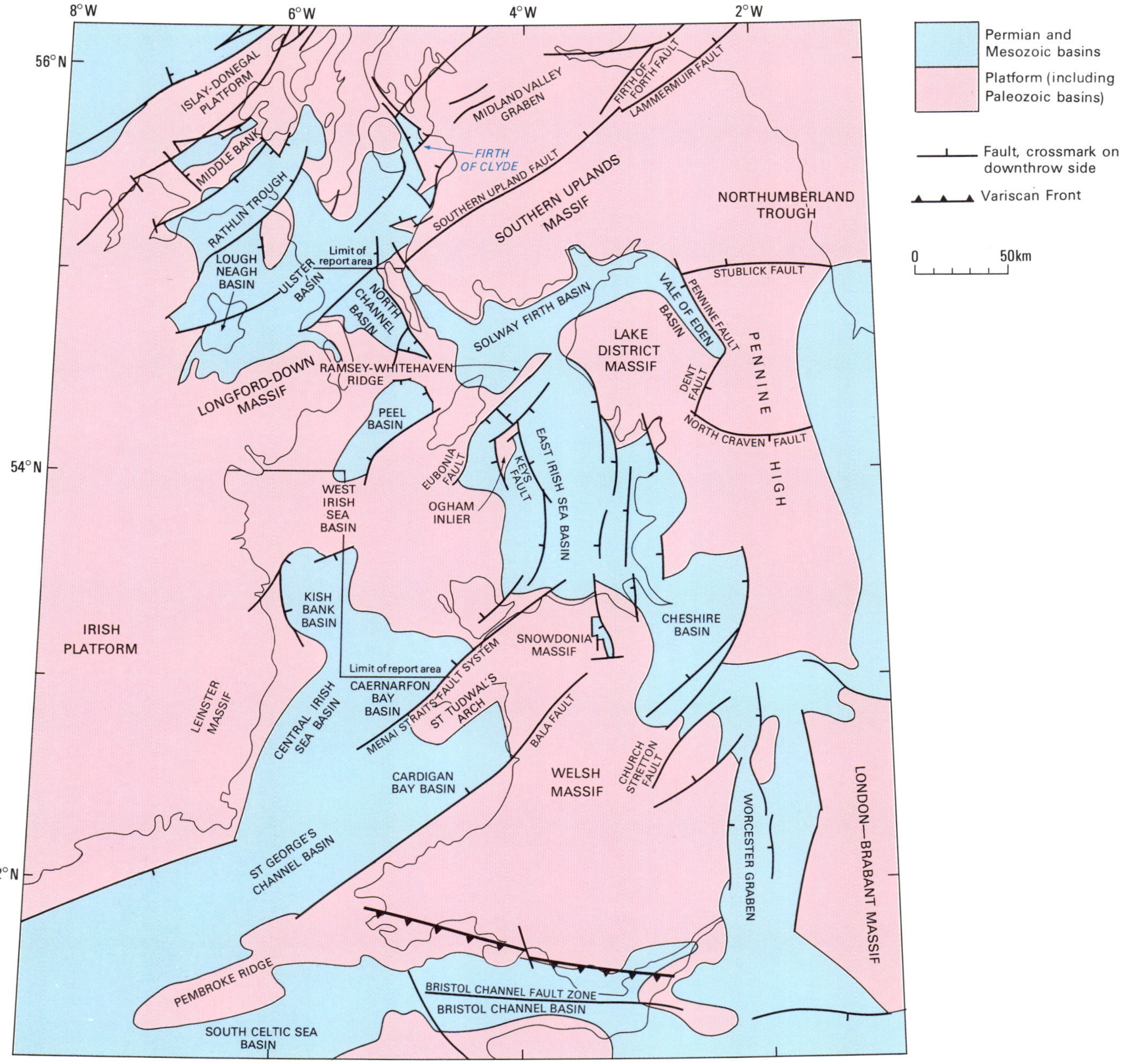

Figure 15 Simplified map of the main structural features in the Irish Sea and surrounding areas. Based on 1:1 000 000 Geology of the United Kingdom, Ireland and the adjacent Continental Shelf maps.

Permo-Triassic. There have probably been repeated posthumous movements on many major faults, including those of Caledonian origin. Seismic-reflection records also suggest possible reversed movement on some syndepositional faults, with the production of adjacent compressional anticlines.

There is evidence in the East Irish Sea Basin that a substantial thickness of the Mercia Mudstone Group has been removed by erosion (Colter, 1978), yet in basinal areas in Northern Ireland, the unconformity at the base of the Upper Cretaceous appears slight (Fletcher, 1977). This suggests that the major erosive episode may postdate the Late Cretaceous, and that late-Cimmerian (Late Jurassic to Aptian) movements may have been relatively slight in the Irish Sea. The major uplift and inversion of the East Irish Sea Basin may thus be Paleocene, rather than pre-Cretaceous and post-Early Jurassic in age, that is, Laramide rather than late Cimmerian (Jackson and Mulholland, 1993).

Around the time of the uplift, a major phase of igneous activity occurred over a wide area of what is now the northern Atlantic, as the north-east Atlantic opened (Morton and Parson, 1988). In Northern Ireland, extrusion of the Antrim Lava Group is thought to have spanned the period from 61 to 58 Ma (Mussett et al., 1988); all the lavas are reversely magnetised within a single magnetic epoch. A phase of central vent eruption followed, producing the shield volcanoes of Slieve Gullion and Carlingford (58 to 56 Ma); the Mourne Granites (56 to 51 Ma) were probably the last episode of this phase of igneous activity. Dyke swarms are the main evidence of Tertiary igneous activity offshore (Chapter 9); one dyke in the East Irish Sea Basin has been dated as between 65 and 61 Ma in age (Arter and Fagin, 1993).

A period of erosion followed this igneous activity, and the topmost part of the Upper Basalt Formation in Northern Ireland was removed before late Oligocene deposition in the Lough Neagh Basin (Wilkinson et al., 1980). Offshore, Paleogene sediments are mapped only in the Kish Bank, Central Irish Sea and St George's Channel basins, and no

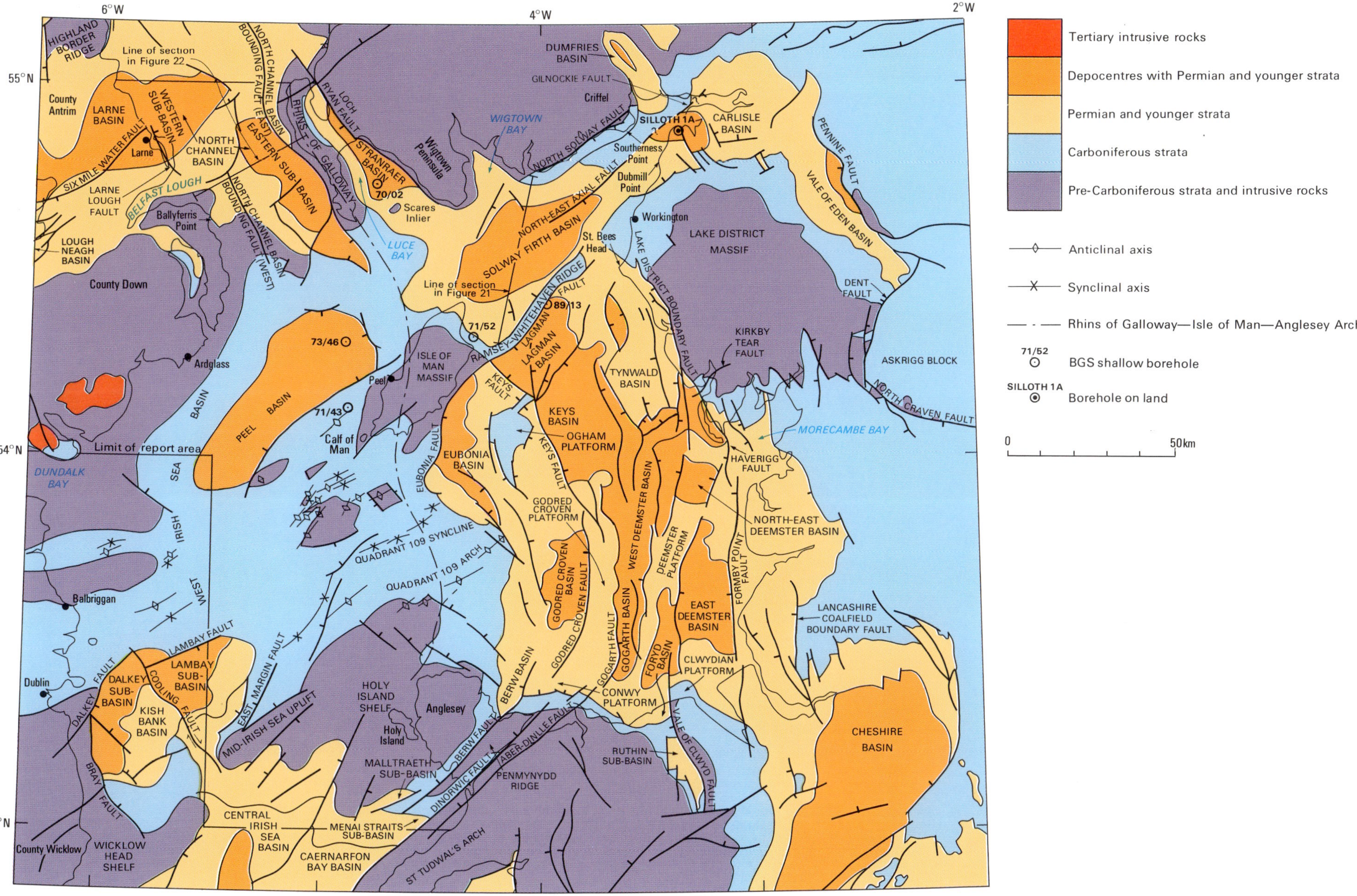

Figure 16 The main structural features of the Irish Sea region.

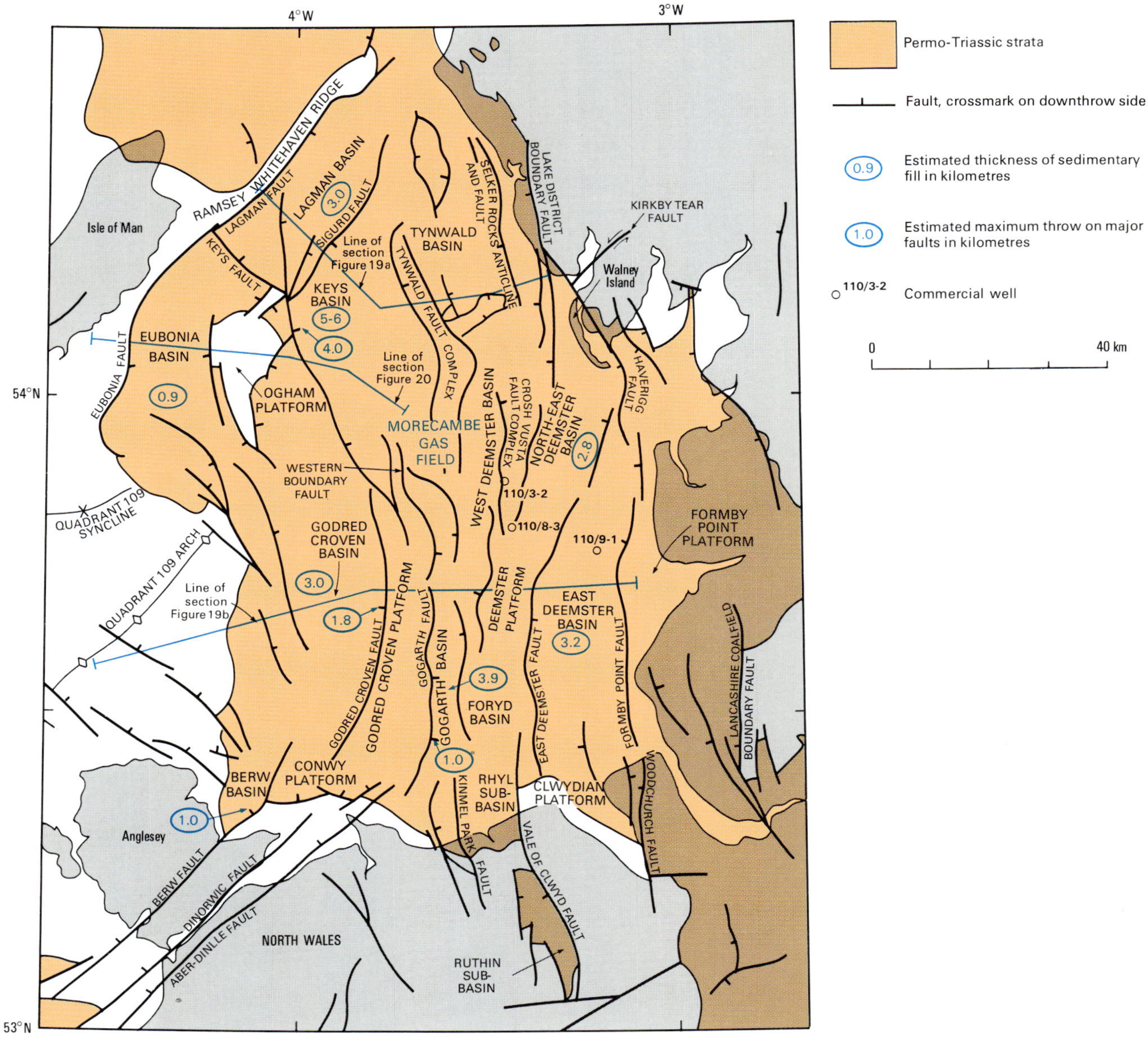

Figure 17 The main structural features of the East Irish Sea Basin.

Neogene sediments have been proved (BGS Solid Geology sheets).

The Irish Sea and northern England appear to be characterised by a series of broad, north–south-aligned, low amplitude, large wavelength warps of Tertiary age. These include the Pennine High and the Rhins of Galloway–Isle of Man–Anglesey Arch which passes southwards to the St Tudwal's Arch (Figure 16). Triassic rocks were probably once deposited over this latter ridge, although the successions in both the Kish Bank and East Irish Sea basins thin towards it. The ridge may have influenced sedimentation patterns from end-Carboniferous times onwards, on the evidence of Permian isopachs and palaeogeographical reconstructions. Enhanced periods of uplift occurred at the start of the Paleocene and during the Eocene, but the final uplift on this structure is post-Oligocene in age, judging by the separation of the Tertiary deposits of the Cardigan Bay and St George's Channel basins over the St Tudwal's Arch and its southerly extension (Figures 15 and 16).

Reactivation of Caledonoid faults occurred before, during, and after the late Oligocene in a north-east to south-west tensional regime. This is demonstrated by the faulted margins of the Central Irish Sea Basin, Kish Bank Basin (Lambay and Dalkey faults), and Lough Neagh Basin (e.g. Sixmilewater Fault). Dextral strike-slip movement of about 6 km, probably in Eocene and early Oligocene times, occurred along the north-westerly trending Codling Fault (Jenner 1981), and a similar sense of movement may also have taken place on the Keys Fault (Figures 15 and 16).

Further warping and normal faulting has resulted in the preservation of probably largely Oligocene continental-facies beds offshore in the Kish Bank, Central Irish Sea and Cardigan Bay basins (BGS Anglesey and Cardigan Bay Solid Geology sheets). Possibly there was an imposition of a southerly tilt to the entire area; this may have occurred in post-Pliocene times, partly connected with changes during the Pleistocene.

EAST IRISH SEA BASIN

The East Irish Sea Basin is one of the largest and deepest post-Carboniferous depocentres of western Britain. It is subdivided into several sub-basins and highs, and contains an estimated 5000 to 6000 m of Permian to Lower Jurassic strata, with Carboniferous rocks extending to a depth of nearly 10 km. The main sub-basins of the East Irish Sea Basin are shown in Figure 17, which also shows an estimate of the maximum thickness of the Permo-Triassic strata, and the maximum present-day throw of the major syndepositional faults. The component basins were formed by the development of grabens or half grabens, hence the maximum thickness of Permo-Triassic sediment is generally found adjacent to the bounding faults. Carboniferous strata occur widely in the East Irish Sea Basin; Westphalian strata are preserved at or near the sea bed to the west of the basin in the core of the Quadrant 109 Syncline (Figure 16). Elsewhere, the base of the Permo-Triassic overlaps Westphalian strata to rest on Namurian and Dinantian rocks.

The East Irish Sea Basin is associated with a relatively low Bouguer gravity anomaly (Figure 18), with smaller, deeper lows centred over individual sub-basins. The basin is bounded to the south and west by the erosional base of the Permo-Triassic, and to the north by the Lagman Fault (Jackson et al., 1987) which defines the southern limit of the Ramsey–Whitehaven Ridge (Figure 17). The Lake District Boundary and Haverigg faults are major syndepositional faults which mark, or guide, its north-eastern margin. In the south-east, the Formby Point Fault and its stepped onshore counterpart, the Boundary Fault of the Lancashire Coalfield, are major synthetic faults influencing the limit of the thick Permo-Triassic strata. Permo-Triassic rocks are absent from the Ogham Platform (Figures 2 and 17), where Carboniferous strata subcrop at the sea bed.

The basin can be divided into two distinct structural domains lying to the north and south of about 53° 50'N. In the northern domain (Figure 17), there are large westerly tilted fault blocks separated by four north-north-westerly trending fault complexes. Fault planes dip at relatively low angles; at depth the movement of some of the faults is taken up by a shallow plane of detachment in Upper Permian halites.

In the southern domain, Permian halites are absent, and the structure is characterised by parallel arrays of sequentially tilted, stepped, fault blocks. The bounding faults are normal, steeply dipping, and predominantly north–south trending. Two, irregular, broad upwarps cross the southern domain: the Conwy–Godred Croven Platform, and the Clwydian–Deemster Platform (Figure 17). These are separated by broad downwarps: the Berw–Godred Croven Basin, the Vale of Clwyd–Foryd–Gogarth–West Deemster Basin, and the East Deemster Basin. Northwards towards the centre of the East Irish Sea Basin, the highs diminish, and the basins expand in areal extent.

The contrast between the two domains is paralleled onshore by the north–south structures of the Lancashire Coalfield, and the north-westerly trending, post-Caledonian structures in the Lake District; these may reflect deep-seated control in the basement (Jackson and Mulholland, 1993). The line between the two domains has a north-easterly trend, and extends from the Quadrant 109 Arch north-eastwards into Morecambe Bay, and possibly as far as the Dent Fault which separates the Askrigg Block from the Lake District Massif (Figure 16).

Northern domain of the East Irish Sea Basin

East–west sections through the northern domain of the East Irish Sea Basin (Figures 19 and 20) show that in the west, the Eubonia Basin (named from a synonym of the Isle of Man) is an asymmetrical, shallow graben. It is bounded by north-easterly to northerly trending faults (Figure 17), but traversed by north-westerly aligned faults. It preserves a maximum thickness of 900 m of Permo-Triassic strata which thin eastwards within the basin, whereas the preserved thickness of underlying Carboniferous rocks thickens eastwards towards the Ogham Platform.

The Ogham Platform (Jackson et al., 1987), named from the Celtic alphabet, is a fault-bounded Carboniferous inlier (Figures 2, 16 and 20) in which subcropping Namurian and Westphalian rocks dip to the north-east. Carboniferous rocks are surrounded by Permo-Triassic sediments that dip radially away from the platform.

The Lagman Basin, named after an eleventh century Lord of Man, lies to the north of the Ogham Platform and the Keys Basin (Figure 17); it is a half-graben bounded to the north-west by the Lagman Fault, a major listric fault which defines the south-eastern margin of the Ramsey–Whitehaven Ridge for much of its length. The major fault movements are post-Carboniferous in age, but the structure may have been initiated during the Carboniferous. The basin contains a marked, north–south, symmetrical syncline. In the south-west, the axes of a number of subsidiary north–south synclines are repeatedly displaced by north–south faults.

The Keys Basin (Jackson et al., 1987) is a major half-graben, active during Permo-Triassic times, that contains an estimated 5000 to 6000 m of Permo-Triassic and Lower Jurassic strata (Figure 17). Up to 4500 m of Carboniferous strata lie beneath the Permian. In the west it is bounded by the Keys Fault (Smith, 1985), a syndepositional fault towards which, in the south, there is a significant westward thickening of the Triassic Sherwood Sandstone Group. The basin is bounded to the north by the north-easterly trending Sigurd Fault (also named after an eleventh century Lord of Man). To the south, a set of north–south-trending antithetic faults include the Western Boundary Fault (see Figure 79) of the Morecambe Gasfield (Ebbern, 1981).

Halokinetic structures are well developed in the Mercia Mudstone Group of the Keys Basin, where both the thickness of individual halites, and the depth of burial, are close to the maximum for the East Irish Sea Basin. Salt pillows and salt diapirs commonly have their source in the Rossall Halite (Jackson et al., 1987), the earliest Triassic halite. Younger salts, especially the Preesall Halite, provide detachment horizons for low-angle listric faults.

The Tynwald Basin is structurally similar to the Keys Basin. Both the Carboniferous and Triassic sections thin eastwards (Figure 19), continuing the trend of the Keys Basin. Towards the Cumbrian coast, the Westphalian and Namurian are lost as a result of pre-Permian erosion (Smith, 1985). The Tynwald Basin is separated from the Keys Basin by the Tynwald Fault Complex, which has a north-north-westerly trend, and denotes a change in the seismic character and thickness of the Sherwood Sandstone Group. The Tynwald Fault Complex is a high-level graben which dies out southwards (see Figure 79) within the Morecambe South Gasfield (Ebbern, 1981). At top-Carboniferous level, the overall throw across the fault complex is small and down to the east (Figure 19), whereas for higher reflectors the overall throw is westwards in the north and eastwards in the south, implying some component of strike-slip or scissor movement. However, the limited north–south extent of the fault complex, and the consistent pattern of the Permian facies and isopach values adjacent to it (see Chapter 6), preclude any major strike-slip movement.

The Sherwood Sandstone Group is absent from the Tynwald Fault Complex, which is interpreted as the root zone of a postdepositional collapse graben. Together with the Crosh Vusta Fault Complex, it shares many characteristics with similar large features in the southern North Sea, in

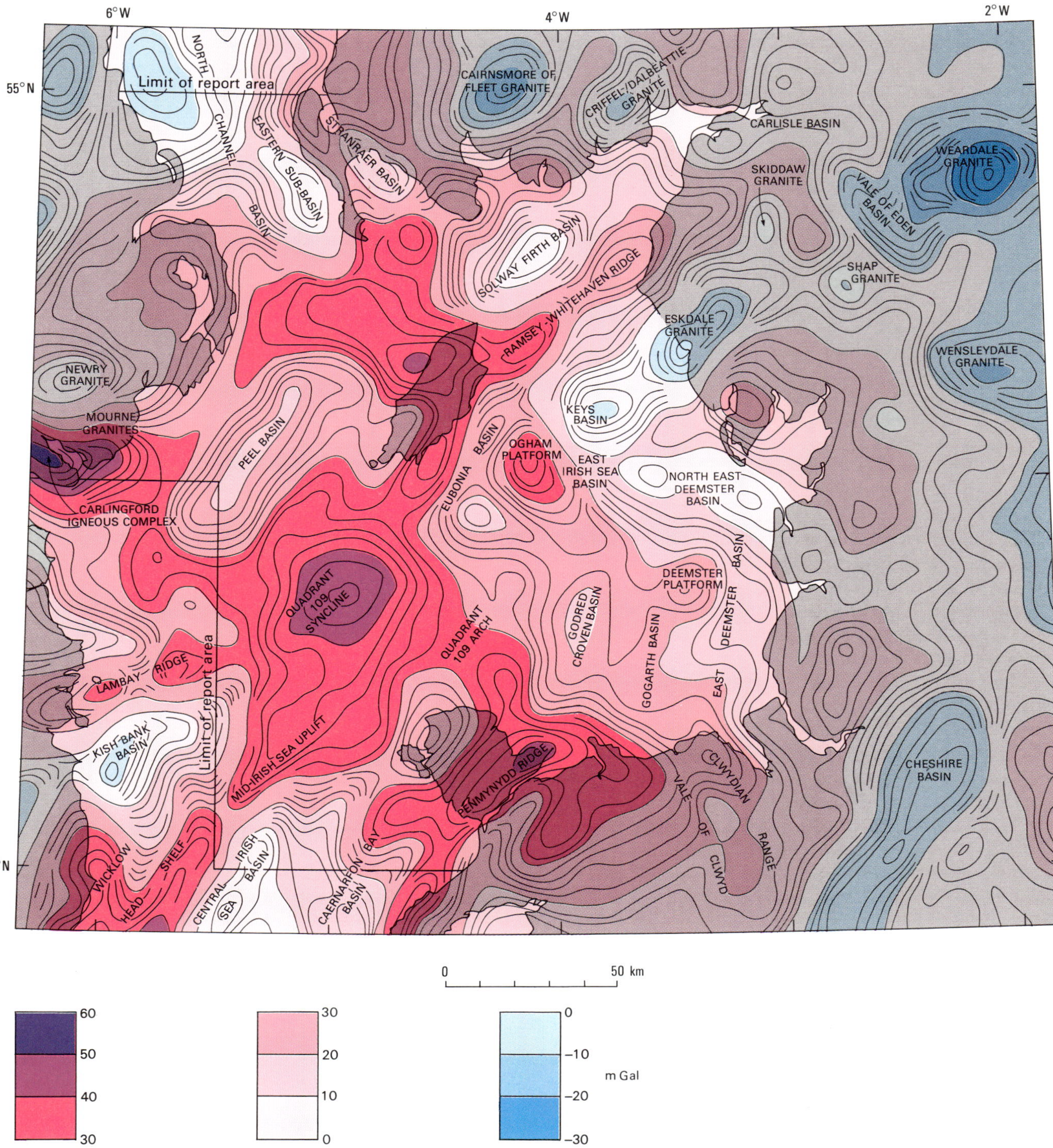

Figure 18 Bouguer gravity anomaly map of the Irish Sea and adjacent areas.

which the Bacton Group (the correlative of the Sherwood Sandstone Group) is absent (Jenyon, 1985; Walker and Cooper, 1987).

In the East Irish Sea Basin, such fault complexes are confined to those areas where the Rossall Halite (lower leaf) and halite of the St Bees Evaporite (Upper Permian) provide stacked detachments. Both broad and tight anticlines, with axes subparallel to the fault margins, are a feature of the Mercia Mudstone Group, and may reflect minor Tertiary inversion and halokinetic modification developed above a Rossall Halite *décollement*.

The Selker Rocks Anticline is a densely faulted, southward-plunging fold; it is poorly developed to the south of a north-easterly trending dislocation (Figure 17; Jackson et al., 1987) which runs from the Tynwald Fault Complex and continues onshore as the Kirkby Tear Fault (Rose and Dunham, 1977). Minor step faults are imaged, with keystone arches over the fold crest. Reflector packages, especially the Sherwood Sandstone Group, thicken eastwards across the anticline towards the Lake District Boundary Fault (Figures 17 and 19), which is the major, synsedimentary bounding fault at the eastern margin of the East Irish Sea Basin (Figure 17). However, within each fault block, there is minor westward thickening (Jackson et al., 1987), suggesting contemporaneous movement on antithetic faults.

To the south of the Selker Rocks Anticline, the Crosh Vusta Fault Complex (Figure 17) is defined by a pair of north–south-trending, listric faults which meet at depth in a prominent cusp; the eastern fault of this pair was penetrated in well 110/8-3. A prominent anticline, developed only in the

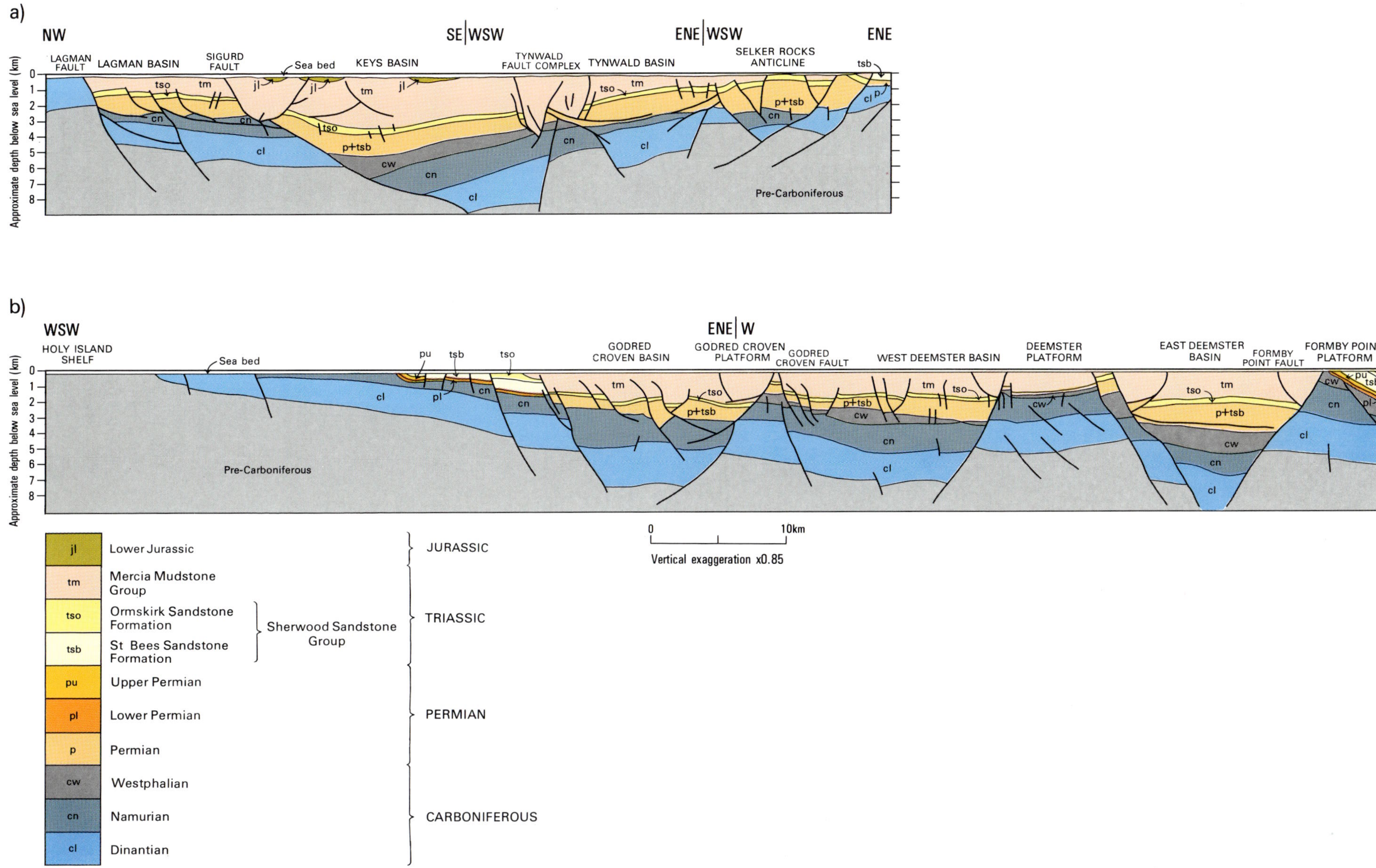

Figure 19 Cross-sections across the East Irish Sea Basin. For locations see Figure 17.

Mercia Mudstone Group, runs axially along the southern part of the fault complex, and was penetrated on the fold axis by well 110/3-2 (Figure 17). In the north, the fault complex is terminated by a poorly defined, west-north-westerly trending cross fault (not shown in Figure 17) with numerous minor splinter faults; these are north-dipping listric faults which sole out, partly along a *décollement* in Permian halites, and partly near the top of the Carboniferous.

The North-East Deemster Basin lies to the east of the Crosh Vusta Fault Complex (Figure 17). It is a fault-bounded centroclinal fold basin, except in the north where reflectors dip consistently southwards from Walney Island.

Southern domain of the East Irish Sea Basin

In the southern domain of East Irish Sea Basin (Figures 17 and 19), the Permo-Triassic strata show growth thickening eastwards across the basin from a present-day erosional feather-edge in the west. The thickness of Carboniferous is more uniform than to the north; the Permo-Triassic rests largely on Namurian strata or on Dinantian strata in the extreme south, with Westphalian preserved mainly in synclines in the West and East Deemster basins and on the Deemster Platform.

In the west, the Godred Croven Basin and the smaller Berw Basin are controlled on their eastern margins by major syndepositional faults that trend north-east or north-north-east (Figures 17 and 19). In the north, the Godred Croven Fault downthrows to the west, and splays near its termination in the south; it may be a major *en-échelon* continuation of the Berw Fault on Anglesey. To the north, the Godred Croven Fault dies out towards the Keys Fault. In the western part of the Godred Croven Basin, a well-developed suite of antithetic faults trends obliquely north-west to south-east, and downthrows to the east, creating a series of small grabens.

In the Berw Basin there may be local inversion near the Berw Fault, reflecting Tertiary uplift. The western depositional limit of the Lower Permian Collyhurst Sandstone appears to have been within the Berw Basin, where it is overstepped westwards by the Upper Permian Manchester Marls (Jackson and Mulholland, 1993).

The Permo-Triassic basin becomes progressively more shallow to the north towards the Carboniferous Ogham Platform. The beds dip eastwards and south-eastwards from the erosional feather edge, and show thickening towards the major eastern bounding faults (Figure 19). Towards the north of the Godred Croven Basin, however, the regional dip swings to the south-east adjacent to a major anticline in Carboniferous strata, termed the Quadrant 109 Arch, which extends north-eastwards beneath the Permo-Triassic (Figure 16).

The Godred Croven Platform is an ill-defined, easterly tilted fault block in which the Permo-Triassic units are some 300 m shallower on average than in the surrounding area (Figures 17 and 19; Jackson et al., 1987). It is characterised by numerous, minor, north-trending step-faults that downthrown to the east.

In the Gogarth Basin (named after Pen-y-Gogarth or Great Ormes Head — see front cover), the Permo-Triassic sediments are the thickest in the southern domain of the East Irish Sea Basin, and show internal westwards expansion towards the Gogarth Fault; there is an overall westerly tilt, with additional minor westerly thickening within each individual fault block. The Lower Permian succession appears to reach a maximum thickness of 1220 m adjacent to the Gogarth Fault, and the Sherwood Sandstone Group also thickens westwards. A marked change in dip occurs across the Gogarth Fault, and a hanging-wall syncline is developed above this basement-involved, listric fault, which may be the offshore continuation of the Aber-Dinlle wrench fault. It appears to have been particularly active during deposition of the Sherwood Sandstone Group, rather than the Collyhurst Sandstone.

The West Deemster Basin (Figures 17 and 19) is an elongate north–south-aligned basin with a north-north-westerly oriented syncline oblique to the bounding faults. There is marked eastward thickening of both Permian strata and the Sherwood Sandstone Group towards the eastern bounding fault. Pronounced rollover with steep dips is seen adjacent to the Tynwald Fault Complex in the north-west.

The Deemster Platform (Figures 17 and 19) is an irregular horst over which the Mercia Mudstone Group has been removed by erosion. It possesses an overall northward tilt in the Sherwood Sandstone Group, and unlike the other inter-basinal platforms of the East Irish Sea Basin, Upper Permian reflectors can be identified clearly.

The East Deemster Basin (Figure 17) forms the south-eastern portion of the East Irish Sea Basin, where it passes from a highly faulted graben in the south, to an unfaulted half-graben in the north, reflecting the influence of Permian halite distribution. At top-Ormskirk Sandstone Formation level, the tilt is slightly westward, but at top-Carboniferous level the tilt is to the east (Figure 19). These characteristics may be due to a combination of local Permian salt withdrawal and syndepositional, Triassic thickening in directions opposed to Permian depositional thickening. There is synsedimentary thickening of the Permian and the Sherwood Sandstone Group eastwards towards the Formby Point Fault (Jackson et al., 1987). The Sherwood Sandstone Group also thickens westwards across the Formby Point Fault from about 800 m in the Formby boreholes on the Formby Point Platform (Falcon and Kent, 1960), to 1286 m in well 110/9-1.

The East Deemster Fault downthrows to the east, and dies out southwards where the movement has been taken up on parallel, *en-échelon* and relay faults to the east and west. Intra-basinal faulting in the East Deemster Basin is dominantly north–south oriented, whereas an east–west alignment appears to have been imparted on later warping. In the south, minor faults splay outwards radially from the Clwydian Platform.

The Formby Point Fault exerts a considerable influence on the basin-margin geology. It is replaced to the north by the *en-échelon* Haverigg–Lake District Boundary Fault, and in the south by a series of smaller *en-échelon* faults. During deposition of the Mercia Mudstone Group, the fault probably separated a halite-dominated depositional area to the west from a zone of mudstone deposition to the east (Wilson, 1990; Jackson and Mulholland, 1993). The Formby Point Platform to the east (Figure 19) is a north–south-faulted pericline.

The North Wales coastal shelf, from which strata dip northwards, forms the southern limit of the East Irish Sea Basin (Figure 17). It is divided into two, broad, faulted anticlines, termed the Conwy and Clwydian platforms, that are separated by the Rhyl–Foryd Basin. The fold amplitudes of the Clwydian Platform and Foryd Basin die out northwards as the effects of the post-Jurassic uplift of North Wales declines. The amplitude of each anticline also decreases upwards through the sequence from the top-Carboniferous reflector to the top-Ormskirk Sandstone Formation reflector.

The Conwy Platform (Figure 17) is an asymmetrical anticline with a steeper eastern limb, a central keystone block, and a set of step-faults dipping westwards towards the main

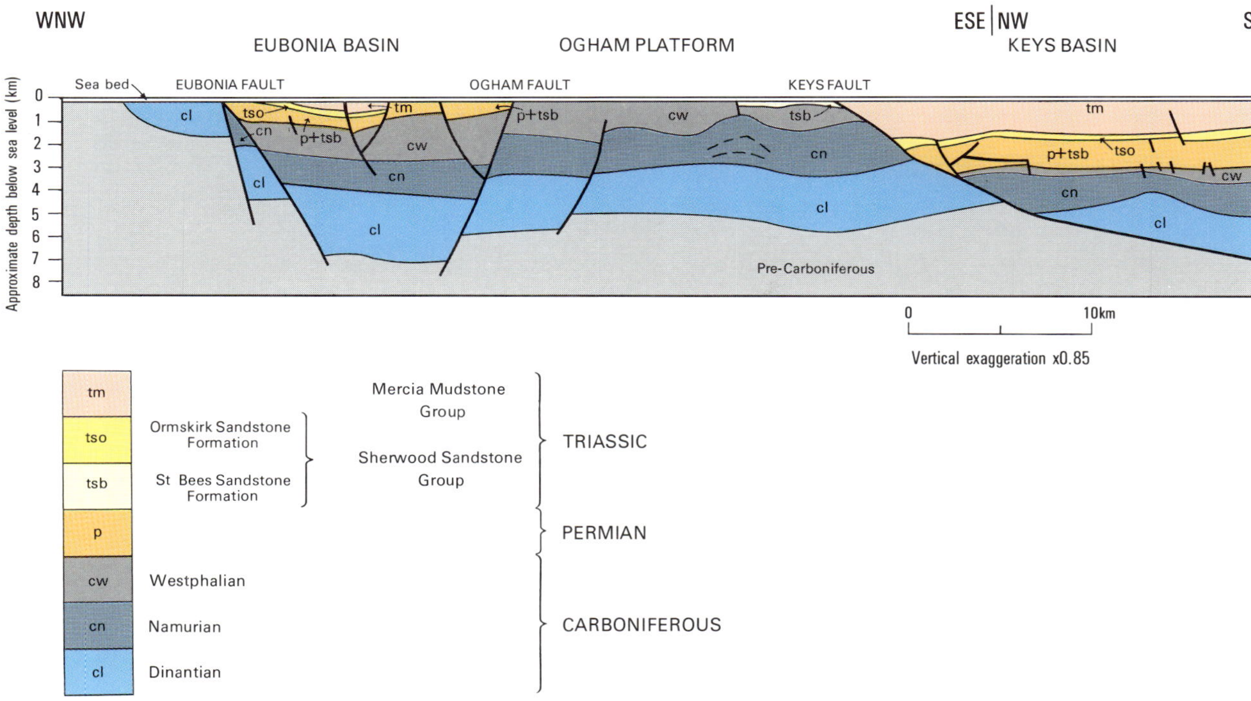

Figure 20 Cross-section through the Eubonia Basin, Ogham Platform and Keys Basin. For location see Figure 17.

boundary fault. The western boundary fault of the Conwy Platform appears to be transverse to the central anticlinal axis, and the fault zone becomes broader to the north. On the eastern limb, the faults have a greater hade; they trend to the north-east in the south, but swing to a more northerly trend in the north. These faults may be northward splays from the Dinorwic Fault, which runs along the Menai Straits, but direct continuity cannot be demonstrated. On the western limb, the faults are north-easterly trending throughout. The Conwy Platform may be the result of inversion of a depocentre, and reversal of movement on the controlling fault.

The Berw, Dinorwic, and Aber-Dinlle faults (Figure 17) are probably normal, down-to-the-north faults with a strong Palaeozoic wrench component. They have possibly had several phases of later movement, although their influence waned considerably during Permo-Triassic times. The highly faulted nature of the area to the north-east of Anglesey poses difficulties in tracing these faults into the East Irish Sea Basin. Their continuation is coincident, or near coincident, with the trace of the Godred Croven and Gogarth normal faults, which either originate from them, or are *en-échelon* continuations.

The Foryd Basin dies out northwards towards the Deemster Platform, as individual tilt blocks become subhorizontal and less clearly differentiated. To the south, it extends onshore as the Rhyl Sub-Basin (Powell, 1956; Wilson, 1959; Collar, 1974), which, together with the Ruthin Sub-Basin, is interpreted as an *en-échelon* sag basin in Lower Permian deposits within the larger Vale of Clwyd half-graben (Figure 17). The Vale of Clwyd Fault to the east is the major synthetic fault, with a suite of lesser antithetic step-faults generally dipping uniformly to the east (Warren et al., 1984). In the south, the Kinmel Park Fault defines the western side of the Foryd Basin, marking the central sag axis of Collyhurst Sandstone deposition.

The anticlinal crest of the Clwydian Platform is marked by a small central horst. A series of small horsts occurs to the west of the axis, and to the east there is an array of step-faults downthrowing dominantly to the west, with individual fault blocks tilted to the east. The progressive reduction in number of faults northwards may be due to the increasing thickness of Permo-Triassic cover above a brittle pre-Carboniferous basement, and to the ability of the Mercia Mudstone Group to absorb minor fault movements.

Slight eastward sedimentary thickening towards the Vale of Clwyd Fault is apparent in preserved Westphalian strata. However, in the Permian Collyhurst Sandstone the trend is reversed, with basinal sagging into the Foryd Basin indicating the influence of Variscan uplift of the Clwydian Platform, which has remained a high since that time.

RAMSEY–WHITEHAVEN RIDGE

The Ramsey–Whitehaven Ridge separates the Solway Firth and East Irish Sea basins (Bott, 1964); it is a long-lived Caledonoid-trending feature marked by relatively high Bouguer gravity anomaly values (Figures 16 and 18). Seismic-reflection data indicate horizontal or gently northerly dipping Permo-Triassic rocks to the north of the ridge, partly overlapping a highly faulted, northward-dipping and relatively thin Carboniferous succession above Lower Palaeozoic strata. Much of the south-eastern margin of the ridge is marked by the Lagman Fault. The appearance of the Ramsey–Whitehaven Ridge is thus of a northerly tilted escarpment with a steep, south-facing scarp. On the northern flank of the ridge, the Permo-Triassic is markedly unconformable on the Carboniferous, and may rest on Dinantian rocks, as encountered in BGS boreholes 71/52 and 89/13 (Figure 16 and see Figure 26).

The ridge appears to have acted as a hinge throughout most of Carboniferous times, when it was either exposed, for example during Namurian times, or only lightly buried. Both Permian and the Lower Triassic sediments thin towards the ridge, and post-Triassic uplift and erosion has been more severe over it than in the adjacent basins. Only thin Permian

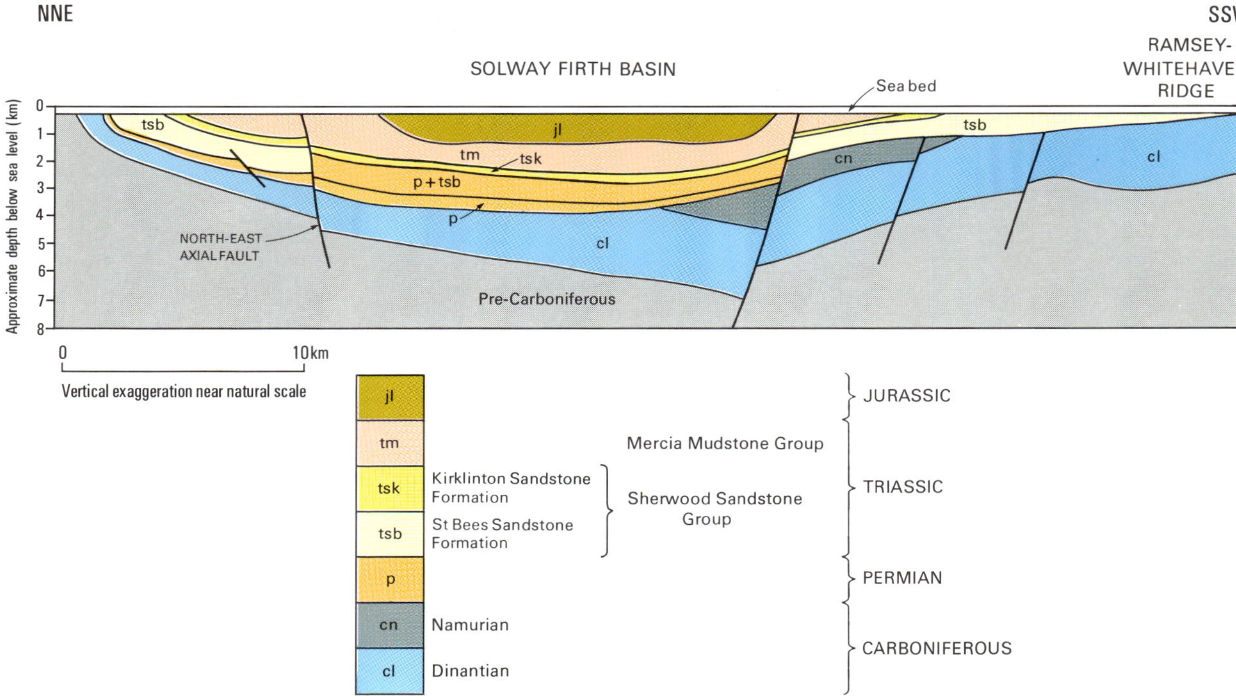

Figure 21 Cross-section through the Solway Firth Basin. For location see Figure 16.

strata and the lower beds of the Sherwood Sandstone Group are now preserved on the north-eastern part of the ridge.

SOLWAY FIRTH BASIN

The Solway Firth Basin is a relatively unfaulted, symmetrical, extensional sag which lies at the south-western end of the Northumberland Trough (Figure 15); its Permo-Triassic fill is only partly linked to that of the East Irish Sea Basin over the Ramsey–Whitehaven Ridge. To the west-south-west, the margin is the erosional base of the Permo-Triassic against the Rhins of Galloway–Isle of Man Arch. The basin axis shows a set of right-stepping, *en-échelon,* axial synclines. The Solway Firth Basin is a discrete structural basin; its separation from the Carlisle Basin is marked by a north-north-westerly trending ridge on the Bouguer gravity anomaly contour map. This feature runs from the Westphalian outcrop at Workington to the Dinantian of the Southerness–Kirkbean area (Figures 16 and 18). It is postulated that the Gilnockie Fault of the Canonbie Coalfield (Picken, 1988) separates thicker Lower Permian deposits of the Dumfries Basin from thinner developments on the Cumbrian coast and extends offshore to join the North-East Axial Fault.

A conformable succession of Permian and Mesozoic deposits is found throughout the basin, reaching an estimated maximum thickness of over 3000 m. It rests unconformably on Carboniferous strata, overstepping Westphalian and Namurian to rest mainly on Dinantian strata both in the north-west (Figure 21) and on the crest of the Ramsey–Whitehaven Ridge. Although there is limited thinning of Permo-Triassic deposits towards the present-day margins, the Solway Firth Basin is evidently an erosional remnant of a larger basin whose bounding Permo-Triassic faults are not seen, but whose axis of subsidence is assumed to be coincident with the preservational axis.

The northern margin of the basin is marked, in part, by the North Solway Fault, which was active during the Carboniferous and probably in later times, although contacts with younger rocks are not preserved. The mapped trace of the North Solway Fault (Deegan, 1973; Ord et al., 1988) dies out along strike both to the east-north-east and west-south-west of the Criffel Granodiorite; the growth of this important depositional fault is intimately related to the buoyancy effects of the granite. The basin may owe its preserved shape not so much to the Permo-Triassic palaeogeography, as to post-Triassic reactivation of Caledonian crustal contrasts between the granite-cored and isostatically buoyant masses of the Lake District and Southern Uplands, and the intervening ground (Bott, 1968).

Faults are less common than elsewhere in the Permo-Triassic sequences of the Irish Sea, and there is little evidence of growth or syndepositional faulting. However, in the main basin, the Sherwood Sandstone Group appears to be thicker on the downthrow side of the North-East Axial Fault (Figure 21), although this may be partly due to halokinetic effects. The North-East Axial Fault passes into a fault complex to the north-east, where a collapse zone is developed in the Mercia Mudstone Group; this is similar to the Tynwald and Crosh Vusta fault complexes of the East Irish Sea Basin. Faults bounding small, tilted fault blocks in the Sherwood Sandstone Group are seen to sole out in a detachment at the base of assumed Upper Permian halites, above a prominent reflector thought to be that of the Upper Permian basal carbonate or anhydrite.

Offshore, from Workington to St Bees Head (Figure 16), high-angle thrust faults directed outwards from the Lake District, and subparallel with the coast, occur within Carboniferous strata. Thrusting may also have taken place in the collapse zone in the Mercia Mudstone Group on the North-East Axial Fault, with movement directed towards the Lake District.

The generally unfaulted nature of the basin, and its similarity in structural behaviour to the northern part of the East Irish Sea Basin, suggests that Permian halite is present locally at depth offshore. Halokinetic effects, low-angle faults in the Mercia Mudstone Group, and dip discordance across the detachment between the Triassic Kirklinton Sandstone

Formation and the Mercia Mudstone Group, are however all less marked than in the East Irish Sea Basin. Although this may be due in part to less-deep burial, these features, along with the evidence from the Silloth 1A borehole (Figure 16), suggest that the proportion of Triassic halite is much less than that in the East Irish Sea Basin.

An ubiquitous disconformity is seen on seismic records in the highest strata near the main basin axis; the relationship is similar to that seen within the youngest deposits of the Keys Basin (Jackson et al., 1987), where Lower Jurassic strata overlie the disconformity.

STRANRAER BASIN

The Stranraer Basin, a satellite basin of the Solway Firth Basin, is a classic half-graben tilted north-eastwards towards the Loch Ryan Fault (Figure 16). Onshore, the throw of the Loch Ryan Fault is estimated at 1525 to 1700 m (Kelling and Welsh, 1970). The fault has governed the depositional history of the basin throughout Triassic, Permian and possibly Carboniferous times, and the preservational history of the basin in post-Triassic times; it appears to have utilised an existing, first-order, dextral, wrench fracture in the Caledonian basement (Kelling and Welsh, 1970).

Sandstones and breccio-conglomerate are postulated to comprise most of the basin fill, and the base of the Permo-Triassic is estimated to reach about 1350 m depth beneath the north shore of Luce Bay (Mansfield and Kennett, 1963). Carboniferous rocks (Fuller, 1958) appear to be relatively thin throughout the basin, and are absent from the Scares Inlier (Figure 16; Barnes, 1989). Lower Permian breccias crop out onshore in the west (Brookfield, 1978; 1989), and interpretation of seismic profiles implies that the Sherwood Sandstone Group, rather than thick Lower Permian sandstones, crop out beneath Pleistocene deposits over large parts of Luce Bay. The red-brown sandstone drilled in BGS borehole 70/02 (Figure 16) can thus probably be assigned to the Sherwood Sandstone Group. A small hanging-wall outlier of Mercia Mudstone Group may be preserved adjacent to the Loch Ryan Fault offshore.

NORTH CHANNEL BASIN

The Permo-Triassic of the North Channel Basin is a southward continuation (Figures 15 and 16) of the Rathlin Trough and the basins of the Firth of Clyde (McLean and Deegan, 1978; Fyfe et al., 1993), and which deepens towards the south-east. The structural trend of the basin is perpendicular to that of the Lower Palaeozoic basement (McLean, 1978), and folding is largely restricted to footwall anticlines and hanging-wall synclines. The major stratigraphical units of the Permo-Triassic (Wilson, 1981; Griffith, 1983) increase in thickness offshore from County Antrim towards the North Channel Basin Bounding Fault (East), to a maximum estimated thickness of 1800 m above almost 1000 m of Carboniferous (Figures 18 and 22).

The North Channel Basin is subdivided, possibly along an anticlinal hinge, into eastern and western sub-basins by mid-channel faults which trend approximately north-north-westwards. Each sub-basin is marked by a pronounced, low, gravity anomaly (Figure 18). In the internally faulted western sub-basin, strata on average dip gently westwards towards the contiguous Larne Basin (Bennett, 1983). The eastern sub-basin, a half-graben with few internal faults, is tilted steeply eastwards towards the controlling North Channel Basin Bounding Fault (East), with concomitant thickening of the Permo-Triassic strata in that direction.

Belfast Lough is the topographic expression of a half-graben trending east-north-east that is a satellite to the North Channel Basin. The half-graben shows substantial thickening to the north-east and north-west, chiefly due to the increase in halite thickness within both the Permian succession and the Mercia Mudstone Group (Griffith and Wilson, 1982). It is traversed by an array of north-north-westerly trending cross-faults, including the Larne Lough Fault which downthrows to the east and is associated with minor synsedimentary thickening.

Cretaceous and Jurassic rocks are very largely absent offshore, despite the north-easterly regional dip; this may be due to the absence of a protective covering of Tertiary basalts. However, it could also be attributed to the dramatic eastward increase in thickness of Permo-Triassic sediments, possibly due to thick halite development, such that the base of the Jurassic to Cretaceous sequence projected eastwards would lie above sea level (Griffith and Wilson, 1982).

WEST IRISH SEA BASIN

The term West Irish Sea Basin is introduced here to refer to the area of largely post-Silurian rocks which lies south of the North Channel, west of the Rhins of Galloway–Isle of Man–Anglesey Arch, and north of the Kish Bank and Central Irish Sea basins (Figure 16). The West Irish Sea Basin consists largely of Carboniferous strata cropping out at or near the sea bed (Figure 2). The basin includes a series of anticlines and Lower Palaeozoic inliers running south-westwards from the Calf of Man towards the Balbriggan Inlier of the Republic of Ireland. The inliers are a continuation of the line of the Ramsey–Whitehaven Ridge, and are marked by a gravity anomaly high (Figure 18), which was attributed by Bott (1964) to basement rocks at or near the surface.

In the northern part of the West Irish Sea Basin lies a major sub-basin that is termed the Peel Basin (Bott, 1964, 1965; Bott and Young, 1971). It trends north-east to south-west, continuing the line of the Solway Firth Basin, and is marked by both a gravity anomaly low and a magnetic low (Figures 18 and 23; Bott, 1965; Bacon and McQuillin, 1972). From gravity modelling, it has been estimated that the sedimentary basin is up to 2.5 km deep (BGS Isle of Man Solid Geology sheet); this interpretation assumed the infill to be of Carboniferous age. However, the geology is not fully resolved, in part because of sparse seismic data. A post-Namurian fill on the basin margin is indicated by the evidence of BGS borehole 71/43 (Figure 16 and see Figure 28), and numerous high-amplitude seismic reflectors have been interpreted as Westphalian coals. Recent reconnaissance BGS seismic-interpretation suggests that the youngest fill comprises a full, but thin, Permo-Triassic succession, gently warped within a shallow graben. The red mudstone smear and cuttings in BGS borehole 73/46 (Parkin and Crosby, 1982) are regarded here as evidence for the Mercia Mudstone Group near the centre of the basin (Figure 2).

South of the Peel Basin and north-east of the Kish Bank Basin, a series of highly faulted synclines and anticlines, with dips varying between 4° and 35°, show a strong Caledonian trend (Figure 16; BGS Anglesey Solid Geology sheet). The boat-shaped Lower Palaeozoic inliers (Figure 2) depicted by Wright et al. (1971) appear in some cases to be Lower Palaeozoic horsts, either exposed or lightly buried by Dinantian strata. However, the possibility that the inliers may comprise highly folded Dinantian of basinal facies cannot be

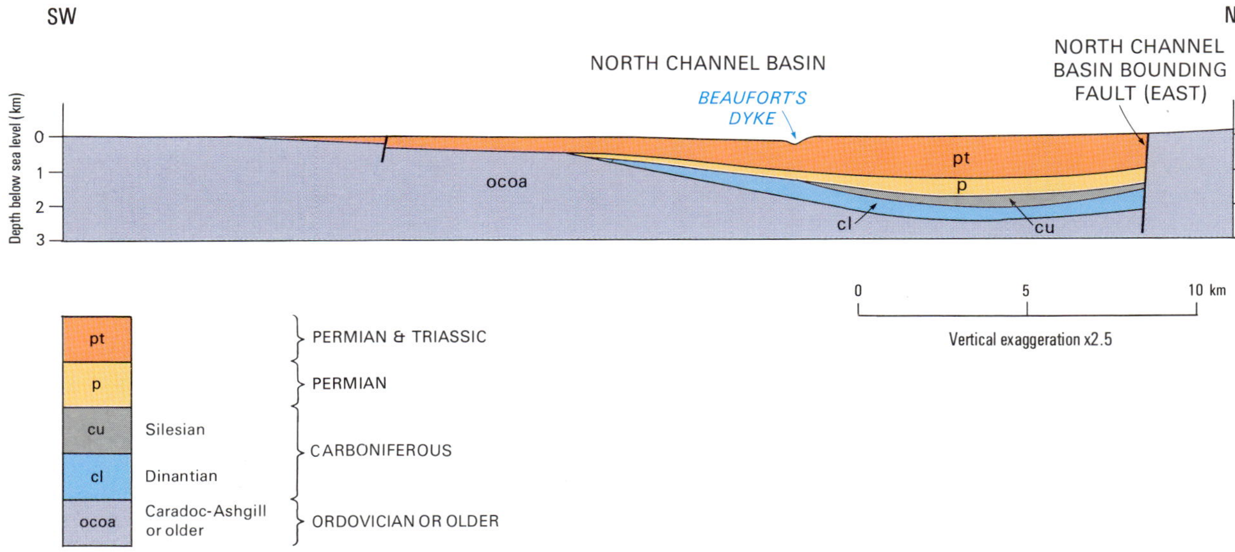

Figure 22 Cross-section through the North Channel Basin. For location see Figure 16. Taken from BGS Isle of Man Solid Geology sheet.

ruled out, for there is no evidence of an unconformity between the rocks in the cores of these anticlines and adjacent strata. Nevertheless, a conformable relationship between Carboniferous and Lower Palaeozoic rocks is at variance with the structural relationships seen elsewhere in the Irish Sea or adjacent onshore areas (Sevastopulo, 1981a, c; Moseley, 1972).

In the south-eastern part of the West Irish Sea Basin, the Quadrant 109 Syncline (Figure 16) is a major, broad, open syncline or centroclinal fold. Westphalian rocks have been proved by BGS drilling on the southern limb at boreholes 70/01, 71/88, 73/68 and 75/06 (Figure 1). Namurian rocks crop out beneath Quaternary on both the southern and northern limbs. In the complementary Quadrant 109 Arch to the south-east, a complex pattern of smaller-scale, north-easterly trending anticlines and synclines (BGS Anglesey Solid Geology sheet) are displaced by cross faults which may indicate wrench movement (Warren et al., 1984).

KISH BANK BASIN

The Kish Bank Basin, which largely lies outside the report area, is a half-graben in which Tertiary strata rest unconformably on Carboniferous, Liassic and Triassic rocks. The half-graben is tilted to the north-west (Figure 16), where it is bounded by the Dalkey and Lambay faults (Jenner, 1981). It is hinged on the south-east along the Wicklow Head Shelf and Mid-Irish Sea Uplift (Dobson and Whittington, 1979), although the impersistent East Margin Fault hints at the development of a local graben. In the south-west, the basin is bounded by the Bray Fault, which has a throw of up to 4000 m. A low gravity anomaly is centred on the western part of the basin (Figure 18), and throws of 3000 to 5000 m are estimated on the bounding faults. The Dalkey–Lambay Fault may have been a Carboniferous syndepositional fault (Smith, 1985) along a similar trend to that of the Eubonia and Lagman faults in the East Irish Sea Basin. A small, compactional, structural-sag syncline within the Mercia Mudstone Group occurs on the downthrow side of the Dalkey Fault (Jenner, 1981).

The Kish Bank Basin is split into the Dalkey and Lambay sub-basins by the north-west-trending Codling Fault, a complex that is 1 km wide and interpreted as a dextral strike-slip fault (Jenner, 1981). An Eocene or early Oligocene age has been suggested for the latest movement on the fault, which may be compared in trend and timing with the Sticklepath–Lustleigh Fault of south-west England (Holloway and Chadwick, 1986).

The basin is preserved as an isolated erosional remnant that may once have been linked with the Eubonia Basin to the north-east, and with the Central Irish Sea and Caernarfon Bay basins to the south-east. Tertiary uplift along both the Anglesey–Isle of Man Arch and the Mid-Irish Sea Uplift has resulted in the removal of intervening Permo-Triassic rocks.

MID-IRISH SEA UPLIFT

The Mid-Irish Sea Uplift is a pronounced, north-easterly trending horst separating the Kish Bank Basin from the Central Irish Sea Basin, with throws of up to 5000 m on the bounding faults (Figure 16). Dobson et al. (1973) concluded that the Mid-Irish Sea Uplift consists of a 900 m-thick succession of nonmagnetic, Lower Palaeozoic rocks, probably Ordovician, that overlie a magnetic core (Al-Shaikh, 1970), that may be equivalent to the Ordovician basic-igneous rocks of County Wicklow. The Wicklow Head Shelf to the south-west (Dobson et al., 1973) is composed of Ordovician and Cambrian sedimentary rocks with some interbedded volcanic rocks.

CENTRAL IRISH SEA BASIN

Only the northern part of the Central Irish Sea Basin (Dobson et al., 1973) is included in the report area (Figure 15). The basin gives rise to a low gravity anomaly (Figure 18), has a north-easterly trend, and plunges to the south-west. The basin is a graben floored by Palaeozoic rocks, and infilled with Mesozoic strata overlain by thin Tertiary deposits. The present-day basin limits are erosional; thin Tertiary deposits occupy a shallow, north-easterly trending, partially fault-bounded sag.

The Central Irish Sea Basin consists of two half-grabens of opposed polarity, separated by a buried anticlinal ridge of

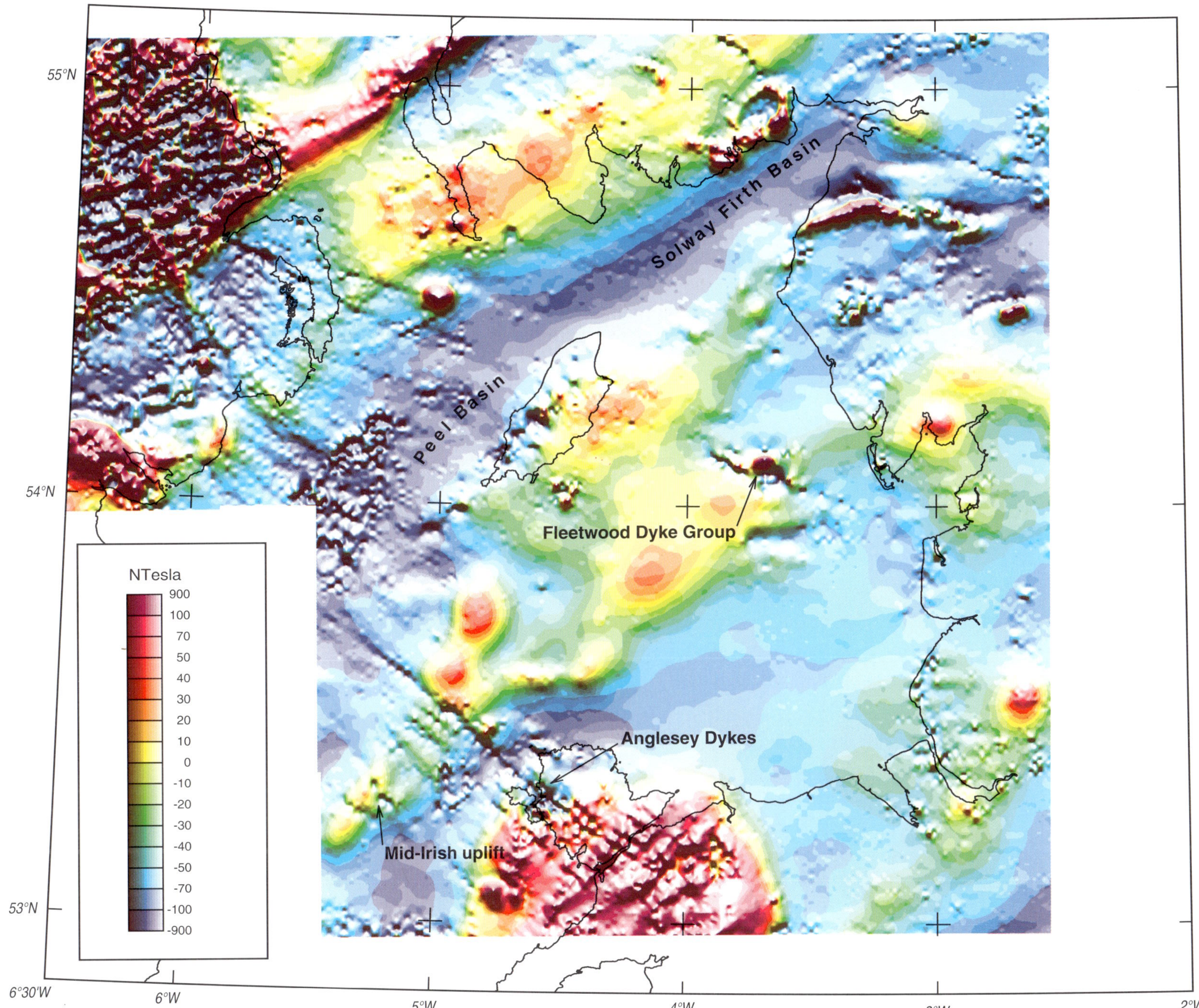

Figure 23 Shaded relief aero-magnetic anomaly map of the Irish Sea.

Lower Palaeozoic rocks plunging to the south-west (Al-Shaikh, 1969; 1970). In the larger, south-eastern half-graben, an estimated 1400 m of Permo-Triassic strata overlie 700 m of Carboniferous with slight unconformity. Beds dip south-eastwards towards a hanging-wall syncline and the bounding fault of the Holy Island Shelf. In the smaller north-western half-graben, about 1600 m of Permo-Triassic strata rest unconformably on some 500 m of Carboniferous, and dip north-westwards towards the boundary fault with the Mid-Irish Sea Uplift/Wicklow Head Shelf.

CAERNARFON BAY BASIN

The Caernarfon Bay Basin (Dobson et al., 1973) is the eastern extension of the Central Irish Sea Basin (Figure 16). It consists of two, Caledonoid-trending, stepped half-grabens, the Malltreath and Menai Straits sub-basins, both tilted to the south-east and with hanging-wall synclines adjacent to their controlling faults. Both synclines die out south-westwards towards the southern margin of the report area, where thin Permo-Triassic rocks rest unconformably on Carboniferous, giving a total sedimentary fill of about 2500 m (Al-Shaikh, 1969). The half grabens are separated by a north-east-trending central ridge which extends onshore as the Penmynydd Ridge (Figure 16; George, 1974).

The Malltreath Sub-Basin in the north-west (Dobson et al., 1973) is bounded by a splay of the Berw Fault, and floored by Westphalian rocks resting unconformably on Monian. The Menai Straits Sub-Basin to the south is bounded by the Dinorwic Fault, and passes north-eastwards into the Menai Straits Syncline in which Westphalian D strata rest unconformably on about 300 m of Dinantian rocks (BGS Anglesey Solid Geology sheet).

HOLY ISLAND SHELF

The Holy Island Shelf (McQuillin et al., 1969; Dobson et al., 1973) is an inlier of Precambrian metamorphic rocks with a fringe of Lower Palaeozoic strata; it straddles the Anglesey–Isle of Man Arch of Tertiary uplift (Figure 16). Its detailed structure is poorly known, but it may incorporate infolded outliers of Carboniferous strata. The shelf is topographically divided into western and eastern parts by the Holyhead Deep (see Figure 66). The smaller western part is characterised by a smooth sea bed, thought to be mostly made up of Lower Palaeozoic rocks. The larger eastern part is characterised by a rugged sea bed, and consists mostly of Precambrian rocks (BGS Anglesey Solid Geology sheet), although Lower Palaeozoic rocks may be present in a magnetically quiet zone south-east of Holy Island (Al Shaikh, 1970; Wright et al., 1971).

The Holy Island Shelf is traversed by the north-westerly trending Anglesey dykes of Tertiary age, which give rise to magnetic anomalies (Figure 23; Al Shaikh, 1970; Wright et al., 1971; Kirton and Donato, 1985). Dyke injection during a period of regional extension in the Paleocene may have been facilitated at the intersection of a major north–south uplift with a dyke swarm radiating from Northern Ireland.

5 Carboniferous

Carboniferous strata subcrop extensively beneath Quaternary deposits in the western part of the Irish Sea, and lie beneath a thick Permo-Triassic cover over most of the remainder of the area (Figure 24). Onshore, they crop out in Northern Ireland, the Republic of Ireland, southern Scotland, the Lake District, the Pennines, North Wales, and on the Isle of Man. Over most of the Irish Sea, the Carboniferous rests unconformably on folded Lower Palaeozoic rocks. Seismic records indicate a maximum thickness of about 6250 m between Anglesey and the Isle of Man (Jackson et al., 1987); this is made up approximately of 2050 m of Dinantian, 2900 m of Namurian, and 1300 m of Westphalian strata. Mudstone lithologies make up most of the Carboniferous, although locally the Dinantian is dominated by limestones, the Namurian by interbedded sandstones, and the Westphalian by interbedded sandstones, seatearths and coals. The Dinantian and Namurian shales and Westphalian Coal Measures have proved to be important sources for oil and gas (Colter, 1978).

During the Carboniferous Period, the British Isles passed through equatorial latitudes and endured a hot, wet climate, although conditions were more arid in Dinantian and late Westphalian to Stephanian times (Johnson and Tarling, 1985). In the Early Carboniferous times, the region lay to the south of the Equator, but movement related to the collision of the Eurasia and Gondwana continental plates resulted in northerly drift, so that by the beginning of the Permian, the region lay in the northern hemisphere under the influence of desert conditions.

Throughout Carboniferous times, large parts of the major Lower Palaeozoic massifs of Longford–Down and the Southern Uplands in the north, and St George's Land in the south (Figure 25), remained above sea level. Between these massifs there were two independently subsiding troughs: in the north was the Peel–Solway-Firth–Northumberland Trough, and in the south was the more-complex Dublin–Craven Basin or Central Province Trough (George, 1958). These were separated by a ridge of 'highs' or islands including the Balbriggan Block, the Manx Massif, the Ramsey–Whitehaven Ridge, the Lake District Massif and the Alston Block. The Central Province Trough includes the East Leinster Basin, the southern parts of the East and West Irish Sea basins, and the Craven Basin. During the Dinantian, a complex of independently subsiding grabens with intervening horsts (Lee, 1988) developed in the Central Province Trough, in response to north-north-west to south-south-east regional extension. Within each graben there was asymmetrical subsidence towards a major bounding fault or growth fault, and a network of smaller faults accommodated extension and settling (Gawthorpe et al., 1989). The sedimentary fills therefore thicken towards the major growth faults and onlap the trailing margins, producing characteristic wedge-shaped profiles. Condensed, fine-grained sequences mark the intervening horsts, and local unconformities may occur both on the horsts and towards the trailing margins of the hanging walls.

The evidence suggests that rifting had largely ceased by the late Dinantian (Leeder, 1982), but subsidence continued during Namurian and Westphalian times as a broad crustal downwarp related to thermal sag. It has been postulated that both rifting and basin sag occurred in response to lithospheric stretching (Leeder, 1982). Contemporaneous volcanic rocks, mainly of basaltic and andesitic composition, reflect this lithospheric stretching. Lavas of Courceyan age occur in the Solway Firth Basin, the Northumberland Trough, and nearby localities; submarine tuffs and lavas in the southern part of the Isle of Man are probably of Brigantian age (Figures 26 and 27; Dickson et al., 1987).

A wide variety of facies is present in the Carboniferous of the region, indicating sedimentary environments ranging from terrestrial, fluvial, tidal and shallow marine to deep water. This variety has given rise to problems in the correlation of diverse, facies-controlled faunas. While these problems still exist, the use of microfossils and spores has overcome some problems, although the biozonations derived from them do not have the resolution of those provided by the macrofossils.

DINANTIAN

During Dinantian times, a number of major facies belts crossed the Irish Sea (Figure 25). In the north, a fluviodeltaic and marginal-marine facies now fringes the Solway Firth Basin and Northumberland Trough on the southern flanks of the Southern Uplands Massif. An extensive area of shelf facies, probably stretching from the east coast of Ireland to the Isle of Man and the Lake District, is separated from the shelf facies of North Wales by the basinal facies of the Central Province Trough (Figure 25), in which a thick sequence of alternating limestone and mudstone was deposited. Data on thickness offshore are sparse, but onshore indications imply considerable variation (Figure 27). Little is known of the Carboniferous strata of the Caernarfon Bay Basin, but onshore, shelf limestones cropping out on Anglesey provide evidence of a breach in the Wales–Brabant landmass (Figure 25).

Northern and western areas

At the basin margin of the Ulster–Midland Valley Trough (Figure 25), the Dinantian is represented at Cultra, north-east of Belfast (Figure 27), by the Craigavad Sandstone and Ballycultra formations, which consist of fluvial and tidal-flat deposits of late Courceyan age (Griffith and Wilson, 1982). Late Courceyan (Smith et al., 1991) rocks were also proved in the nearby Belfast Harbour No. 1 borehole (Smith, 1986). At Castle Espie at the head of Strangford Lough (Figure 26), a small outlier of pink-stained limestone of Brigantian age at the feather edge of a much larger subcrop underlying Strangford Lough indicates that the barrier created by the Lower Palaeozoic rocks of the Longford–Down Massif (Figure 25) had been breached by late Dinantian times.

South-east of Strangford Lough, borehole 73/67 proved 3.6 m of pale grey, bioclastic limestone containing crinoidal debris and early Asbian foraminifera; to the south-west, borehole 73/65 proved 0.5 m of massive, grey, bioclastic limestone with late Arundian foraminifera (N J Riley, written communication, 1990).

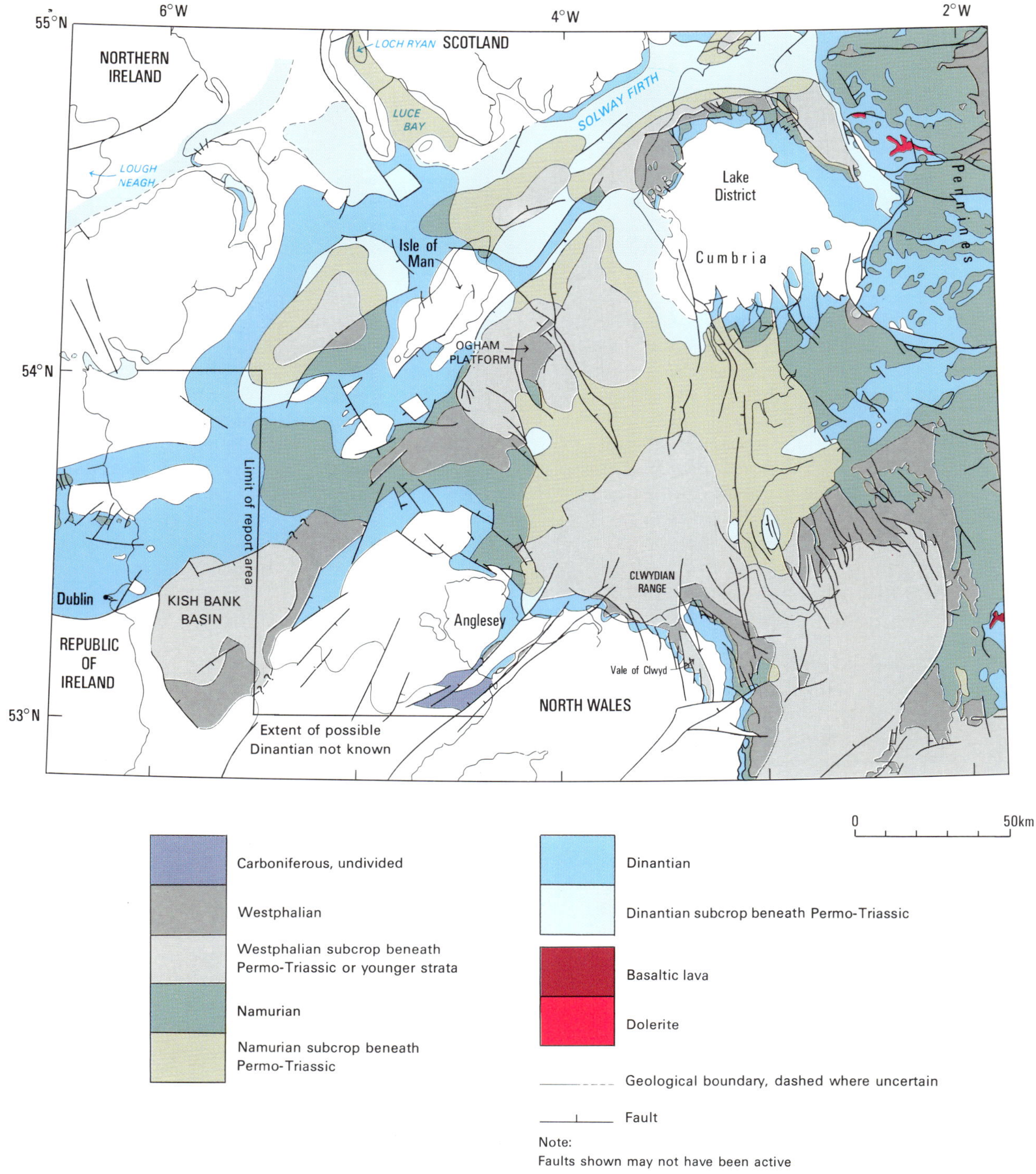

Figure 24 Distribution of Carboniferous strata in and around the report area. Note that revised Permo-Triassic boundaries in the Solway Firth and Stranraer basins are not shown in figures in this chapter.

Largely terrestrial, fluviodeltaic and marginal-marine facies occur onshore in the fault-bounded Solway Firth Basin and Northumberland Trough (Figure 25), and similar lithologies might be expected offshore. The Lower Border Group of the Northumberland Trough is interpreted as fluviatile and deltaic deposits separated by areas of nonclastic, marginal-marine deposition (Leeder, 1973). The clastic material was carried by a number of river systems draining the Southern Uplands Massif. Clastic deposits are preserved adjacent to the North Solway Fault, notably around Rascarrel Bay (Figure 26; Deegan, 1973; Ord et al., 1988), and marginal-gulf carbonates occur a few kilometres to the north-east around Kirkbean (Craig, 1956). The beds range in age from Courceyan to Holkerian (Figure 27). North of the Isle of Man, BGS borehole 73/53 proved 9 m of reddish-brown-stained, grey sandstone and mudstone which yielded palynomorphs of late Viséan to early Namurian age (Wilkinson and Halliwell, 1979). Similar lithologies were seen in borehole 73/52 to the west; both may be facies equivalents of the Scottish 'Calciferous Sandstone'.

On the northern Isle of Man, there are strata representing the Solway Firth Basin succession (Figures 25 and 26). The beds, concealed under thick drift, comprise about 250 m of limestones with sparse shale interbeds resting on 30 m of Basal Conglomerate (Taylor et al., 1971). The succession is similar to the continental and shallow-shelf succession seen

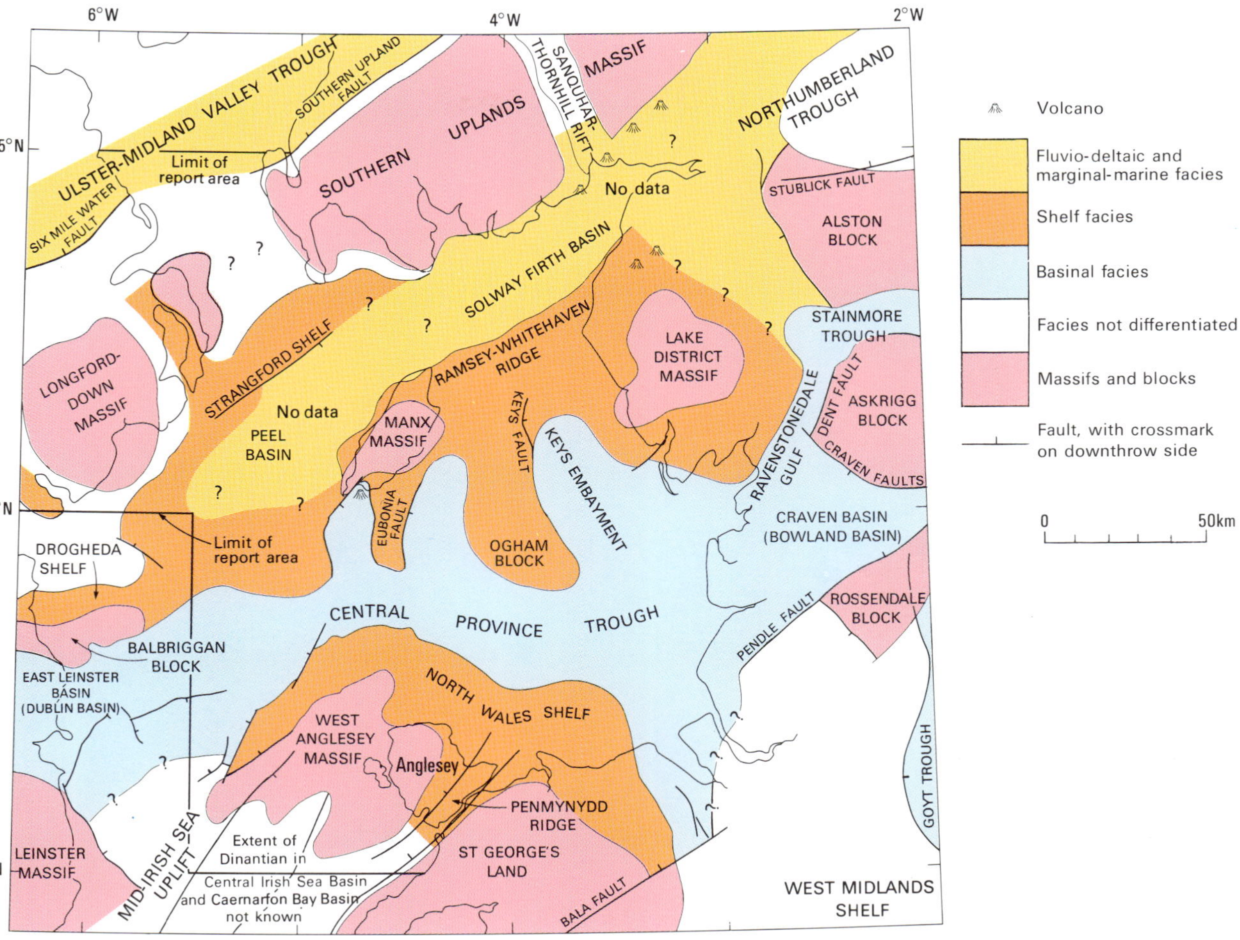

Figure 25 Generalised late Dinantian palaeogeography of the Irish Sea and surrounding areas. Based on sources quoted in text.

north-west of the Lake District, where the deltaic Yoredale influence appears in late Brigantian sediments, and becomes increasingly dominant eastwards (Mitchell et al., 1978; Arthurton and Wadge, 1981). A Dinantian age has been suggested, on lithological grounds, for a sucrosic and coarse-grained dolomite in BGS borehole 71/52 near the north-east coast of the Isle of Man (N J Riley, written communication, 1990).

On the south of the Isle of Man (Figure 27), the basal Langness Conglomerate (Dickson et al., 1987), deposited as an alluvial-fan, is overlain by dark grey, thinly bedded, shallow-shelf limestones of the Derbyhaven, Ballasalla and Castletown formations. These are in turn overlain by calcite mudstones of the Balladoole and Poyllvaaish formations. The latter are interpreted as 'reef' or mud-mound facies analogous to those developed at the margin of the Craven Basin farther east in the Central Province Trough (Miller and Grayson, 1982). Dark grey mudstone of Dinantian age was recovered from the adjacent Ramsey–Whitehaven Ridge by BGS borehole 89/13 (Figure 26). The sedimentary sequence on the southern Isle of Man is capped by the Scarlett Volcanic Formation, of early to mid-Brigantian age, that comprises submarine basaltic lavas, tuffs and agglomerates with thin beds of black limestone.

Carboniferous rocks are thought to occur widely in the West Irish Sea Basin. By analogy with strata at the southern tip of the Isle of Man and in the area to the north of Dublin (Sevastopulo, 1981a), much of this sequence is likely to be of Early Carboniferous age. In the Peel Basin, geophysical evidence suggests that a thick sequence of Carboniferous rocks underlies a prominent unconformity, and may rest on a Caledonian granite (Figure 3; Wright et al., 1971; Bacon and McQuillin, 1972).

East Irish Sea Basin and adjacent areas

The Basement Beds of continental origin are widespread around the Lake District (Figure 27), where they rest on a red-stained surface, and represent infill of the hollows and valleys of an irregular landscape. They consist of conglomerates with locally derived pebbles, interbedded with sandstones, mudstones, and some limestones and gypsum that indicate occasional marine influence. The beds thin northwards on to the Lake District Massif (Rose and Dunham, 1977). In the Cockermouth area (Figures 26 and 27), where deposition was interrupted by olivine basalt of the 100 m-thick Cockermouth Lavas (MacDonald and Walker, 1985), their age has been determined as Courceyan on the basis of miospore examination (Butterworth and Butcher, 1983; Mitchell et al., 1978).

The Basement Beds are succeeded by thickly bedded limestones which are typical of the Dinantian around the Lake District, and are indicative of shallow-shelf conditions. Numerous transgressions and regressions are reflected in the complex of tidal and shallow-marine facies, but are less obvious in the sediments deposited in deeper water (Ramsbottom, 1973). These fluctuations have been attributed to changes in sea level, possibly caused by the growth and deterioration of contemporary ice caps in the southern hemisphere, but tectonic activity also played an important role (Horbury, 1989).

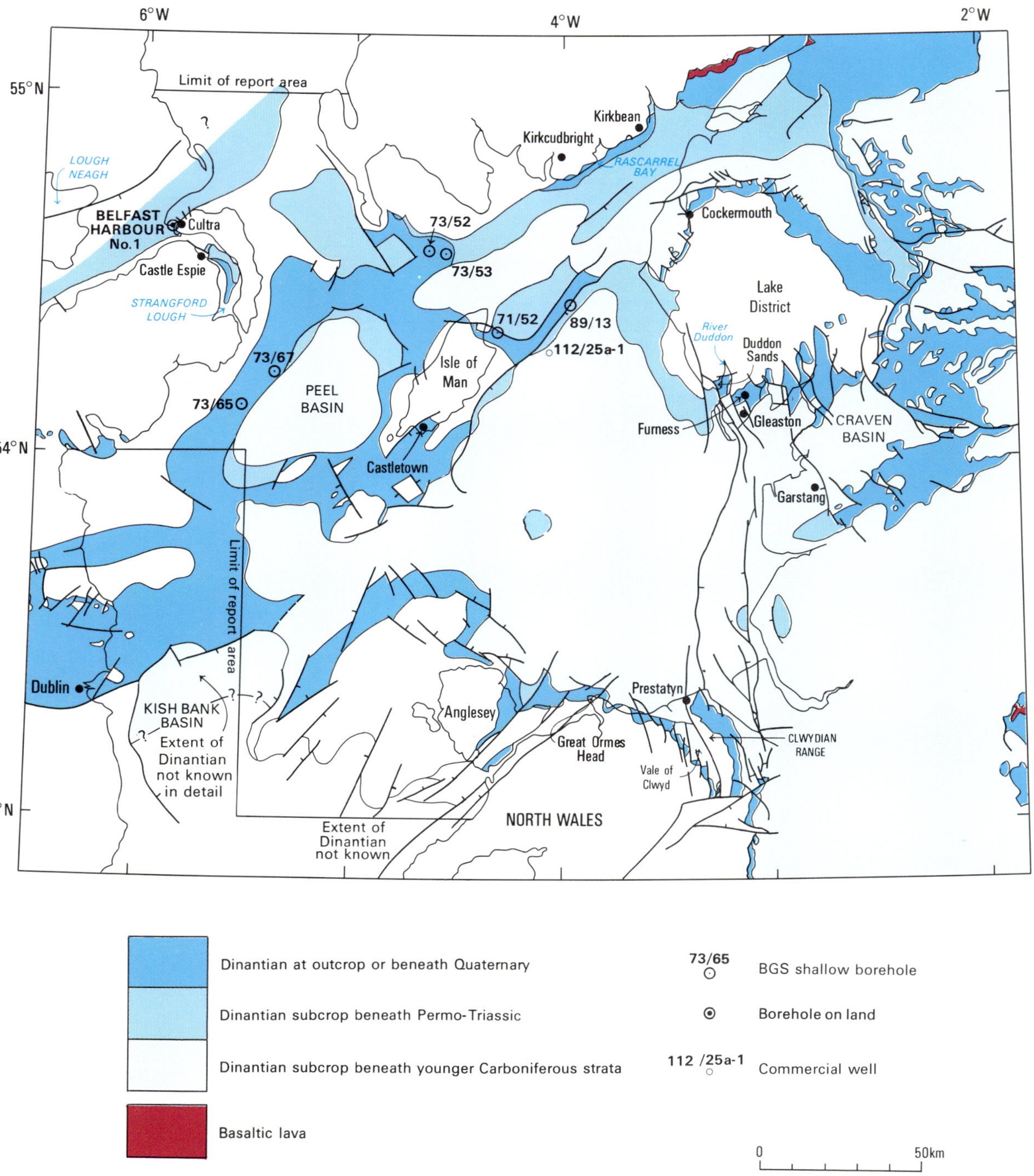

Figure 26 Distribution of Dinantian strata.

In general, the regressions are represented by algal and dolomitic, fine-grained limestones and shales with some sandstones; emergence is indicated by breccia beds, unconformities, and palaeokarstic features (Mitchell et al., 1978). The transgressions are marked by marine bioclastic limestone with a fauna indicative of the open seas. To the south of the Lake District, shallow-shelf conditions were established by the late Courceyan (Rose and Dunham, 1977), but to the north-west, Courceyan and Chadian rocks of marine origin have not been recognised (Mitchell et al., 1978).

At the close of the Asbian, a major change in deposition occurred; the Brigantian and overlying Namurian strata around the Lake District belong to the Yoredale facies (Mitchell et al., 1978). Large quantities of clastic material were supplied from a major delta in the north-east, resulting in thickly bedded limestones separated by variable thicknesses of terrigenous material. Onshore, the clastic content decreases westwards, so that the terrigenous influence may be limited offshore. The Gleaston Formation, of Yoredale facies, is lithologically highly variable and includes black mudstones, thin sandstones and limestones, as well as dark, cherty, argillaceous limestone, and pale-coloured, shelf limestones (Rose and Dunham, 1977). The beds thicken towards the south and west, and at Gleaston (Figure 26) the formation is transitional between shelf and basinal facies. The formation passes south-westwards and offshore into the basinal facies of the Roosecote Limestone, a uniform, dark, bituminous, cherty limestone with regular mudstone partings. About 108 m of grey, argillaceous limestone proved in the East Irish Sea Basin in well 112/25a-1 is equated with the Gleaston Formation (Figure 27).

The shelf sequence at the southern margin of the Central Province Trough mirrors that of the northern margin. Progressive southerly onlap onto the Lower Palaeozoic massif

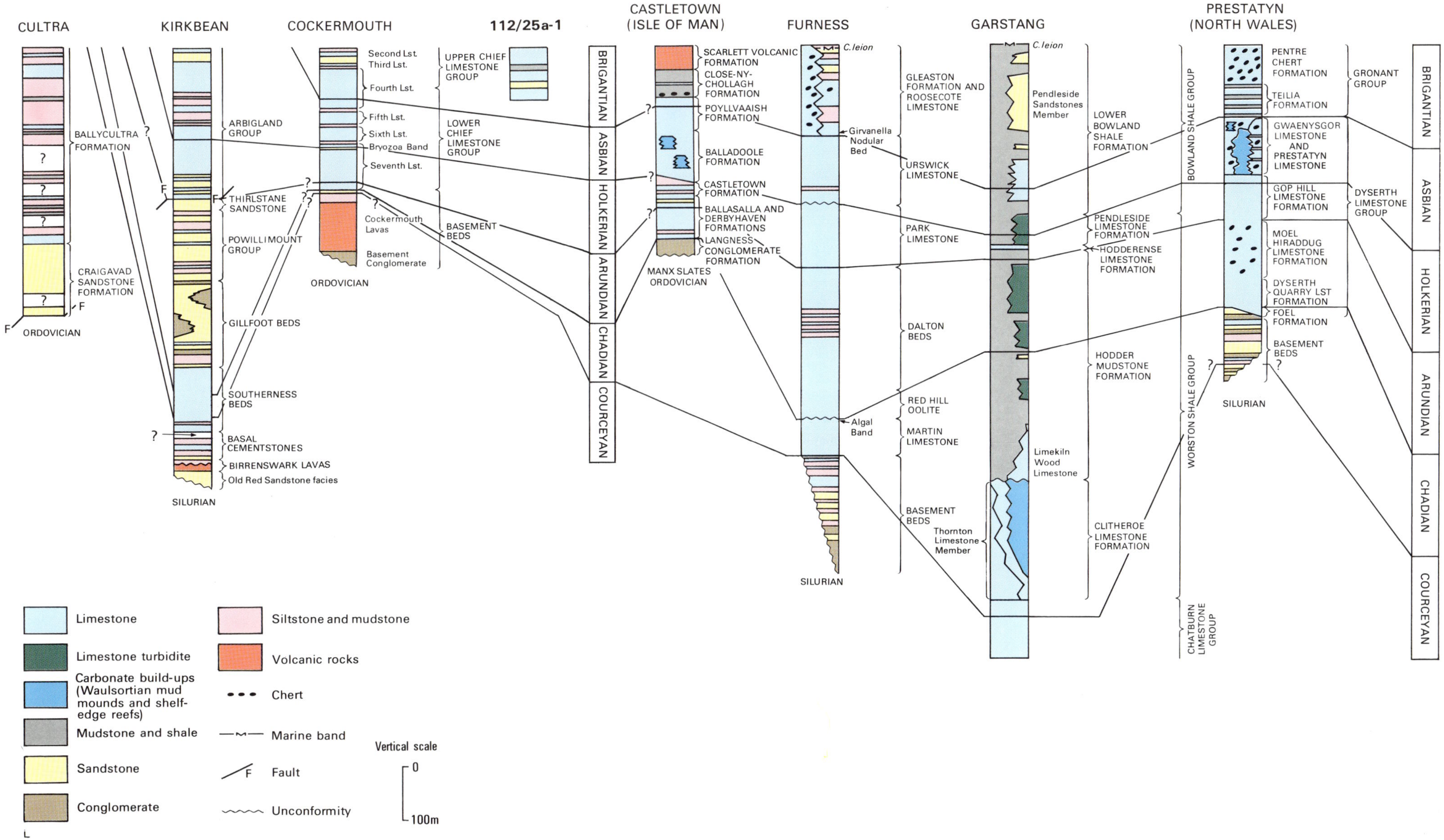

Figure 27 Correlation of Dinantian strata in and around the Irish Sea. For locations see Figure 26. Land sections based on authors quoted in text.

of North Wales allowed carbonate sediments to be deposited on a broad, shallow shelf lying to the north of St George's Land. Onshore, the presence of the Asbian and Brigantian stages has long been recognised (Warren et al., 1984), but older strata of Chadian and Arundian age have also been proved (Somerville et al., 1986, 1989; Davies et al., 1989). It therefore appears that the main transgression took place in Chadian times, and it is likely that a complete sequence of Dinantian rocks exists offshore.

The Central Province Trough is marked by a thick, variable sequence of limestones and mudstones with minor turbiditic sandstones. Onshore, in the Craven Basin (Figure 26), an estimated thickness of over 4000 m of Dinantian sediments is indicated by gravity modelling (Lee and Gawthorpe in Gawthorpe et al., 1989). Courceyan sediments alone are estimated to be up to 1720 m thick, indicating that major subsidence began early in the Dinantian.

The earliest Dinantian beds at Garstang (Figure 27), in the Craven Basin, comprise the Chatburn Limestone Group (late Courceyan to early Chadian), a carbonate-ramp sequence of shallow-water limestones and fine-grained terrigenous clastics (Charsley, 1984; Riley in Aitkenhead et al., 1992). Differential subsidence of adjacent fault blocks within an asymmetric graben is apparent in the overlying Worston Shale Group, which ranges in age from early Chadian to Asbian (Riley, 1990). The lowest formation of the group, the Clitheroe Limestone Formation of early Chadian age, consists of shallow-water, bioclastic limestones (Thornton Limestone Member) in the north, and deeper-water, Waulsortian lime mounds in the south; this indicates tilting of the graben floor. Following a period of submarine erosion in the late Chadian, probably as a result of the retreat of fault scarps away from active faults, a conspicuous limestone-conglomerate sequence (Limekiln Wood Limestone), derived from the underlying Clitheroe Limestone Formation, was deposited. Interbedded crinoidal-packstone turbidites mark the profuse production of crinoidal debris on intrabasinal highs.

Subsequently, a hemipelagic, dysaerobic depositional regime was established, and persisted throughout the remainder of the Dinantian (Riley, 1990). Carbonate sediment supply switched from intrabasinal to extrabasinal as adjacent platforms became the main sources. Sediment supply varied in type and volume depending on changes in sea level on adjacent platforms; sandstones are few, but limestone turbidites make up a conspicuous part of the sequence. The thickest accumulations built up in depressions on the basin floor adjacent to active faults.

The black, organic-rich shales of the Bowland Shale Group (Asbian to Pendleian) mark a diminution in the supply of carbonate sediment to the basin (Aitkenhead et al., 1992). This was initiated by the growth of fringing shelf-edge reefs between the Craven Basin and the surrounding carbonate platform, and then maintained by uplift of the Askrigg Block and southern Lake District high, isolating the basin from their influence. The first major influx of terrigenous sand from the encroaching deltas to the north occurred in mid-Brigantian times, in the form of the Pendleside Sandstones Member (Figure 27).

Reef knolls or lime mud-banks (Lees, 1964) are a major feature of the Central Province Trough. Two types of limestone build-up are distinguished: firstly, the Waulsortian facies of lime mounds and lime mudbanks (formerly termed reefs) of Courceyan to early Chadian age, formed in a submarine-slope environment in water depths ranging from less than 120 m to over 300 m (Lees and Miller, 1985). Secondly, shallow-water, coral/algal, shelf-edge and platform-margin reefs of Asbian to early Brigantian age which fringe the shelf areas (Miller and Grayson, 1982; Ramsbottom, 1969b). The reefs occur along the Craven fault system at the southern margin of the Askrigg Block (Figure 25); they reach their maximum development in Ireland, where they are associated with lead-zinc mineralisation (Sevastopulo, 1981d). They are therefore likely to be present in the offshore part of the Central Province Trough.

NAMURIAN

Namurian strata form the backbone of northern England, for they crop out extensively on the Pennines and their foothills (Figures 15, 24 and 28). A narrow crop occurs north and west of the Lake District, and there is also a small outcrop west of Loch Ryan in southern Scotland. In North Wales, rocks of this age occur to the east of the Clwydian Range, but are absent onshore to the west. In the western part of the Irish Sea, rocks of Namurian age subcrop beneath Quaternary sediments, notably on the limbs of broad folds (Figure 28). Elsewhere in the report area, they occur beneath Permo-Triassic rocks. Namurian strata are apparently absent from the Kish Bank Basin (Jenner, 1981), and only remnant outliers occur to the north and north-west of Dublin, including a full Namurian succession near Kingscourt (Jackson, 1965; Sevastopulo, 1981b). Offshore, the Namurian sediments are thickest between the Isle of Man and Anglesey, where they reach 2900 m (Jackson et al., 1987). Calculation of original offshore thicknesses for the Namurian is tentative; nevertheless, an isopach map (Figure 29) highlights the contrast between basins, shelves and adjacent landmasses.

In general, the Namurian rests conformably on the underlying Dinantian strata, but the lowest beds are absent from large parts of the basin margins and towards Lower Palaeozoic highs. A later episode of intra-Namurian erosion or nondeposition is also indicated over contemporaneous topographic highs (Ramsbottom et al., 1978); this disconformity is most clearly detected in the north-west of the Lake District and over the Ramsey–Whitehaven Ridge, where only the Pendleian to Arnsbergian and Yeadonian stages are represented.

Facies variations are clearly displayed throughout the Namurian; they involve the pulsed advance of a deltaic facies encroaching from a broadly northerly direction across an intermittently subsiding shelf. Accumulation of basinal mudstones continued in most of the former Dinantian depocentres in the Central Province Trough (Figure 25); these are oil-prone source rocks that show high gamma-ray values locally. Three major facies are present, the Yoredale, Millstone Grit and Basinal (mudstone) facies (Figure 30). The Yoredale and Millstone Grit facies are mainly clastic deposits of shallow-marine and deltaic origin that form a transitional sequence between the marine conditions of the Dinantian and the fluviodeltaic/alluvial-floodplain swamp conditions of the Westphalian.

The Yoredale facies is found over, and to the north of, the Askrigg Block (Figure 30), forming the characteristic terraced topography of the northern Pennines. It consists of rhythmic units comprising, in ascending order: limestone, shale, sandstone, seatearth and coal. Each unit represents a cycle of marine transgression and regression, although part of the cycle, particularly the coal, may be missing in any given unit. The marine fauna of the limestones and shales includes crinoids, brachiopods and foraminifera, but ammonoids (goniatites) are rare (Edwards and Trotter, 1978). The lithologies are similar to those of the late Dinantian, but the limestone beds are generally thinner, and terrigenous sediments become increasingly dominant.

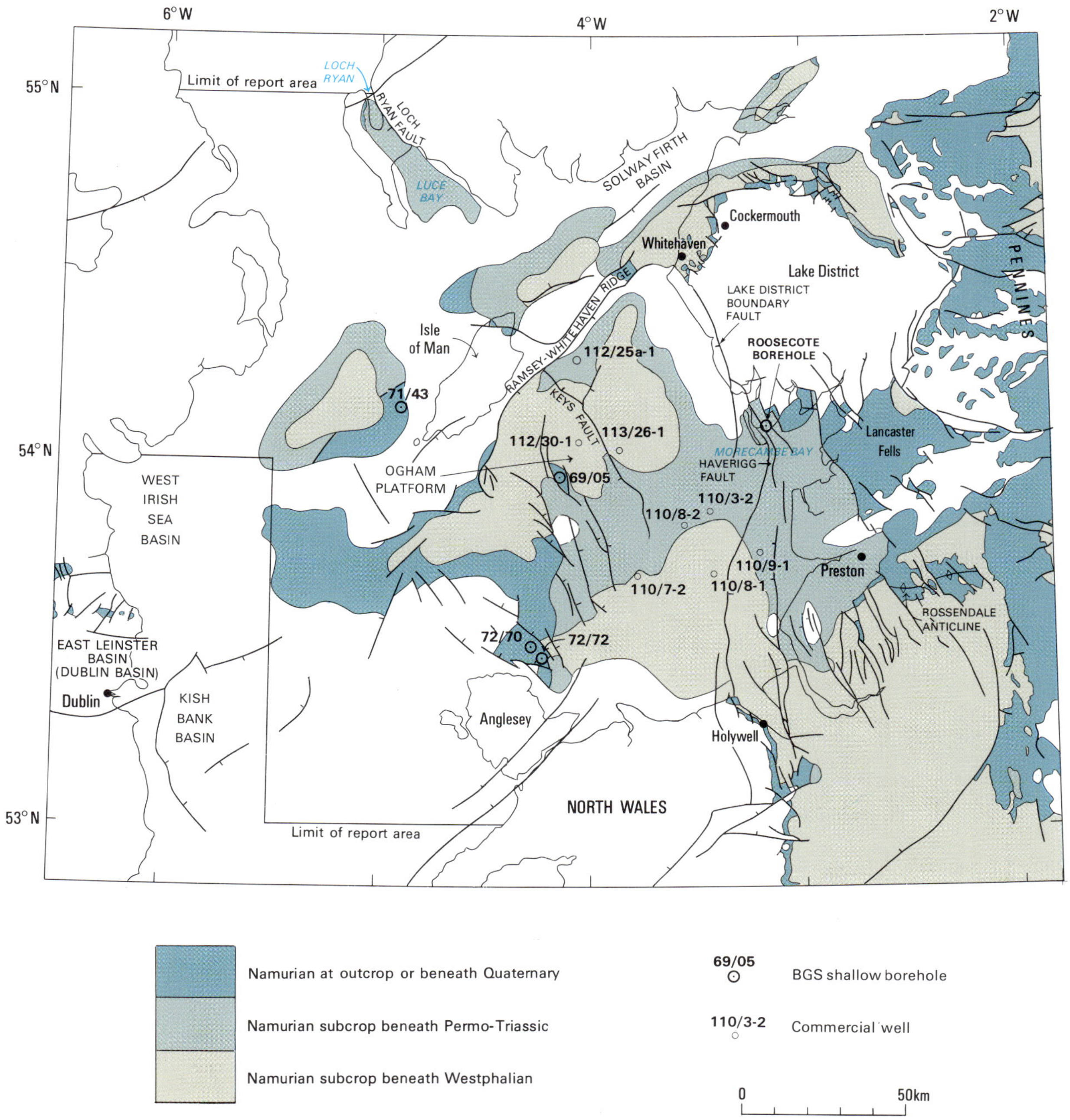

Figure 28 Distribution of Namurian strata.

The Millstone Grit facies is found to the south of the Askrigg Block (Figure 30), where it interfingers with the Basinal facies. The Millstone Grit is a rhythmically bedded sequence of deltaic origin, that generally lacks the limestone bands of the Yoredale facies. A complete rhythm consists of mudstone, siltstone, sandstone, seatearth and coal, although again parts of the cycle may be missing. The sandstones are commonly thick and coarse-grained, and the numerous named 'grits' can be traced over large areas (Ramsbottom, 1977; Aitkenhead et al., 1985). The lithology indicates derivation from two distinct sources: protoquartzites, transported into the basin by turbidity currents flowing from St George's Land to the south, are most common in the Pendleian and Arnsbergian but persist locally into the Marsdenian. Feldspathic sands were largely derived from the north, where thick deposits were laid down early in the Pendleian; they did not reach some parts of the basin centre until the Marsdenian, and are represented there initially as fine-grained turbidites.

Progradation of deltaic sediments into the basin has been charted by detailed analysis of the fauna and flora in marine bands (Ramsbottom et al., 1978; Aitkenhead et al., 1985; Collinson, 1988); Pendleian grits are thickest in the Craven Basin, whereas the later grits are thickest farther south. Yeadonian sandstones, such as the Rough Rock (Figure 31) are fairly uniform in their thickness, and of almost basinwide extent; seat earth and coal are developed locally. The Yeadonian Haslingden Flags of south-east Lancashire show a derivation from the west, possibly from a source under the Irish Sea.

The Basinal facies is a thick, monotonous, mudstone sequence, primarily of Pendleian to Kinderscoutian age; it has been proved in a number of offshore wells (Figure 31). This marine facies is also the dominant lithology in the onshore Roosecote Mudstone, the Edale, Upper Bowland and Holywell Shales, and the Leinster Shale Formation. It consists mainly of grey to dark grey mudstones and fine-grained siltstones deposited fairly rapidly from suspension, and minor, distal, turbiditic sandstones; in places there is a high

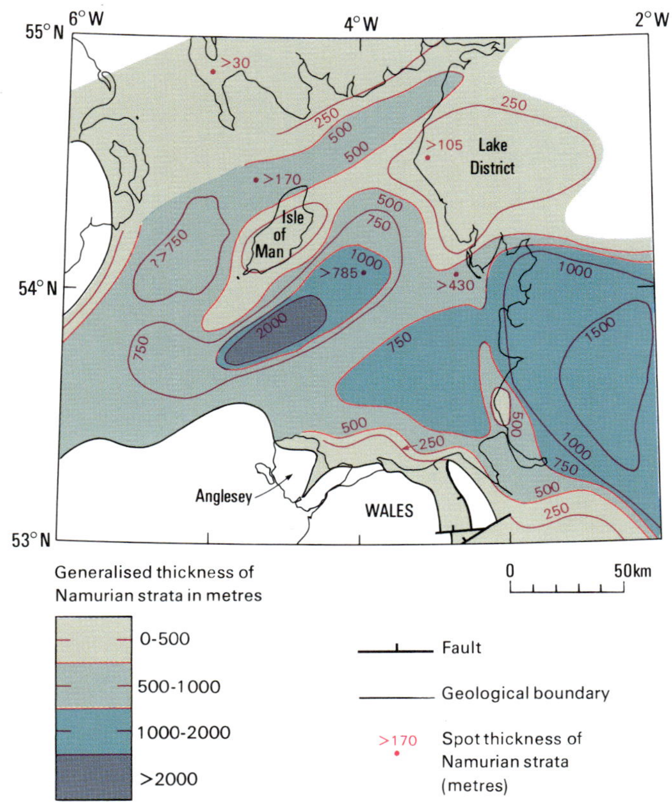

Figure 29 Depositional thickness of Namurian strata. Onshore data taken from Ramsbottom (1969a).

proportion of slumped beds (Johnson, 1981). Thin, dark grey to black, marker bands with distinctive ammonoid and bivalve faunas are slightly finer grained, and indicate periods of higher salinity with good connections to the open ocean (Collinson, 1988), at least for pelagic faunas, although dysaerobic conditions prevented the development of a diverse benthos. The condensed equivalents of the mudstone sequence on local, sediment-starved highs, were deposited during periods of sediment starvation, and are characterised by numerous, closely spaced ammonoid bands, thin argillaceous limestones, cherts and phosphates.

Northern part of the report area

In the western and northern parts of the report area, the deposits of the West Irish Sea and Solway Firth basins are not well known, although it is likely that strata of Namurian age are widespread in the West Irish Sea Basin (Figure 28). They have been proved in BGS borehole 71/43, west of the Isle of Man, where laminated grey mudstone with dolomite and very fine-grained sandstone up to about 15 cm thick yielded palynomorphs of Kinderscoutian to Marsdenian age (Wilkinson and Halliwell, 1979). It is possible that a complete Namurian succession, resting unconformably on the Dinantian, occurs locally to the north of the Ramsey–Whitehaven Ridge (Smith, 1985), where the effect of the intra-Namurian disconformity is considerably reduced.

In the Solway Firth and Stranraer basins (Figure 21), some extrapolation may be made from adjacent onshore sequences. A narrow outcrop west of Loch Ryan (Figure 28), formerly classified as Dinantian, is now tentatively included with the Namurian on the basis of lithological similarity with the Passage Beds of Ayrshire (BGS Stranraer 1:50 000 Solid Sheet). The sequence consists of grey, secondarily reddened and mottled sandstone, shale and thin seatearths, and becomes more arenaceous towards the top; it also contains a thin basaltic lava. Deposition occurred in a narrow half-graben controlled on its eastern boundary by the contemporaneous Loch Ryan Fault.

At the northern end of the Isle of Man, the succession is closely comparable with that north of the Lake District and in the Midland Valley. The Pendleian is about 142 m thick and dominantly argillaceous, consisting of three to four Yoredale cyclothems in which the top 90 m of strata are secondarily reddened beneath the base Permian unconformity. The overlying dark shales contain thin limestones and sandstones, some of which have been correlated with the Cumbrian sequence (Taylor et al., 1971).

Southern part of the report area

To the south of the Ramsey–Whitehaven Ridge, the sequence is known incompletely from a number of released wells. As data are too meagre to permit identification of individual

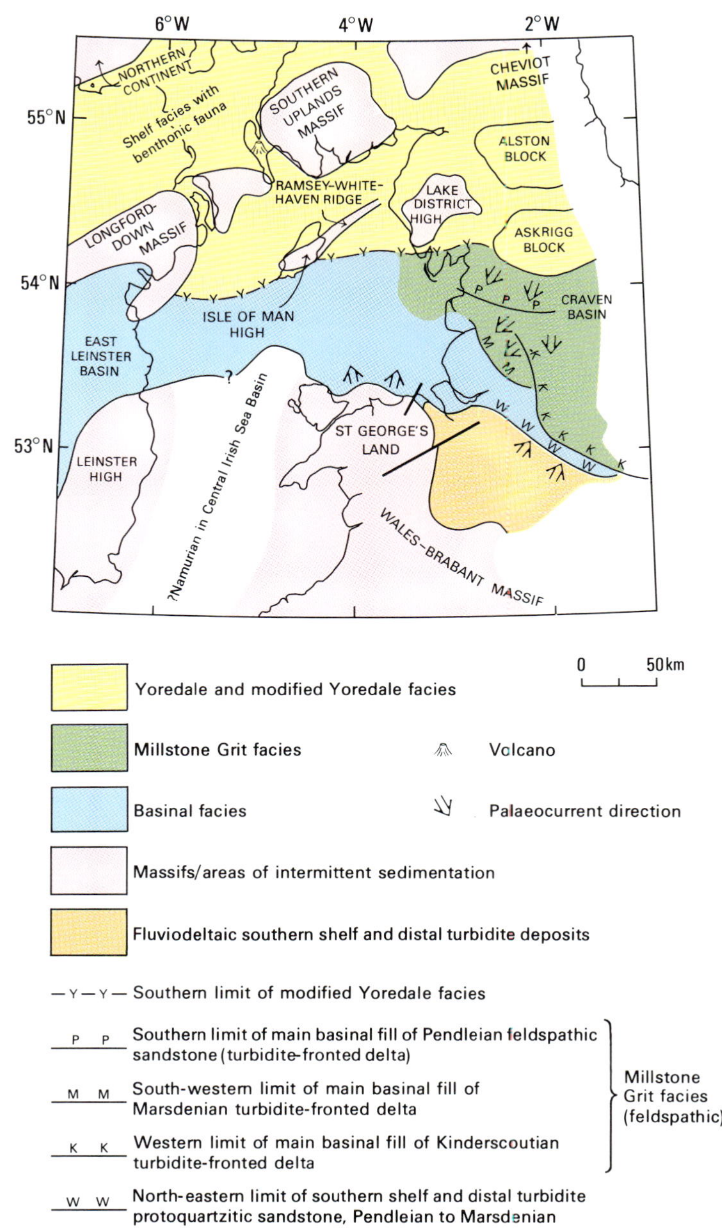

Figure 30 Generalised Namurian palaeogeography, showing the encroachment of Millstone Grit facies into the basin. Based on sources quoted in the text.

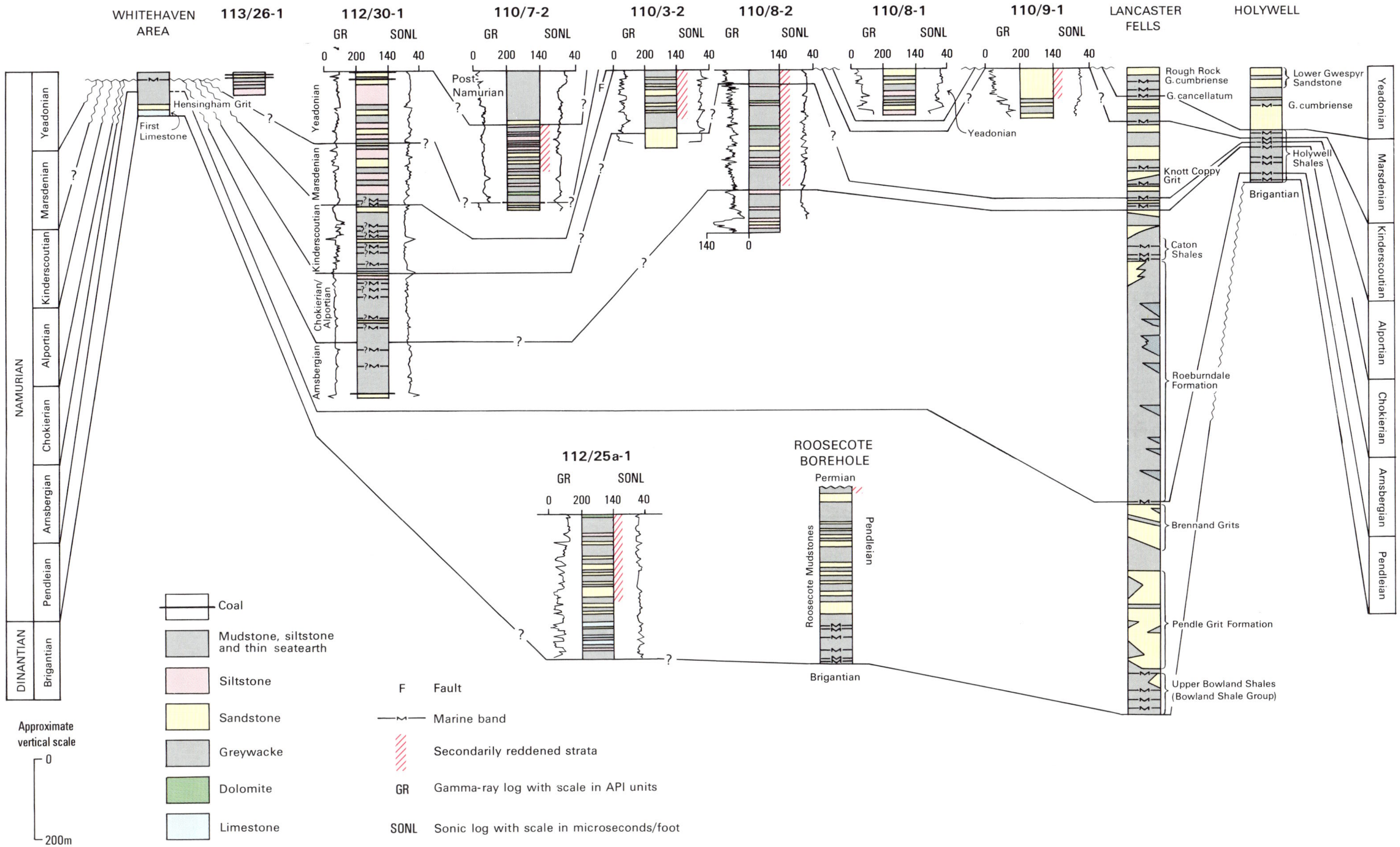

Figure 31 Correlation of Namurian strata in and around the Irish Sea. For locations see Figure 28. Land sections based on sources quoted in text.

marine bands, all stage boundaries in the following account are approximate and taken largely from composite logs. The Namurian of both the West and East Irish Sea basins is probably largely of basinal facies; the maximum proven thickness is 784 m in well 112/30-1 (Figure 31), which shows an unbroken succession of Arnsbergian to Yeadonian age (Jackson et al., 1987). Several other wells have each proved parts of the Namurian sequence.

In the main depocentre between the Isle of Man and Anglesey, a thickness of about 2900 m is postulated from seismic evidence. The depositional axis of the basin extended east–west from the Rossendale Anticline of mid-Lancashire, across the East Irish Sea Basin, and through the Dublin Basin (Figure 28) to the Shannon Trough (Sevastopulo, 1981b).

Beds thought to be of Pendleian age have been proved offshore only in the northern part of the basin, in well 112/25a-1 (Figures 28 and 31). The lithology is a modified Yoredale facies, and the succession can be divided into three parts. A lower cyclical division of alternating limestone and siltstone becomes coarser grained upwards, with sandstone at the top. A thick middle division, over half of which is secondarily reddened, comprises siltstone and silty sandstone with thin ribs of turbiditic sandstone; a prominent sandstone rib marks the top. The upper division of 'hot shales' is entirely secondarily reddened, and is probably a deeper-water marine shale.

On the northern shore of Morecambe Bay, the BGS Roosecote borehole (Figures 28 and 31) proved 455 m of Pendleian strata, named the Roosecote Mudstones by Rose and Dunham (1977), resting conformably on Brigantian limestone. The lower 122 m consist of dark, weakly calcareous mudstone with marine bands, and are comparable with the Upper Bowland Shales of the region to the south-east, indicating that this district lay within the northern part of the Craven Basin province. The upper 333 m consist of a sandstone-siltstone-mudstone turbidite sequence that correlates with the Pendle Grit Formation (Sims in Aitkenhead et al., 1992). The Upper Bowland Shales, the Pendle Grit Formation and Brennand Grits of the Lancaster Fells (Figure 31) have not yet been proved offshore, but are expected to occur in the eastern part of the East Irish Sea Basin.

Arnsbergian sediments are thought to occur in wells 112/30-1 and 110/8-2 (Figure 31), where they consist of uniform, dark grey to black, carbonaceous shales and mudstones (Jackson et al., 1987); these are rarely calcareous or pyritic, and are thus analogous to the Holywell Shales/Lower Sabden Shales. Two, thin, more-radioactive layers overlying siltstone partings in well 112/30-1 may mark uranium-rich ammonoid bands at the base of marine-mudstone cyclothems. Scarce, pale grey, sandstone ribs up to 5 m thick, possibly equivalent to the protoquartzitic turbidites seen onshore, occur in both wells. In the absence of faunal confirmation, the top of one such sandstone rib is taken as the succeeding stage boundary in well 110/8-2. A thin coal resting on a sandstone rib in well 112/30-1 (Figure 31) indicates the presence of a modified Yoredale-type cyclothem in the northern part of the area.

North of Anglesey (Figure 28), BGS borehole 72/70 proved dark grey to black, pyritic, Arnsbergian mudstone with scattered, thin, grey limestone and sandstone bands. Nearby, BGS borehole 72/72 recovered grey mudstone and siltstone, locally with laminae and bands of sandstone, which contain plant fragments and sparse miospores of late Arnsbergian age or younger (B Owens, written communication, 1988).

Beds of Chokierian and Alportian age are poorly represented offshore, probably as a result of the intra-Namurian disconformity, although they are believed to have been penetrated in wells 112/30-1, 110/8-2 and possibly 110/3-2 (Figure 31; Jackson et al., 1987). The dark grey to black, marine mudstones contain sporadic siltstones and scarce, siliceous, sandstone ribs, including prominent sandstones marking the tops of the successions. Crinoid debris was noted in well 112/30-1, in addition to thin dark brown limestones; this is an approximate facies equivalent of the Lower to Middle Sabden Shales. Two thin radioactive shale beds were noted towards the base in well 112/30-1, where a marked change in dipmeter readings is suggestive of an unconformity between the Arnsbergian and Chokierian sequences.

Kinderscoutian rocks are most clearly displayed in well 112/30-1 (Figure 31), where dark grey to black shale has a lower velocity, higher background radiation, and a more varied and spiky gamma-ray log response than the underlying Arnsbergian to Alportian strata. Clearly differentiated sandstone ribs up to 13 m thick occur in the middle and at the top of the succession; scattered, argillaceous, dark brown limestone interbeds occur towards the base.

Possible lateral equivalents are believed to have been drilled in wells 110/9-1 and 110/3-2 (Figure 31), both largely in secondarily reddened beds with scant evidence of age. Siltstones and silty calcareous mudstones produce a varied, spiky and erratic gamma-ray log response, with evidence of both upward-fining and upward-coarsening cycles. Isolated, high, gamma-ray peaks recorded in dark grey mudstones probably indicate ammonoid bands; numerous ammonoid bands have been recorded in the onshore correlative, the Upper Sabden Shales (Price et al., 1963). Low-velocity spikes on the sonic record in well 110/3-2 may indicate the former existence of 6 thin coals which are now totally oxidised by pre-Permian secondary reddening. In well 110/3-2, it may be significant that the beds are more arenaceous than the succession to the west, with thicker, dark grey, shaly sandstone possibly equivalent to the Knott Coppy Grit (Arthurton et al., 1988). This may indicate an approach to the western margin of the deltaic Millstone Grit of the Pennines.

Marsdenian rocks were penetrated in well 112/30-1, and possibly in well 110/7-2 (Figure 31). The deposits are largely cyclothemic, but show an upward change from thinly bedded deltaic deposits with strong marine influences in the form of ammonoid bands, to thicker deltaic cycles with plant debris and nonmarine bivalves at the top. In well 110/7-2, above a basal black shale with high gamma-ray values, the succession comprises thickly interbedded units of micaceous siltstone and silty mudstone, with scarce, thin, pale grey, silty sandstones, some with a coal capping. BGS borehole 69/05, on the Ogham Platform (Figure 28; Wilkinson and Halliwell, 1979), proved black, sandy mudstone upon fossiliferous pyritic shale from the base of the Marsdenian (Wright et al., 1971).

Yeadonian rocks closely resemble those of the paralic, coal-bearing Westphalian sequence, and consist of two upward-coarsening cyclothems. They have been proved in wells, and show some lithological diversity (Figure 31). In well 113/26-1, siltstone is subordinate to mudstone, and four coals are reported. Siltstones predominate in other wells, which show small-scale variation throughout on the gamma-ray logs. Radioactive peaks tentatively identified with the Cancellatum and Cumbriense marine bands are noted. In addition, two thin coals, or former coals now secondarily reddened and converted to dolomite (cf. Mykura, 1960), are interpreted from sonic and compensated-density logs in wells 110/7-2 and 112/30-1. The entire Yeadonian succession appears to be less arenaceous than the equivalents in North Wales and Lancashire, though a 25 m-thick sandstone in well 112/30-1,

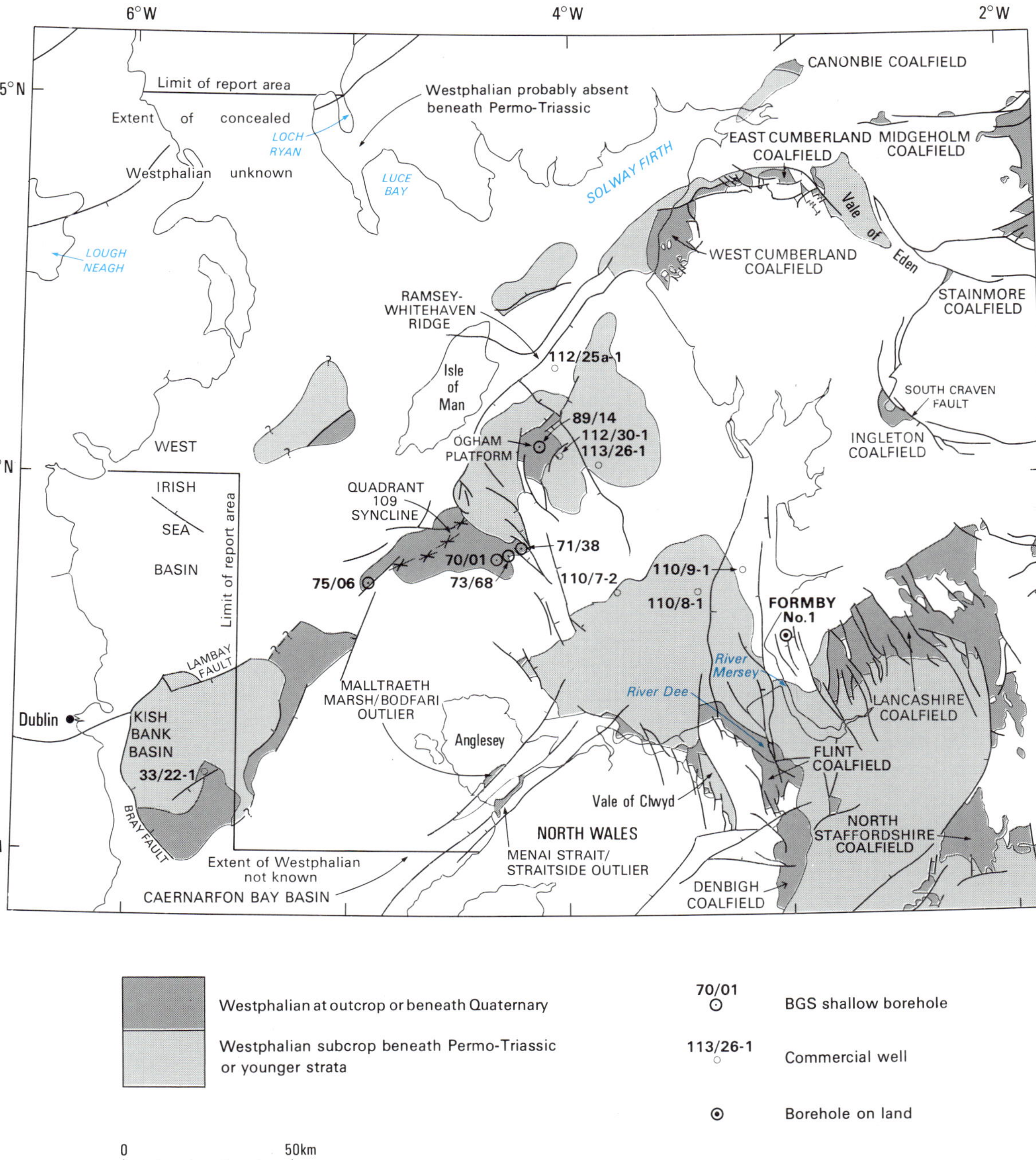

Figure 32 Distribution of Westphalian strata.

associated with two to three coals, may be the local equivalent of the Rough Rock.

WESTPHALIAN

The Westphalian succession is restricted to erosional remnants of a formerly much thicker sequence. On land these occur to the north-west of the Lake District, in southern Lancashire, and in North Wales. Offshore, the Westphalian is largely preserved beneath the Permo-Triassic unconformity (Figures 24 and 32). Nevertheless it crops out in the core of the Quadrant 109 Syncline between the Isle of Man and Anglesey (Wright et al., 1971; Smith, 1985), over much of the Ogham Platform, possibly in the West Irish Sea Basin, and south-west of Anglesey in the Caernarfon Bay Basin. Westphalian beds also crop out in the Kish Bank Basin (Jenner, 1981; McArdle and Keary, 1986). No Stephanian rocks have been proved to date in the report area.

The complete Westphalian sequence has not been drilled in the Irish Sea, but estimates from seismic evidence south of the Isle of Man suggest a minimum thickness of 4000 m (Jackson and Mulholland, 1993), although the top is not seen. In basinal areas, a complete Westphalian sequence was probably deposited prior to uplift and erosion during the Variscan orogeny. It is possible that a full succession was also deposited over parts of Ireland (Sevastopulo, 1981b), for the small inliers which occur there all record Langsettian (Westphalian A) successions comparable in thickness with that of the Lancashire Coalfield, and thermal maturity levels imply the former presence of a thick overburden. Youngest Westphalian strata (Westphalian D) have been proved south of Wexford (Clayton et al., 1986), in the Kish Bank Basin (Jenner, 1981) and are also known in the St George's Channel south-west of Pembroke (Barr et al., 1981).

In basinal areas, Westphalian strata rest conformably on Namurian deposits, but progressively overlap the Namurian towards the basin margin, and the Dinantian near the crests of the Lower Palaeozoic massifs such as Anglesey and the Ramsey–Whitehaven Ridge. In the Kish Bank Basin to the west of the report area, presumed Duckmantian (Westphalian B) strata rest on probable Cambro-Ordovician rocks (Naylor and Shannon, 1982) in well 33/22-1 (Figures 32 and 33). On Anglesey, Bolsovian (Westphalian C) and Westphalian D directly overlie the Precambrian locally.

Folding, and erosion of the Lower Palaeozoic massifs, occurred in mid-Bolsovian times; this is recorded by the Symon Unconformity (Ramsbottom et al., 1978; Tubb et al., 1986), which in the southern North Sea marks the base of the primary redbed sequence. The unconformity has been linked with Variscan inversion (Leeder and Hardman, 1990). In the UK, primary redbeds overlie grey Westphalian Coal Measures; these include the Ruabon Marl and Erbistock Beds of North Wales, and the Etruria Marl and Enville formations of the English Midlands. The base of the redbeds is diachronous, but mainly Bolsovian in age. Strong evidence for the Symon Unconformity occurs in coalfields adjacent to the report area; in the Lancashire Coalfield, the unconformity increases in magnitude westwards. To date it has not been identified offshore, although Jenner (1981) refers to a minor unconformity in mid-Bolsovian deposits in the Kish Bank Basin.

The sedimentation pattern of the late Namurian continued into the early Westphalian, as major deltaic progradation continued from the north. The rate of subsidence kept pace with deposition, such that alluvial, shallow-water and paralic conditions were maintained. The Westphalian Coal Measures were deposited in a large, paralic, swampy area stretching eastwards from western Ireland to Germany and beyond. The depositional environment has been compared to the present-day Atchafalaya swamps west of the Mississippi Delta (Elliot, 1986). The Lower Palaeozoic massifs continued to exert an influence (Figure 34), probably as low-lying, forested hills standing above the level of the peaty swamps. Marine influence increased towards the axis of subsidence (Calver, 1968).

In the Pennines, Guion and Fielding (1988) have shown that lower delta-plain and shallow-water deltaic depositional environments were dominant in early Langsettian times, although occasional marine transgressions drowned the delta lobes. By late Langsettian times, there was a gradual transition to an upper delta-plain environment, and marine influences were minimal throughout late Langsettian and Duckmantian times, with the notable exception of two major transgressions that deposited the Vanderbeckei and Aegiranum marine bands (Figure 33). Beds are mostly cyclothemic and represent deposition in distributary channels, crevasse-splay systems, shallow lakes and inland peat swamps. The latter environments are responsible for most of the economically exploitable coal and constitute the richest gas-prone source rocks. In the absence of direct sedimentological evidence in the Irish Sea, it is assumed that similar depositional environments prevailed here also.

A considerable change in environment occurred after mid-Bolsovian times, for humid, tropical conditions were replaced by more arid conditions. In most of England and Wales, the grey Coal Measures are overlain by primary redbeds (Besly, 1988); these are diachronous, appearing earlier in the south adjacent to the Wales–Brabant Massif. The red colour is regarded as a syndepositional feature, indicating low water-table conditions which contrast markedly with the water-logged swampy environments which prevailed during the accumulation of the grey Coal Measures. The redbeds are interpreted as well-drained, alluvial-fan and alluvial-plain deposits, and are composed of sandstones and breccias alternating with upward-fining units of mottled green and yellow siltstone and red or purple silty mudstone, and capped by a lateritic palaeosol.

An episode of reddening, termed secondary reddening (Trotter, 1939), has affected many rocks, but is most noticeable in Namurian and Westphalian strata (Kent, 1948). It probably occurred mainly during Stephanian times (Mykura, 1960; Wagner, 1983), as both reddened and grey Dinantian limestone clasts occur in the Permian breccias of the Vale of Eden (see Chapter 6). A thick zone of secondary reddening occurs in all wells in the East Irish Sea Basin (Figure 33), except where it is cut out by faulting, and it can be traced both offshore and onshore as a swathe 6 to 7 km wide beyond the feather edge of the Permo-Triassic outcrop (Trotter, 1953; 1954; Jackson et al., 1987).

In the past, the base of the Permian offshore has commonly been taken as the base of the reddened strata (e.g. Colter, 1978; see Chapter 6), leading to inconsistent and variable isopach and palaeogeographic maps for the Lower Permian. A revised boundary between the Carboniferous and the Permian is placed in this report at the junction in well-log signatures between spiky fluctuations in reddened Carboniferous, and the more regularly patterned Lower Permian (Jackson et al., 1987). In some wells, this choice is substantiated by density and dipmeter logs.

In addition to the reddening of grey shales and argillaceous sandstone, the effects of this arid weathering include the decomposition of basalts and the conversion of coal to dolomite and ironstone (Strahan, 1901; Mykura, 1960). The depth of reddening (Figure 33) is believed to represent the lowest level of the contemporary water table, and thus to present a muted profile of the landscape at that time. The depth (Jackson et al., 1987) is greatest (546 m in well 112/30-1) over the structural highs, and least in basinal settings (69 m in well 110/9-1 — Figure 32). The zone is 339 m thick in Formby No. 1 borehole on a major structural high (Kent, 1948).

Duricrusts or soil horizons may have formed immediately beneath the sub-Permian unconformity, and incipient silicification is seen (Kent, 1948) in the highest 7 m of Formby No. 1 (Figure 32). The dolomites which mark the top of the reddened zone in wells 112/25a-1 and 112/30-1 (Figures 32 and 33) might be interpreted as caliches. Both the soil horizons and the reddening indicate semiarid conditions, for they require high seasonal temperatures and ground-water circulation for their formation (Walker, 1967).

Stratigraphical details

Patterns similar to those of the Westphalian stratigraphy onshore can be recognised in the sparse data available from offshore wells, and BGS boreholes such as 75/06 which drilled probable Westphalian grey sandstone with mudstone and coal (Figures 32 and 33). With the exceptions of well 33/22-1 in the Kish Bank Basin, much of well 113/26-1, and the basal Westphalian of 112/30-1, all other beds in offshore wells classified in this account as Westphalian are secondarily reddened (Jackson et al., 1987). On seismic sections, the Westphalian package shows subparallel, high-amplitude, high-frequency, low-continuity reflections that are interpreted as coals.

The Langsettian in well 112/30-1 includes two thin mudstone bands with high radioactivity (Figure 33) that are tenta-

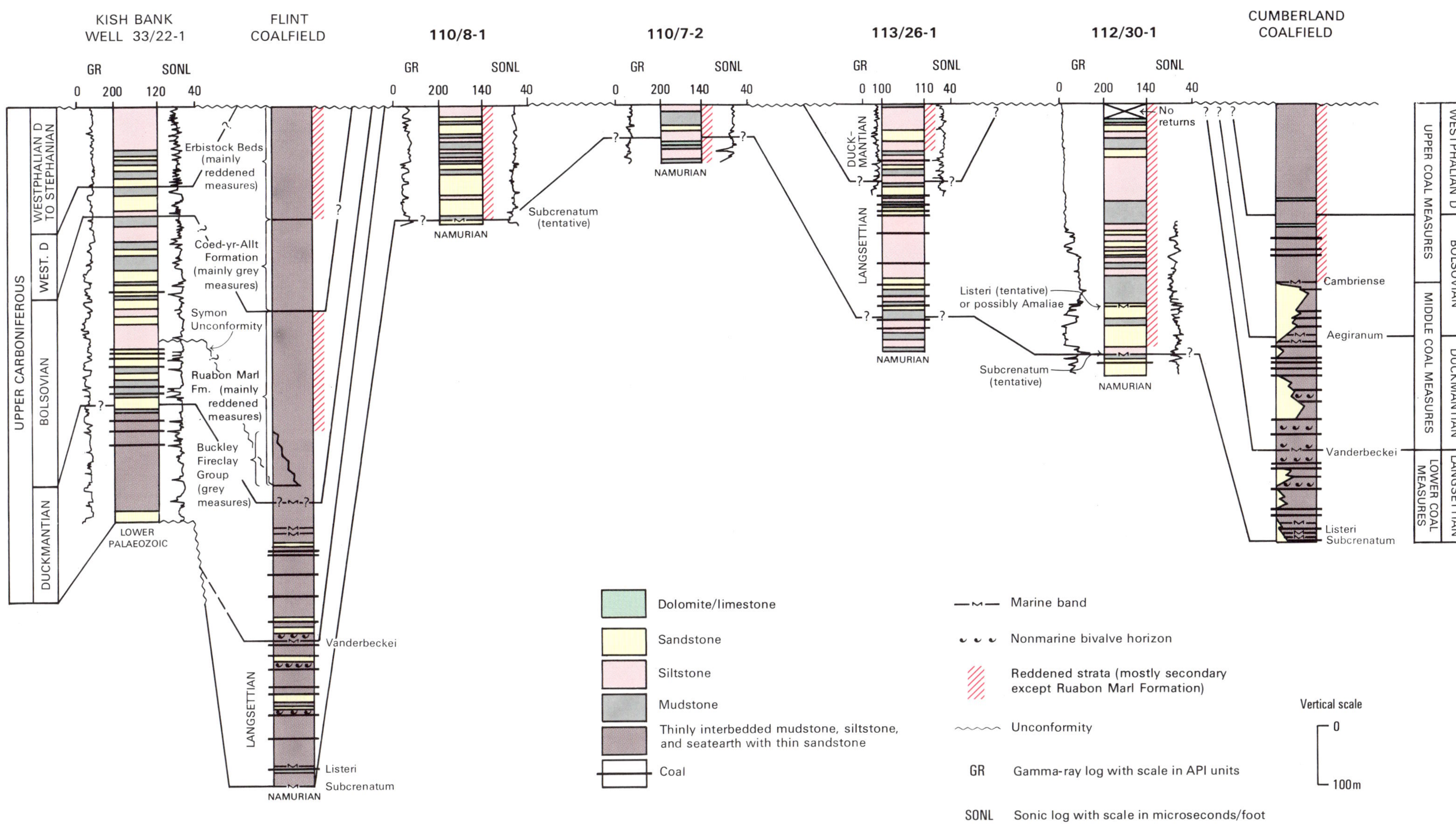

Figure 33 Correlation of Westphalian strata in and around the Irish Sea. For locations see Figure 32. Land sections based on sources quoted in text.

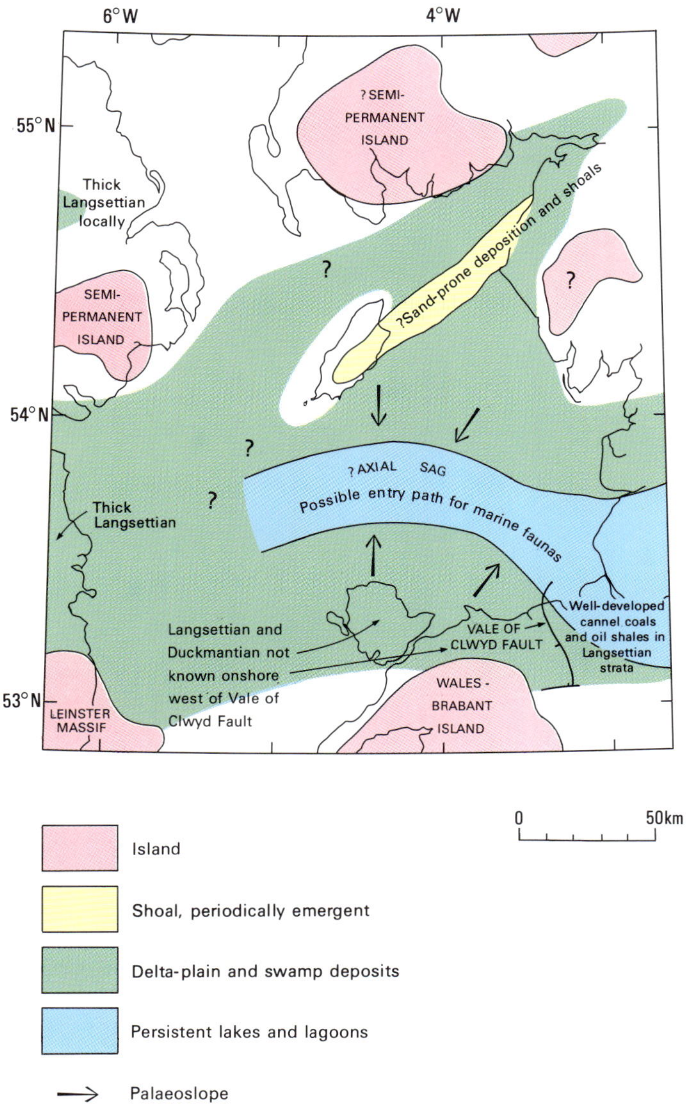

Figure 34 Postulated Westphalian facies and palaeogeography. Adapted from sources quoted in text.

tively identified as the Subcrenatum and Listeri (or possibly Amaliae) marine bands respectively (cf. Whittaker et al., 1985; Ramsbottom et al., 1978). Two, impure, pale grey, massive sandstones (38 m and 21 m thick) with 'box-car'-type gamma-ray signatures separate the radioactive mudstones; these are possibly distributary-channel deposits. In well 110/8-1, the thickest sandstone and the underlying radioactive shale, possibly the Subcrenatum Marine Band, can be correlated on the basis of well-log signature with the basal Westphalian of well 112/30-1.

Beds from near the top of the Langsettian were recovered from BGS borehole 70/01 north of Anglesey (Figure 32; Parkin and Crosby, 1982; Wilkinson and Halliwell, 1979), where 2.4 m of dark grey mudstone contain *Carbonicola* aff. *cristigalli* Wright; boreholes 71/38, 73/68 and 75/06 also penetrated early Westphalian rocks. The Langsettian is about 224 m thick in well 113/26-1 (Figure 33), in which 29 coals, some up to 2 m thick, were logged from analysis of cuttings. Some 80 per cent of the succession comprises mudstone and siltstone, with a 12 m-thick sandstone near the top. Beds of Langsettian to Duckmantian age were recovered in BGS borehole 89/14, south-east of the Isle of Man (Figure 32). These comprise 4 m of fine-grained, well-bedded sandstone with mudstone interbeds and abundant plant and coal fragments, with grey micaceous siltstone.

The overlying 212 m of Duckmantian in well 113/26-1 contain only one coal, and one radioactive mudstone bed, possibly the Vanderbeckei Marine Band, (Figure 33). The measures are again dominantly argillaceous, with three upward-fining, and four upward-coarsening cyclothems; the latter are successively thicker towards the top, and each culminates in a 3 m-thick silty sandstone that is possibly a crevasse-splay sandstone. No other released wells in the report area have drilled the Duckmantian, but 202 m of presumed Duckmantian beds were drilled in well 33/22-1 in the Kish Bank Basin. The sequence there comprises siltstone and sandstone with 7 coals, above a 10 m-thick basal sandstone (McArdle and Keary, 1986).

Bolsovian and Westphalian D deposits are known from boreholes in the offshore parts of the Cumberland and North Flint coalfields (Figure 33; Taylor, 1961; Lane, 1987). In the Dee Estuary (Figure 32), the succession above the productive Coal Measures consists of redbeds of the Ruabon Marl Formation overlain by grey measures of the Coed-yr-Allt Formation, and the reddened Erbistock Formation (Calver and Smith, 1974; Ramsbottom et al., 1978).

Well 33/22-1, in the Kish Bank Basin (Jenner, 1981), proved 329 m of interbedded grey siltstone and pale to dark brown sandstone with 19 coals in the Bolsovian Stage (Figure 33). The aggregate thickness of the coal is 11 m, but individual seams are less than 1 m thick. A minor unconformity, marked by red shales and angular quartz pebbles within the Bolsovian, may be the equivalent of the Symon Unconformity, for vitrinite-reflectance studies on the coals (McArdle and Keary, 1986) confirm a hiatus at about this level. The beds are overlain by 47 m of interbedded grey shale, thin sandstone and a few thin coals, possibly of Westphalian D age. The sequence is completed by 144 m of barren, interbedded grey and red shales with grey-brown sandstones, alleged to be Westphalian D to Stephanian in age, but taken to be reddened measures of Westphalian D age (Higgs in McArdle and Keary, 1986).

6 Permian

Permian rocks crop out sporadically around the Irish Sea (Figure 35), and where the sequence in the East Irish Sea Basin is not faulted, have been proved in all those wells which penetrate the conformable base of the Triassic. They have also been proved at or near the sea bed in BGS shallow boreholes (see Figure 41). Seismic evidence suggests that Permian strata are also present beneath the Triassic in the North Channel, Solway Firth and Kish Bank basins. In the Caernarfon Bay and Central Irish Sea basins, their presence beneath younger beds is more speculative (Barr et al., 1981), and they are thought to have been removed by erosion over a wide area of the West Irish Sea Basin. The Permian succession in the Irish Sea is subdivided into Lower Permian, largely continental redbeds, and Upper Permian, mainly marginal-marine and evaporitic deposits. The Lower and Upper Permian correspond to the Rotliegend and Zechstein respectively of the southern North Sea (Cameron et al., 1992).

During most of the Carboniferous Period, the area lay to the south of the Equator, but by Permian times, plate movements had carried it into the northern hemisphere (Johnson and Tarling, 1985). The climate changed gradually and damp, subtropical conditions gave way to a desert environment in the latest Carboniferous and Early Permian. The Early Permian climate is envisaged as having been arid or semiarid, with the prevailing wind blowing from the east or east-north-east (Brookfield, 1978; Smith, 1972).

Secondary reddening of strata beneath the Permian is widespread, making it difficult to identify the Permian–Carboniferous contact, especially above Silesian rocks (Trotter, 1953). In this account, the base of the Permian in wells is taken where the more regular pattern and 'box-car' character of gamma-ray and sonic logs of the Lower Permian red sandstone passes down into a spiky signature in the reddened Carboniferous. This designated boundary is considerably higher than that chosen by Colter and Barr (1975), Colter (1978) and Jackson et al. (1987) in some wells in the East Irish Sea Basin.

During the earliest Permian, topography was controlled by north-north-westerly and north-easterly trending active faults. Local volcanism was associated with the earliest sedimentation; lavas and pyroclastic rocks of basic to intermediate composition occur in the Lower Permian of both Northern Ireland, south-west Scotland and possibly in the centre of the East Irish Sea (Hardman, 1992) and south Cumbria. Volcanic activity was confined to the Early Permian, but crustal extension and basin subsidence continued through the Triassic and into the Early Jurassic. Around the Irish Sea, the Lower Permian is restricted to narrow grabens and half-grabens; it invariably rests unconformably on the Carboniferous, and is overlapped by Upper Permian sediments, particularly on the trailing edges of half-grabens and at basin margins (Jackson and Mulholland, 1993).

Occasional flash floods during the Early Permian carried debris flows from adjacent highlands, building up steep alluvial fans, with clasts becoming finer grained away from the sources. The fans pinch out abruptly basinwards and in general, aeolian processes were dominant over large parts of the basins passively infilling large depressions, possibly below OD (Smith, 1972), in the fault-induced topography (cf. Poole and Whiteman, 1955). By the end of the Early Permian, the landscape had been reduced to a gently rolling topography.

At the beginning of the Late Permian, a widespread marine transgression, believed to have entered from the Boreal Sea to the north, flooded the Irish Sea area, reaching the northern part of the Cheshire Basin at its maximum extent (see Figure 42). This sea, known as the Bakevellia Sea, was effectively separated by the Pennine Ridge from the Zechstein Sea (Smith and Taylor, 1992), a major basin which covered the area now occupied by much of the North Sea, Germany, Denmark and Poland. The Bakevellia Sea inundated the North Channel/Ulster, Solway Firth, and East Irish Sea basins, but little is known of its extent between Ireland and the Isle of Man. There may have been a trans-Pennine connection between the Bakevellia and Zechstein seas via the proto-Ribble and Aire valleys (Figure 42, see also Jackson, 1994 for further discussion). Shallow-water conditions prevailed during the marine transgressions, but there is evidence of periodic emergence; at least four main cycles of transgression and regression can be identified (Jackson et al., 1987).

The dry, continental conditions of the Early Permian were not conducive to the preservation of organic remains, although plant fragments and tetrapod footprints have been recorded in some of the southern Scottish basins, and footprints from the Vale of Eden (Smith, 1972; Delair, 1991). The largest fossil collections have come from Upper Permian marine sediments (Pattison in Smith et al., 1974). The macrofauna is restricted to a few species, mostly molluscs of which the bivalve *Bakevellia binneyi* (Brown) is the most abundant, particularly in deposits of the initial transgression. Even this species is rare in the sediments of the second transgression, and no marine faunas have been recorded in subsequent deposits of the Bakevellia Sea, indicating more adverse conditions and restricted circulation, or even isolation from normal marine waters. Plant remains include terrestrial plants washed into nearshore sediments, and marine algae. The terrestrial plants are mostly conifers of which *Ullmannia frumentaria* (Schlotheim) Göppert is the commonest (Pattison et al., 1973).

The greatest thickness of Lower Permian strata is found in the rift basins on and towards the perimeter of the Irish Sea; these deposits appear to thin towards the centre of the East Irish Sea Basin (Figure 36). Over 1000 m of Lower Permian sedimentary and volcanic rocks have been proved in the North Channel Basin at the Larne No. 2 borehole (Figure 35), and over 1400 m of sedimentary rocks are estimated to occur in the Dumfries Basin. A more recent seismic interpretation than that shown in Figure 36 suggests that in an east-north-easterly trending belt between Formby and Anglesey a maximum of 1150 m occurs, with possibly as little as 10 m in the centre of the East Irish Sea Basin (Jackson and Mulholland, 1993).

In contrast, the Upper Permian strata show a regional increase in thickness towards the centre of the East Irish Sea Basin, as well as local increases towards major growth faults (Figure 37). Thick halite sequences in the hanging walls of faults mark the depocentres of the North Channel and East Irish Sea basins, and halite may also be present in the Kish

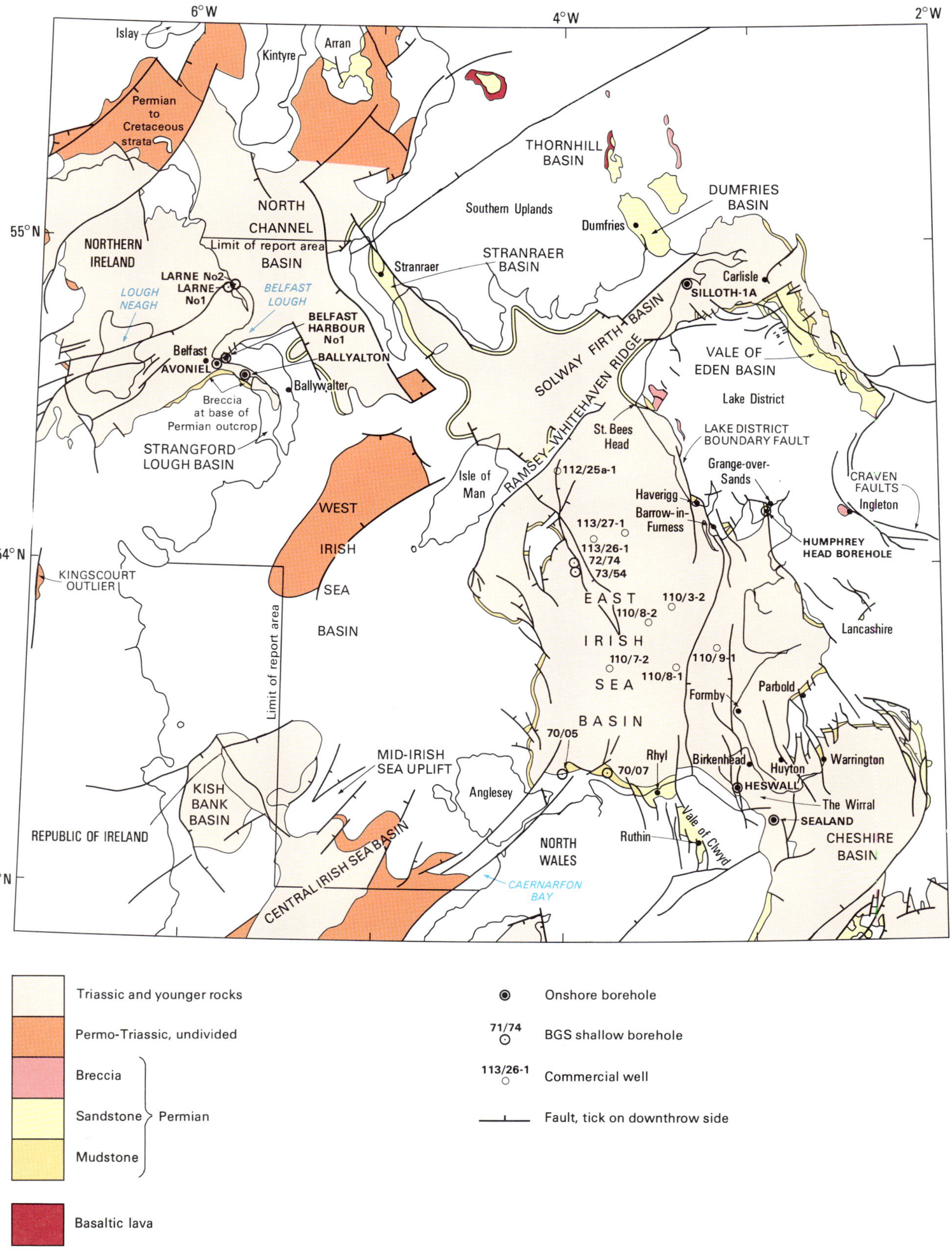

Figure 35 Distribution of Permian strata in and around the report area.

Bank Basin, and locally in the Solway Firth Basin (Figure 35). The halite passes laterally into an attenuated anhydrite and carbonate sequence on the margins of the evaporating basin, and across growth faults into mudstone-dominated sequences in the footwall.

Variations in thickness and lithology demonstrate that the East Irish Sea and Solway Firth basins subsided independently throughout the Permian; the Solway Firth Basin successions are dominated by relatively thick aeolian deposits during the Early Permian, whereas the northern East Irish Sea Basin is characterised by thick evaporites during the Late Permian. Lithological evidence also indicates that the Ramsey–Whitehaven Ridge, which had been active throughout much of the Carboniferous, continued to exert an influ-

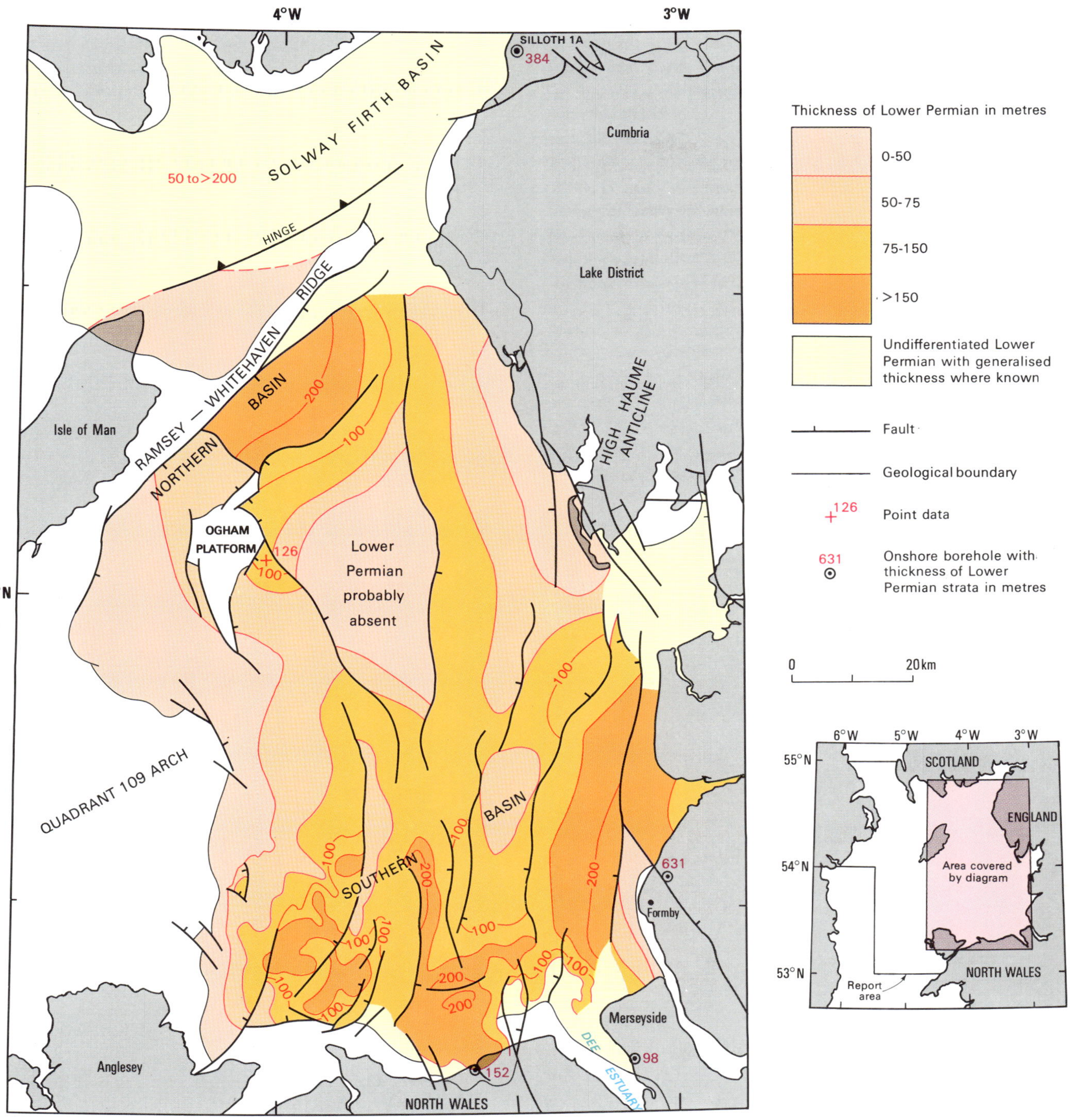

Figure 36 Thickness of Lower Permian rocks in the East Irish Sea Basin. Since this figure was prepared, acquisition of the first seismic data from the southern part of the basin, and recent commercial drilling, now demonstrate that Lower Permian deposits range up to 1150 m in thickness. The deposits are believed to infill a large palaeo-topographic depression, trending ENE–WSW, coincident with a Westphalian cored Variscan syncline, extending from east of Anglesey to Formby. To the north, over a corresponding Caledonoid Variscan anticline, the Quadrant 109 Arch–High Haume Anticline, Lower Permian deposits are thin or absent (Jackson and Mulholland, 1993).

ence on sedimentation by separating the East Irish Sea Basin from the basins to the north.

The lithological contrast between Lower and Upper Permian rocks disappears both southwards and towards the basin margins. The complete Permian succession is represented by a largely aeolian sandstone facies in the south-eastern part of the East Irish Sea Basin and adjacent parts of North Wales and the Cheshire Basin (Figure 42). The eastern margin of the East Irish Sea Basin is marked by alluvial-fan breccias fringing the Lake District Massif; this is the Brockram of Arthurton et al. (1978). On the southern and western margins of the Lake District, these breccias can be shown to span both the Early and Late Permian. Breccias are also present in southern Scotland, and at the head of Strangford Lough, and south of Belfast, in Northern Ireland (Figure 35). All the breccias are believed to be associated with pre-existing fault scarps and penecontemporaneous faults.

Over most central and northern parts of the East Irish Sea Basin, the Permian succession is not easily divided on seismic-reflection records, but in favourable locations and in the south, two packages can be recognised. The lower package shows a few weak events and largely corresponds to the Lower Permian; it oversteps and truncates Carboniferous reflectors. The upper package corresponds approximately to the Upper Permian, and comprises from one to three, persistent, high-amplitude reflections which are correlated with thin carbonate and anhydrite layers identified in wells. Triassic rocks generally succeed the Permian conformably, but overlap the Permian at the basin margins (Jackson and Mulholland, 1993, fig. 9a).

LOWER PERMIAN

The Lower Permian sediments in and around the north of the report area were deposited in narrow, intermontane grabens and half grabens, and are generally coarser grained than those farther south. The strata are now believed to be thin or absent in the centre of the East Irish Sea Basin and to comprise thick aeolian sandstones in the southern East Irish Sea Basin (Jackson and Mulholland, 1993); aeolian sandstones and sandstones with adhesion ripples predominate in North Wales and the Cheshire Basin (Figure 38). The strata are best known from onshore basins; wells penetrating Lower Permian rocks in the East Irish Sea Basin have largely been drilled on structurally complex highs with thin successions.

In the onshore basins of Northern Ireland, southern Scotland and north-west England (Figure 35), the sequences are broadly similar. The basal part of the Lower Permian in Northern Ireland and southern Scotland contains volcanic rocks locally, and there was probably contemporaneous volcanicity near Humphrey Head south of the Lake District (Rose and Dunham, 1977). The sedimentary infill of the half grabens is believed to consist typically of a Lower Breccia division, a Sandstone division and an Upper Breccia division (Figures 39 and 40). These facies are generally sequential, but interfinger to some extent. Although the sequences in individual basins are unlikely to be time equivalent, it is likely that there was broadly contemporaneous evolution in response to regional climatic and tectonic changes (Smith, 1972).

The Lower Breccia division is found mainly in the Scottish basins to the north of Dumfries (Brookfield, 1978), but it may also be present at depth in the Dumfries and Stranraer basins (Figure 35). North-east of the Lake District it is represented by the Penrith Brockram (Burgess, 1965), which is interpreted as both a product of sheetfloods, and as wadi fans deposited in narrow, intermontane basins from fluidised debris flows. Some of the sediments show evidence of aeolian reworking.

The Sandstone division consists of distinctive, well-sorted, quartzitic sandstone which is typically mica free; it forms over half the Lower Permian succession. These beds were laid down in central 'sandsea' areas of the basins termed draas, and consist mainly of dune sand and the deposits of the interdune areas (Brookfield, 1978; 1989). In some sections, the aeolian sandstones are interbedded with sandstones of fluvial origin. In the south of Scotland, the palaeowind direction was uniformly from the east-north-east (Brookfield, 1989).

The Upper Breccia division is well developed in the Stranraer (Stone, 1988) and Dumfries basins (Figures 35 and 40), and also forms a residual deposit generally less than 3 m thick to the west and south-west of the Lake District (Arthurton et al., 1978). East of Belfast, it is represented by a polymict breccia less than 4 m thick. In the Kingscourt Outlier in the Republic of Ireland, beds near the top of the lithologically similar Conglomerate Member have yielded Late Permian pollen (Visscher, 1971). Nevertheless, the larger part of this breccia was regarded as Lower Permian by Smith et al. (1974).

North Channel Basin

At the western margin of the North Channel Basin on the Antrim coast, the Larne No. 2 borehole (Figure 35) penetrated 617 m of Lower Permian trachyte, trachyandesite and pyroclastic rocks (554 m) resting on over 63 m of breccia, conglomerate and sandstone belonging to the Lower Breccia division (Penn et al., 1983). The full thickness of this succession has not been proved. To the south, trachyte pebbles were recorded in a Permian breccia in the Ballyalton borehole (Bazley, 1975). In Larne No. 2, the Sandstone division is represented by 440 m of red, interbedded dune and fluvial sandstones (Figure 40). The fluviatile beds indicate deposition in alluvial fans or braided streams (Penn et al., 1983). The sequence thins towards the south and west; in the Avoniel borehole (Figure 35), the equivalents of the Lower Breccia and Sandstone divisions are only c.28 m and 23 m thick respectively. The Upper Breccia division occurs in the Belfast Harbour borehole, where the grey-green colour of the 3.5 m-thick Coolbeg Basal Breccia (Smith, 1986), and its equivalent nearby at Cultra (Griffith and Wilson, 1982), may be related to chemical reduction associated with the Late Permian marine transgression.

Solway Firth Basin and adjacent basins

No Lower Permian sections have been penetrated offshore in the Solway Firth Basin, but the adjacent land geology (Figure 40) points to its likely nature. In the Stranraer Basin (Figure 35), the Permian strata may be up to 1500 m thick (Mansfield and Kennett, 1963); the Loch Ryan Breccia Formation, equated with the Upper Breccia division, probably conceals an aeolian sandstone (Brookfield, 1978; Stone, 1988). The Dumfries Basin may contain up to 1600 m of Permian strata; about 1400 m of aeolian sandstone of the Locharbriggs Sandstone Formation pass upwards into, and are interbedded with, alluvial-fan deposits of the Doweel Breccia Formation (Brookfield, 1978). North of the Dumfries Basin (Figure 38), earliest Permian lavas, agglomerates and subaerial basic tuffs form part of a line of volcanic vents extending north-westwards from the Thornhill Basin to Ardrossan (Brookfield, 1989), and possibly as far as Glas Eilean near Islay (Upton et al., 1987); other more speculative subparallel lines through sites of Permian volcanicity can be drawn to the south-west, firstly through Ballantrae, Kirkcudbright and Humphrey Head, and secondly through Larne, Ballyalton and block 110/2b-10 (Hardman, 1992).

The Solway Firth Basin extends eastwards into the Vale of Eden Basin (Figure 35), where the Lower Breccia and Sandstone divisions are represented by 100 m of Penrith Brockram and the overlying, 400 m-thick, Penrith Sandstone respectively (Burgess, 1965; Arthurton et al., 1978). North-west of the Lake District on the margin of the Solway Firth Basin, the Lower Breccia division is represented by 40 m of breccia in the Silloth 1A borehole. This passes up into 344 m of Penrith Sandstone, which shows a uniform, cylindrical, log motif. The sandstone is red, coarse grained and friable, with well-rounded and well-sorted grains that show the characteristic bimodal grain distribution of aeolian sandstone. Interbedded fine-grained sandstones, comprising subrounded quartz grains and a cement of dolomite, anhydrite and silica, may be interdune deposits.

At the south-western margin of the Solway Firth Basin, up to 11 m of presumed Lower Permian deposits are known

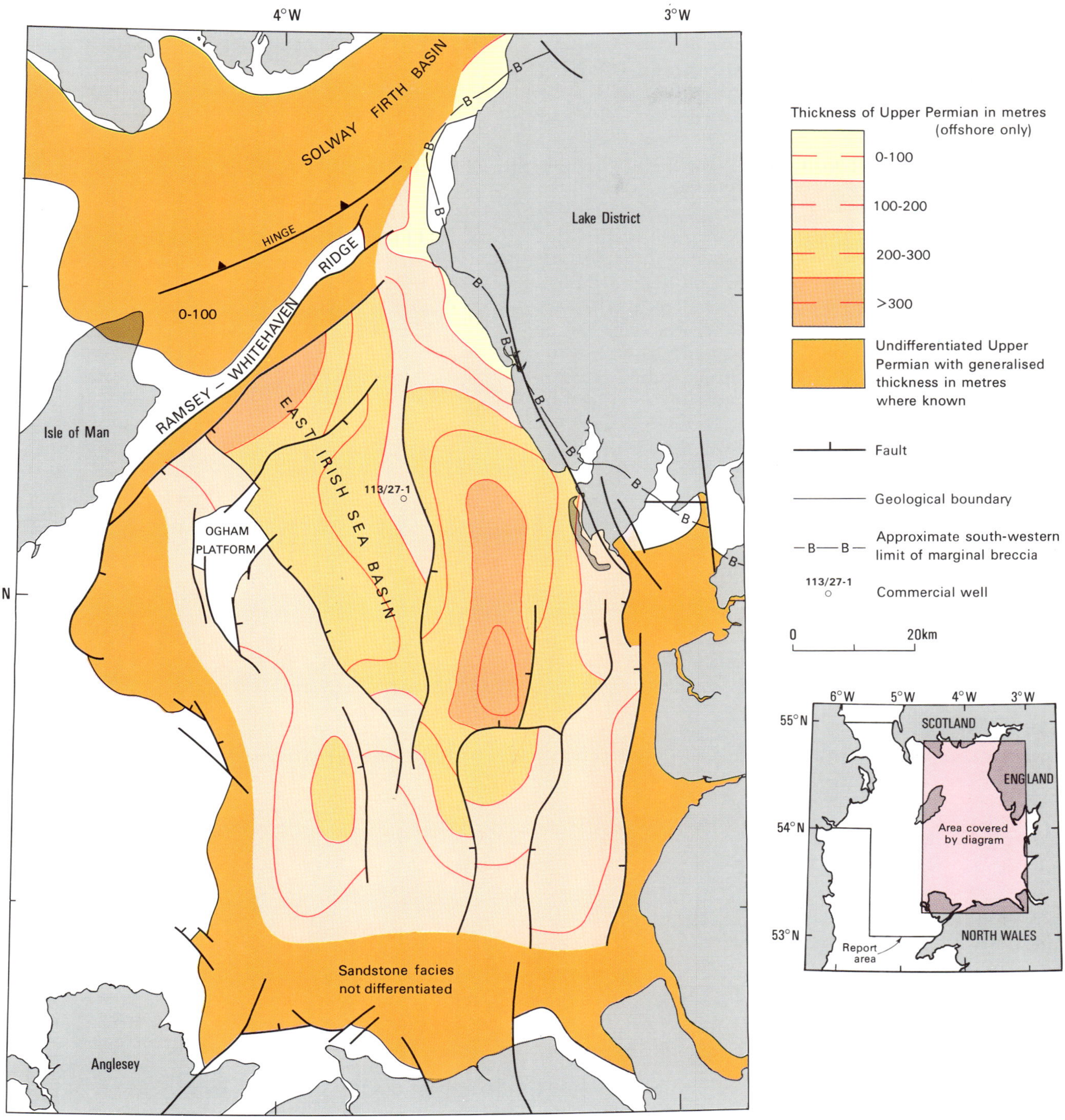

Figure 37 Thickness of Upper Permian sediments in the East Irish Sea Basin. Isopach values need to be modified following recent drilling results and reinterpretation of successions in earlier wells as fault-bounded and halokinetically-disrupted sequences (see Jackson, 1994).

from boreholes on the Isle of Man (Figure 40; Lamplugh, 1903). A lower, calcareous, breccia-rich division containing small pebbles and cobbles of dolomitized Dinantian and Lower Palaeozoic rocks is overlain by coarse-grained, brown, aeolian sandstones or gritty shales.

East Irish Sea Basin and surrounding land areas

On the eastern flank of the East Irish Sea Basin, the Basal Breccia (Arthurton et al., 1978) of south and west Cumbria (Figure 35) is here interpreted as the same facies as the Upper Breccia division. This widespread breccia is generally only 1 to 3 m thick, but exceptionally up to 26 m (Rose and Dunham, 1977); it is separated from Upper Permian deposits by a nonsequence. Locally, the breccia infills hollows and fissures in the underlying Carboniferous rocks; this is interpreted as a residual piedmont gravel, largely or wholly of Early Permian age. Near St Bees Head, the top 5 to 10 cm are reworked, locally pyritic, and penetrated by sand-filled fissures. Chemical reduction of these highest beds possibly occurred as the sediments were inundated by the Bakevellia Sea.

Farther south around Formby, the dominantly aeolian Collyhurst Sandstone ranges up to 715 m in thickness (Figures 35, 36 and 40; Falcon and Kent, 1960), and thickens towards the North Wales coast, possibly indicating the passive infill of an original topographic depression (Jackson and Mulholland, 1993). South of a line through Birkenhead, Huyton and Warrington, the Collyhurst Sandstone becomes increasingly difficult to differentiate as the overlying Manchester Marls pass laterally into a sandstone facies, the Kinnerton Sandstone Formation (see Figure 43; Colter and Barr, 1975; Warrington et al., 1980). It is tentatively suggested that in the Heswall borehole on the Wirral Peninsula (Wade, 1910), the basal 60 m of the 'Kinnerton Sandstone Formation' corresponds to the Collyhurst Sandstone.

The 'Kinnerton Sandstone Formation' of the Vale of Clwyd crops out in two structural depressions, one on the coast around Rhyl, and the other inland around Ruthin (Figure 35); the preserved thicknesses are estimated at 300 m and 525 m respectively (Powell, 1956; Wilson, 1959; Collar, 1974). Friable, red, aeolian sandstones are associated with subordinate, cemented sandstones containing silty bands and mud-flake breccias of fluvial origin (Warren et al., 1984). Interpretation of offshore seismic records from the Dee Estuary (Figure 36), and correlation with the Heswall borehole, support the suggestion of Smith et al. (1974) that the 'Kinnerton Sandstone Formation' preserved in the Vale of Clwyd is entirely of Early Permian age and thus equivalent to the Collyhurst Sandstone.

Offshore in the East Irish Sea Basin, the Lower Permian succession (Figure 41) consists largely of uniform, clean sandstones with cylindrical ('box-car') gamma-ray and sonic-log motifs; these are equivalent to the Collyhurst Sandstone of south Lancashire and the northern Cheshire Basin (Jackson et al., 1987). However, pale grey, fine-grained sandstone proved in BGS borehole 70/05 near the Anglesey coast (Figure 41; Wright et al., 1971) may indicate colour reduction by hydrocarbons. Deposition is now believed to have occurred in two Caledonoid-trending basins to the north and south of the Quadrant 109 Arch–High Haume Anticline over which Lower Permian deposition was thin or non-existent. The deposits proved to date in the northern sub-basin (e.g. 112/25a-1, Figure 41) are much thinner than those in the larger and deeper southern sub-basin (Jackson and Mulholland, 1993).

In the southern part of the East Irish Sea Basin, the red-brown to grey Collyhurst Sandstone is exceptionally thick throughout an east-north-easterly trending belt, or draa, which can be mapped from east of Anglesey to the Formby area; it reaches a maximum thickness of 1150 m, exceeding the 715 m proved in Formby No. 1 (Figure 41). The lithology is predominantly medium- to coarse-grained, friable aeolian sandstone with frosted grains. More scarce very fine- to medium-grained sandstones with subangular to subrounded grains possess a low to moderate porosity, low to moderate bimodal sorting, and a clayey, dolomitic, or more rarely anhydritic, cement. By analogy with other draas and with the southern North Sea, these latter sandstones are tentatively considered to be interdune deposits with thin adhesion sheets infilling damp hollows and wind-shadow areas. Near the basin centre, two upward-fining cycles recognised by Jackson et al. (1987) in wells 113/26-1 and 110/8-2 are now interpreted as secondarily reddened Silesian.

In the centre of the East Irish Sea Basin, the Lower Permian succession is now believed to be much thinner. Here, Smith et al. (1974) and Colter and Barr (1975) interpreted interbedded red mudstones, siltstones and sandstones between Upper Permian deposits and grey-coloured measures of the Carboniferous, as a playa-lake facies of Early Permian age, comparable with the Silverpit Formation of the southern North Sea (Cameron et al., 1992); they are characterised by spiky gamma-ray and sonic-log profiles. However, a problem arises in the East Irish Sea Basin in attempting to identify the Permian–Carboniferous boundary in thinly interbedded red sandstones and mudstones, namely between primary redbeds and secondarily reddened measures of Silesian age. In south Cumbria, Dunham and Rose (1949) and Rose and Dunham (1977) experienced similar problems in placing the Permian–Carboniferous boundary in boreholes; they concluded that in many instances the Lower Permian is absent entirely. Thus, the putative playa-facies, wholly or in part, is now interpreted as secondarily reddened Silesian rather than Lower Permian beds. In this reinterpretation, the Lower Permian succession is taken to comprise merely the 10 m-thick clean sandstones with 'box-car' gamma-ray profiles directly beneath the Upper Permian, as for example in wells 110/8-2 and 113/26-1 (Figure 41).

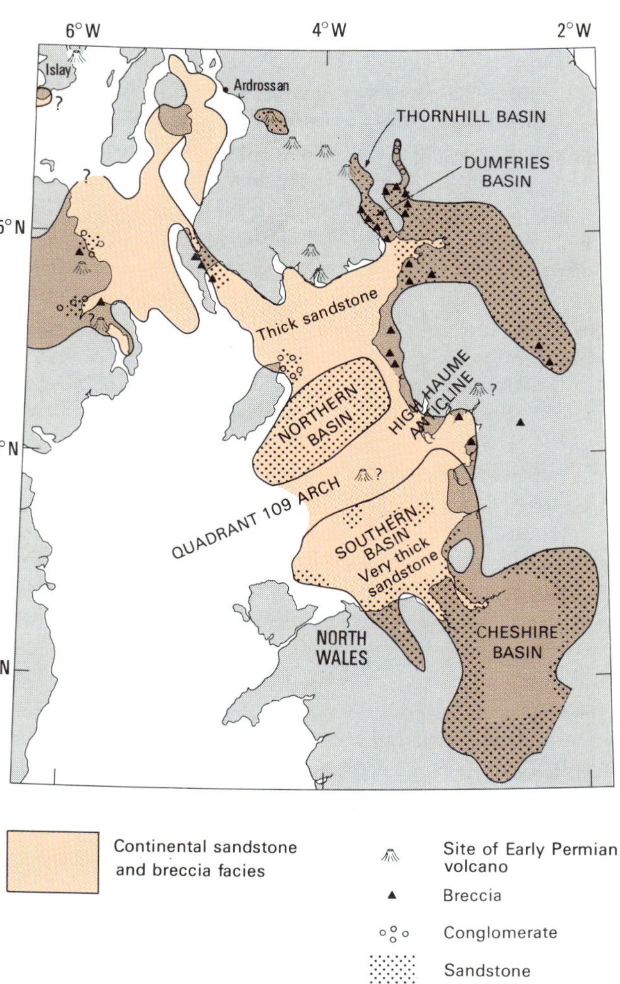

Figure 38 An Early Permian palaeogeography of the Irish Sea. Adapted from sources quoted in text.

UPPER PERMIAN

The Upper Permian deposits in and around the report area consist of a complex interfingering of basinal evaporites surrounded by coastal plain mudstones and fluvial sandstones, and passing outwards to localised marginal breccias and aeolian sandstones. Depositional cycles, named BS (Bakevellia Sea) 1 to 4 (Jackson et al., 1987), can be demonstrated within the evaporitic sequences, and more speculatively elsewhere (Figures 41 and 42); each has been correlated with cycles recognised in a condensed sequence by Arthurton and Hemingway (1972) onshore on the axis of the intra-basinal divide of the Ramsey–Whitehaven Ridge (Figure 35).

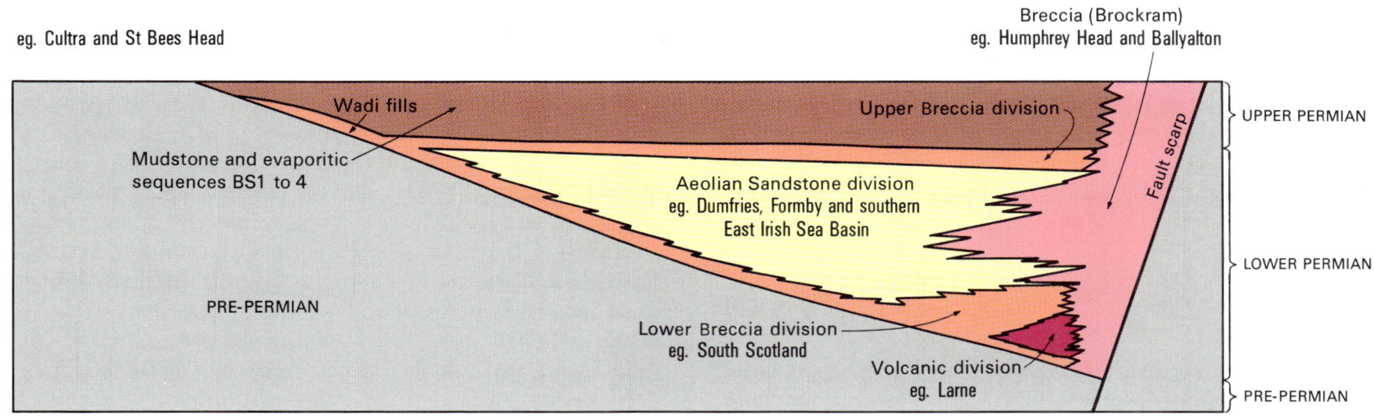

Figure 39 Schematic cross-section of a typical half-graben, showing facies relationships of the Lower Permian deposits.

A tentative correlation with the Zechstein cycles of north-east England and the North Sea is shown in Figure 43, although the cyclical nature of the Bakevellia Sea deposits is less clear, and precise equivalence of Bakevellia Sea and Zechstein cycles should not be assumed; a revised correlation between the Bakevellia and Zechstein cycles has been suggested by Jackson (1994).

Cycles BS1 to BS3 represent a phase of repeated marine transgression and regression. Ideally, the base of a cycle is marked by a breccia pavement, beach deposit or regressive sandstone upon which carbonate sand or mud of the transgressive phase was deposited. During the regressive phase, extensive intertidal flats were established; algal mats covered the surface, trapping carbonate sediments, and periods of emergence allowed the growth of gypsum. Periodic flooding on high, supratidal areas produced characteristically varved anhydrite deposits. Halite deposition occurred in the final stages of evaporation (Hsü, 1972).

Only cycles BS1 and BS2 have yielded marine macrofaunas (Pattison, 1970), and the entire Upper Permian sequence shows a progressive upward increase in terrestrial deposition. Grey siltstones or carbonates of relatively constant thickness (c.26 m) commonly represent the BS1 cycle (Figures 42 and 43). In the centre and north of the East Irish Sea Basin, these are overlain by two, thicker, halite-dominated evaporitic sequences (BS2 and BS3) that are separated by a marker anhydrite, and which pass up into monotonous and ubiquitous red mudstones of the BS4 cycle. A similar, though thinner sequence with two persistent anhydrites occurs in the North Channel Basin, at Kingscourt (Visscher, 1971) and in south and west Cumbria. In west Cumbria at St Bees Head, the deposits of the first three cycles are referred to the St Bees Evaporites, and the fourth to the St Bees Shales.

Where the evaporites are absent, this four-fold division cannot be recognised unequivocably. Adjacent to the basin margin, and especially near contemporaneously active fault margins such as in the Strangford Lough Basin, on the western and southern fringes of the Lake District, and in the Vale of Eden Basin (Figure 35), the Permian successions consist almost entirely of breccias. Towards the southern margins of the Solway Firth Basin, red coastal-plain mudstones (the St Bees and Eden Shales) dominate the sequence (Figure 43), and the basal carbonate is poorly developed or absent; a similar but more arenaceous succession is found on the Isle of Man.

In the more rapidly subsiding areas of the northern and eastern Cheshire Basin, and northwards into west Lancashire, Upper Permian strata consist largely of the Manchester Marls (Figure 43), which are red mudstones, marine in the lower part, with subordinate carbonate beds and siltstones. In the western part of the Cheshire Basin and over the Llŷn–Rossendale Ridge of Jackson and Mulholland (1993), which separates the Cheshire and East Irish Sea basins, subaerial conditions prevailed, and the sequence consists largely of terrestrial sandstones (Kinnerton Sandstone Formation). A transitional zone stretches through north Cheshire and Merseyside (Figure 37), where the Upper Permian consists of sandstones with basal calcareous siltstones and thin limestones, overlain by mudstones.

Evaporite and mudstone sequences

BS1 cycle

In evaporite sequences of the East Irish Sea Basin offshore, deposits of the first cycle (Figures 42 and 43) consist either of coarse-grained shelly dolomite (Gleaston Dolomite and equivalents), or of calcareous siltstone (Saltom Siltstone and equivalents) overlain by fine-grained dolomite (Saltom Dolomite and equivalents). These were equated by Jackson et al. (1987) with the Saltom Cycle of the St Bees Evaporites at St Bees Head (Arthurton and Hemingway, 1972), and lithological equivalents can be recognised onshore in Northern Ireland and at Kingscourt. This initial transgression of the Bakevellia Sea contains the most diverse marine faunas, in beds interpreted as largely shallow marine or littoral. The lower boundary is sharp, and probably represents a nonsequence with the Lower Permian (Arthurton and Hemingway, 1972); the upper boundary, where it can be well defined, appears to occur within a carbonate unit, and corresponds to a supposed nonsequence between the Saltom Dolomite and the Sandwith Dolomite (Arthurton and Hemingway, 1972). The cycle averages 26 m in thickness offshore, and shows little variation.

In Belfast, the BS1 cycle has been identified in the Belfast Harbour borehole (Figure 43; Smith, 1986), where the basal Musgrave Clastic Member of the Belfast Harbour Evaporite Formation comprises pale grey sandstone overlain by laminated siltstone, and is equated here with the Saltom Siltstone. The carbonate facies is represented by the overlying micritic limestone members, consisting of pale grey to pinkish grey micrite with algal-mat dolomite and gypsum nodules comparable with the Saltom Dolomite. In the Larne No. 2 borehole (Penn et al., 1983), the BS1 cycle is represented by the lower part of a well-developed, 22 m-thick, pale grey to white, fine-grained dolomite — the Magnesian Limestone.

	NORTHERN IRELAND (Larne No.2 and Belfast Lough)	STRANRAER BASIN	DUMFRIES BASIN	VALE OF EDEN	SOUTH AND WEST CUMBRIA	ISLE OF MAN	CHESHIRE BASIN AND LANCASHIRE	VALE OF CLWYD AND THE WIRRAL also SW CHESHIRE BASIN
UPPER BRECCIA DIVISION	Polymict breccia (<4m) and Coolbeg Basal Breccia	Loch Ryan Breccia Formation (>150m)	Doweel Breccia Formation ?(up to 1400m)	'Rotten' Brockram (up to 100m)	Basal Breccia (26m)			
SANDSTONE DIVISION	Lower Permian Sandstone, with dune-sand member just above the base (440m)	Sandstone	Locharbriggs Sandstone Formation (up to ?1000m)	Penrith Sandstone (400m)		Aeolian Sandstone (10-15m)	Collyhurst Sandstone (715m)	Kinnerton Sandstone Formation (525m)
LOWER BRECCIA DIVISION		? Breccia present at depth	?Breccia present at depth	Penrith Brockram (up to 150m)		Calcareous breccia (2-7m)		
VOLCANIC DIVISION	Lower Permian volcanic rocks, including breccia-conglomerate at base (617m)							

Figure 40 Nomenclature of onshore Lower Permian strata. Estimated maximum thicknesses are shown in brackets. The dating of most of these strata is problematical, and some, especially the Upper Breccia Division of south-west Scotland, may be younger than Early Permian. Based on sources quoted in text.

To the west at Avoniel (Figure 35), and at Grange in County Tyrone, a faunal break occurs towards the top of the shelly Magnesian Limestone (Pattison, 1970) at a thin, but persistent, barren limestone that may correspond to the barren Saltom Dolomite of west Cumbria (Arthurton and Hemingway, 1972) and to a dolomitic claystone reflected by a prominent log break in well 110/7-2 in the East Irish Sea Basin (Figure 41). Coeval strata of the Kingscourt Gypsum Formation (Figure 43) are the grey mudstones, shales and laminated siltstones of the Lower Mudstone Member, which contains acritarchs, foraminifera and palynomorphs (Visscher, 1971; Wilson, 1972).

In the East Irish Sea Basin, the calcareous siltstone facies of the BS1 cycle occurs as blue-grey shale and siltstone up to 33 m thick in well 110/9-1 (Figure 41), and grey, calcareous siltstone, up to 27 m thick with a rich palynoflora, in well 113/26-1. The siltstone facies was also proved in BGS boreholes 73/54 and 72/74; the latter also penetrated the Saltom Dolomite equivalent (Wilkinson and Halliwell, 1979).

The coarse-grained dolomite facies is seen in well 110/7-2 (Figure 41), where 22 m of limestone resting on argillaceous dolomite are interpreted as a shallow-water shoal deposit on the Godred Croven Platform. This unit is correlated with most of the Gleaston Dolomite of south Cumbria, with the Saltom Dolomite of west Cumbria, and with the lower 'Magnesian Limestone' of other onshore areas. It generally consists of impure dolomite.

In well 110/8-2 in the West Deemster Basin, 8 m of interbedded black dolomite and dolomitic shale overlying 11 m of grey, carbonaceous sandstone are regarded as belonging to the BS1 cycle. The beds were deposited under reducing conditions, in common with the similarly coloured Hilton Plant Beds of the Vale of Eden Basin, although it is possible that the highest 8 m are the basinal carbonates of the BS2 cycle. The carbonaceous sandstone includes dolomite and shale interbeds, and contains incorporated millet-seed grains possibly reworked from the Collyhurst Sandstone.

The Saltom Siltstone passes laterally into the red basal Manchester Marls in west Lancashire, which at Skellow Clough, near Parbold (Figure 35), are 9 m thick and yield a Late Permian fauna including *Bakevellia* (Pattison, 1970). Near the southern limit of the marine Upper Permian in the East Irish Sea Basin, a thin basal dolomite and mudstone beneath the Kinnerton Sandstone Formation in Formby No. 6 (Colter, 1978) yield foraminifera in the lowermost 4 m, and *Bakevellia* 9 m above the base, from either the BS1 or BS2 cycles (Pattison, 1970).

BS2 CYCLE

The deposits of this cycle are about 16–230 m thick in the East Irish Sea Basin (Figure 41) and have been equated by Jackson et al. (1987) with the Sandwith Cycle of west Cumbria (Arthurton and Hemingway, 1972). Halite sequences of this cycle occupy the centres, and areas of most rapid subsidence, of the East Irish Sea and North Channel basins. The halites are considerably thicker (222 m maximum) than the surrounding anhydrite and carbonate sabkha facies, and the crudely subconcentric arrangement of evaporite facies can be compared with the Zechstein basin (Taylor, 1990).

In west Cumbria, the condensed BS2 cycle deposits (Arthurton and Hemingway, 1972) begin with the thin Sandwith Dolomite (1 m), a transgressive, shelly, marine-shoal or beach carbonate containing *Bakevellia* and *Schizodus*, with a capping of intertidal, algal-mat dolomite. It is overlain by the regressive, 15 m-thick Sandwith Anhydrite, which is interpreted as an intertidal- to supratidal-flat deposit. It correlates with the Haverigg Haws Anhydrite of south Cumbria (Smith et al., 1974; Rose and Dunham, 1977), in which a more complete succession comprising the regressive parts of the cycle, with thick sandstones and breccia intercalations, has been preserved. Equivalent anhydrite and dolomite units have been proved offshore in BGS borehole 72/74 (Figure 41; Parkin and Crosby, 1982) near the Ogham Platform.

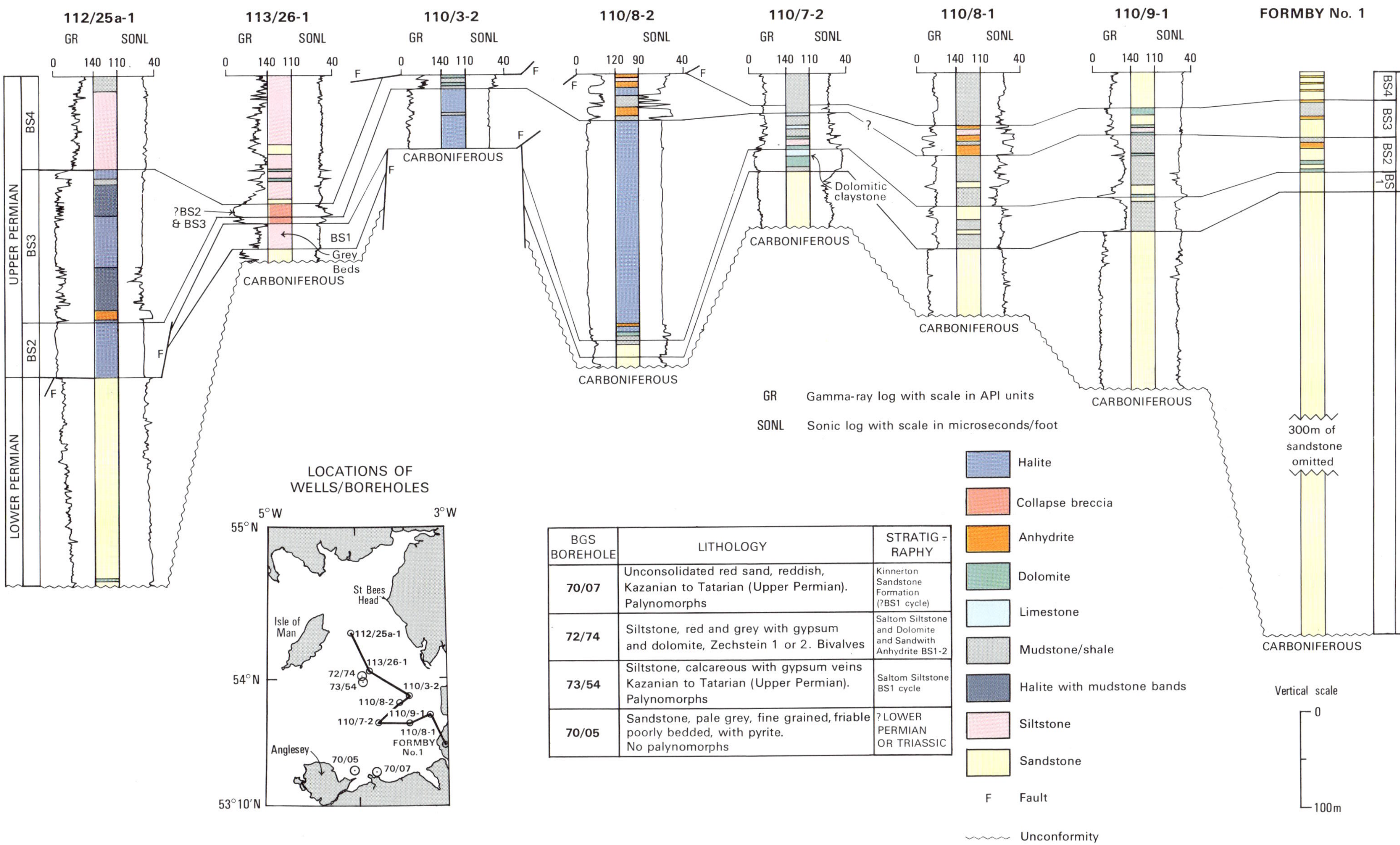

BGS BOREHOLE	LITHOLOGY	STRATIG - RAPHY
70/07	Unconsolidated red sand, reddish, Kazanian to Tatarian (Upper Permian). Palynomorphs	Kinnerton Sandstone Formation (?BS1 cycle)
72/74	Siltstone, red and grey with gypsum and dolomite, Zechstein 1 or 2. Bivalves	Saltom Siltstone and Dolomite and Sandwith Anhydrite BS1-2
73/54	Siltstone, calcareous with gypsum veins Kazanian to Tatarian (Upper Permian). Palynomorphs	Saltom Siltstone BS1 cycle
70/05	Sandstone, pale grey, fine grained, friable poorly bedded, with pyrite. No palynomorphs	? LOWER PERMIAN OR TRIASSIC

Figure 41 Logs of a number of boreholes proving Permian strata in the East Irish Sea Basin. Formby section after Kent (1948).

Deposits of the anhydrite facies of the BS2 cycle of Ireland (Figure 43) include the Lower Gypsum Member of Kingscourt (Visscher, 1971), and the 'A' Anhydrite Member in the Belfast Harbour Evaporite Formation (Smith, 1986), for which a coastal sabkha environment was postulated (Wilson, 1981). Evaporitic sequences may also be present in the Kish Bank Basin, and within a limited axial zone in the Solway Firth Basin (Figure 42).

A 222 m-thick, clean halite occurs in well 110/8-2, and a similar lithology was proved in the Larne No. 2 borehole to be 113 m thick (Penn et al., 1983), in both cases it overlies a thin basal anhydrite (Figures 41 and 43). The latter is also seen in well 113/27-1 (Figure 37), and is tentatively associated with the Basalanhydrit Formation of the southern North Sea (Cameron et al., 1992). In wells 110/3-2 and 112/25a-1 (Figure 41; Jackson et al., 1987), the anhydrite is probably faulted out by low-angle glide planes or *décollements* at the base of the Upper Permian halites.

In well 113/26-1 (Figure 41), a breccia (17 m true vertical thickness) of siltstone clasts in a halite matrix has been reinterpreted in the light of recent drilling, and is now believed to encompass both the BS2 and BS3 cycles. It is regarded here as an attenuated residue, or collapse breccia, formed by halite solution or collapse near the Keys Fault, and perhaps complicated by halite withdrawal from the Keys Basin.

To the east and south of the evaporite sequences in the East Irish Sea Basin, the proportion of mudstone increases sharply towards the Manchester Marls succession (Figure 42). However, in wells 110/8-1 and 110/7-2 on intrabasinal platforms, the transgression is still marked by a basal carbonate. Both wells contain a median sandstone, together with vestigial anhydrite in well 110/8-1. Similarly, at Skellow Clough, near Parbold (Figure 35; Jones et al., 1938), a 2 m-thick 'magnesian limestone' containing *Schizodus obscurus* (J. Sowerby) is equated with the 4 m-thick dolomite in well 110/9-1, which is taken as the base of the cycle (Figure 41). Sandy limestones occur at a similar position in Formby Nos. 3 and 6 (Colter, 1978). The remainder of the Formby succession (Figure 41) largely comprises sandstones of the Kinnerton Sandstone Formation, but anhydrite stringers in Formby No. 1 (Wray and Cope, 1948) probably mark the climax of sulphate deposition on the coastal sabkha of this cycle. A thin (0.15 m) halite bed, 99 m above the base of the Kinnerton Sandstone Formation in the Heswall borehole (Figure 35; Wade, 1910), is similarly taken here to represent the maximum extent of the BS2 evaporites, although the cycle cannot generally be identified south of Formby.

BS3 CYCLE

The BS3 deposits range from 12 to 170 m in thickness in the East Irish Sea Basin, and correlate with the Fleswick Cycle (Figure 43) in the St Bees Evaporites of west Cumbria (Arthurton and Hemingway, 1972; Jackson et al., 1987). The base of the cycle is taken as the base of the Fleswick Breccia and equivalents. The distribution of facies is similar to that of the BS2 cycle, except that the development of evaporite may be slightly more widespread (Figure 42). The only organic remains recorded from beds of this cycle are some poorly preserved pollen grains from Kingscourt (Figure 35; Visscher, 1971).

In west Cumbria, the Fleswick Breccia (2 m thick) marks a nonsequence at the base of the cycle (Figure 43) and is overlain successively by transgressive dolomites, siltstones, and regressive anhydrites (Arthurton and Hemingway, 1972). The Fleswick Dolomite and Fleswick Anhydrite have been correlated with the Roosecote Dolomite and Anhydrite respectively of south Cumbria (Smith et al., 1974). A comparable succession occurs both in the Avoniel borehole (Manning et al., 1970) and at Kingscourt (Visscher, 1971), but in the Belfast Harbour borehole carbonate and mudstone members are absent, and the 'B' Anhydrite Member (Smith, 1986) rests disconformably on the Refinery Breccia Member interpreted here as a lateral correlative of the Fleswick Breccia.

In the basin centres, clear and colourless halites (160 m thick) showing an upward increase in mudstone interbeds comprise the highest parts of the St Bees Evaporite Formation (Figure 41). In basinal successions, the base of the cycle is marked by a thin high gamma shale, overlain by a persistent anhydrite (6–10 m) with a dolomitic lower portion (wells 110/8-2, 112/25a-1 and 113/27-1) the same anhydrite is well developed in the Larne No. 2 borehole and it constitutes an excellent marker unit equated here with the Hauptanhydrit Formation of the southern North Sea (Cameron et al., 1992). On the Deemster Platform in well 110/8-1 (Colter, 1978), a basal siltstone overlain by three anhydrite beds 4 to 8 m thick, with intervening sandy shale, probably marks the position of the sulphate sabkha zone. Thinly interbedded anhydrite and mudstone characterise the lower part of the sequence in wells 110/8-2 and 110/3-1, where the remaining parts of the cycle are faulted out, and in well 113/27-1. In well 110/3-2, the highest Upper Permian deposits comprise black, silty dolomite interbedded with shale now regarded, following recent drilling, as a faulted slice of the basinal limestone facies of the BS3 cycle (Figure 41) rather than an unfaulted BS4 sequence as in Jackson et al. (1987).

The evaporite deposits of this cycle in the East Irish Sea Basin, like those of the BS2 cycle, pass southwards and eastwards through an intermediate belt of interbedded anhydrite and mudstones (well 110/8-1) into platform carbonates and mudstones of the Manchester Marls (Figures 42 and 43). The lateral facies-change southwards into the Kinnerton Sandstone Formation appears to occur farther north than in the BS2 cycle. The beds in well 110/9-1 (Figure 41) are largely sandstone with some interbedded siltstone and dolomite, and in well 110/7-2 (Jackson et al., 1987), a basal, grey, fine-grained sandstone, taken as the equivalent of the Fleswick Breccia, occurs below a limestone and mudstone sequence.

In most boreholes at Formby, cycle boundaries cannot be identified positively within calcareous sandstone that contains millet-seed grains (Kinnerton Sandstone Formation), but thin beds of marl and anhydrite in Formby No. 1 (Figure 41; Wray and Cope, 1948) are taken to indicate the feather edge of evaporitic deposits in this cycle.

BS4 CYCLE

Over large areas, the deposits of this cycle (Figure 42) comprise barren red mudstones equivalent to the St Bees Shales of Cumbria, but southwards and near the Lancashire coast they pass into the upper leaf of the Kinnerton Sandstone Formation. Bedded evaporite deposits are absent from this cycle, which shows fewer facies changes than earlier cycles.

In west Cumbria, the 62 m-thick St Bees Shales contain penecontemporaneous mudstone breccias and can be divided into a lower aeolian division and an upper, water-laid division (Arthurton et al., 1978). A Tomlin depositional cycle was recognised by Arthurton and Hemingway (1972) locally at the base of the St Bees Shales, but it is not clear how this cycle relates to the offshore succession. Minor lenticular breccias (brockram) also interdigitate with the mudstones in south Cumbria (Rose and Dunham, 1977).

Onshore correlatives of BS4 deposits (Figure 43) include the uppermost part of the Eden Shales in north-east

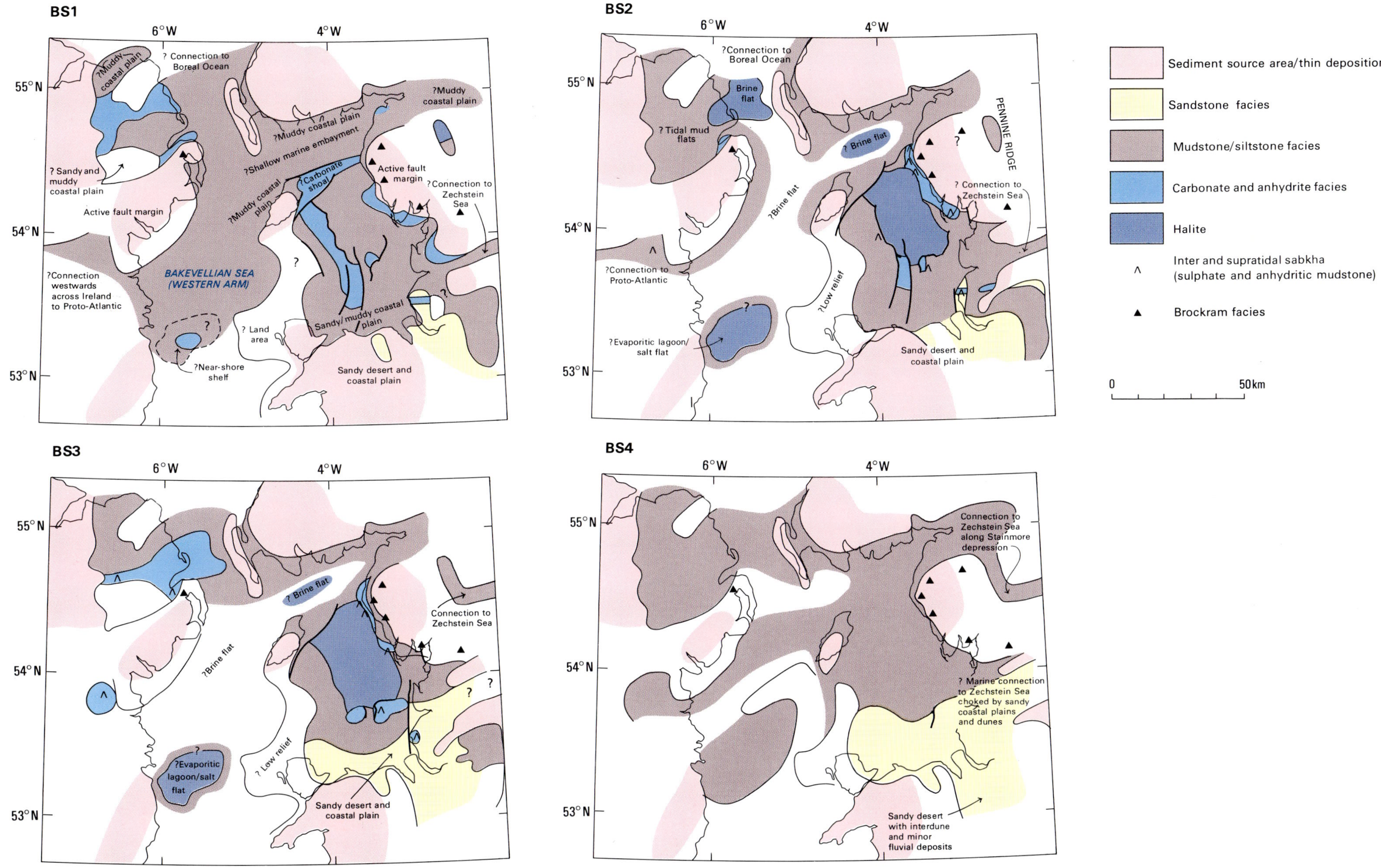

Figure 42 Palaeogeographic sketch maps of the Irish Sea in Late Permian times, showing the depositional patterns associated with the BS1 to BS4 transgressions. Based on sources quoted in text.

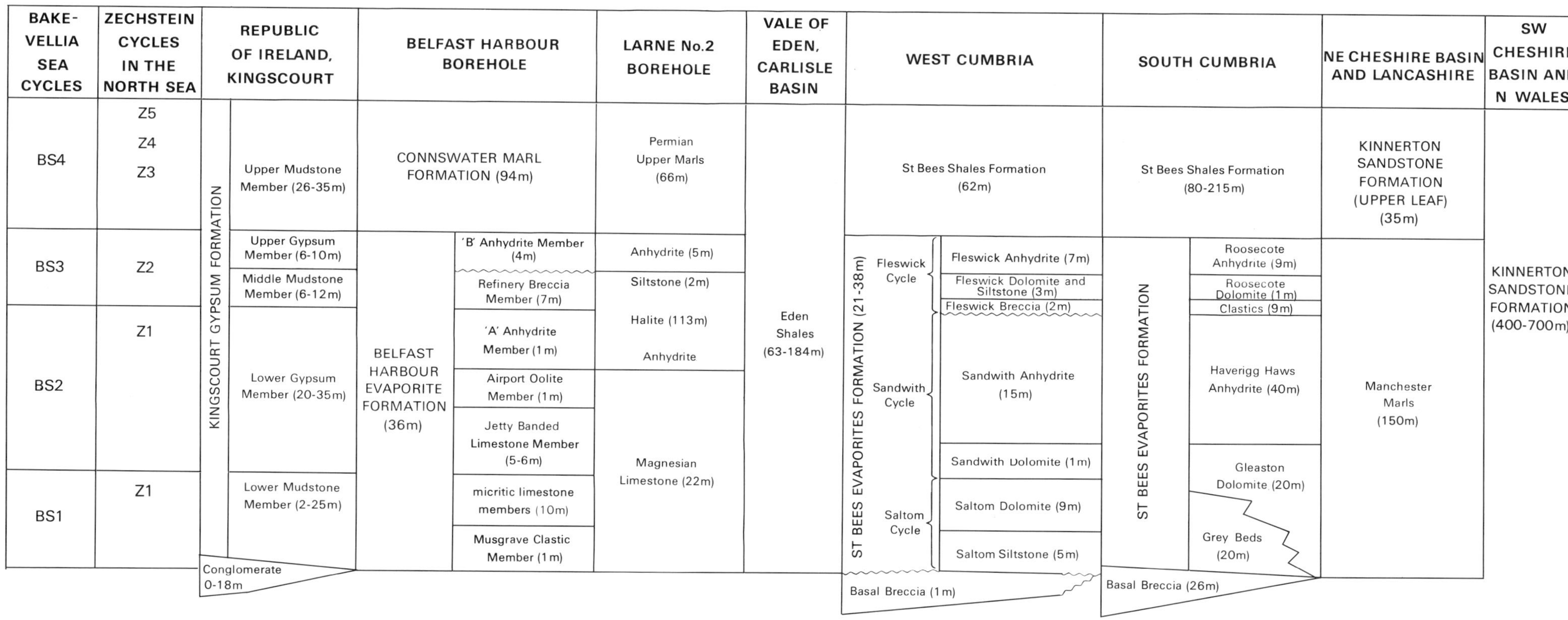

Figure 43 Correlation of onshore Upper Permian strata. This figure was compiled using the established correlations shown in Smith et al. (1974), Arthurton et al. (1978), Smith (1986) and Jackson et al. (1987). A revised correlation of the Bakevellia and Zechstein cycles and cycle boundaries, together with amendments to certain lithostratigraphic correlations, highlighting the importance of nonsequences in basin margin successions, has been made by Jackson (1994). Disconformities and nonsequences omitted.

Cumbria, and the Connswater Marl Formation of Belfast (Smith, 1986). Diagnostic palynological assemblages in BS4 strata have been discovered only at Kingscourt, where Late Permian (Thuringian) spores and pollen were recorded from the Upper Mudstone Member (Visscher, 1971).

The lower boundary of the cycle is taken at the sharp contact of mudstones with the underlying evaporite succession, as defined by Arthurton and Hemingway (1972) at St Bees Head. In offshore wells, the beds appear to thicken towards the north, and also towards the growth faults. This may indicate an increase in tectonic activity, with clastic deposition heralding the regional subsidence of the Early Triassic and the influx of coarser-grained sediment. The upper limit is marked by a decrease in radioactivity as recorded by gamma-ray logs at the sharp contact with the overlying Sherwood Sandstone Group; the boundary probably represents a nonsequence.

The succession is typically developed in well 112/25a-1 (Figure 41), where 104 m of red-brown calcareous mudstone reveal poorly developed, upward-fining sub-cycles that begin with a basal siltstone. In well 113/26-1, where the probably equivalent succession is thicker and more silty, there are two to three upward-fining cycles with thin basal sandstones recognised on the gamma-ray motif. Elsewhere in the East Irish Sea Basin, red-brown shales comprise the succession in wells 110/7-2 (34 m) and 110/8-1 (58 m). The deposits appear to thin to the south of 53° 45'N.

The model of a small evaporite centre with a fringe of carbonates in the BS4 cycle of wells 110/3-2 and 110/8-2 (Jackson et al., 1987) cannot now be sustained in the light of recent drilling. The highest beds of the St Bees Evaporites of these wells are now regarded as fault-bounded slivers and halokinetically disturbed parts of cycle BS3 (Figure 41).

Near the Lancashire coast, sandstones comprising an upper leaf of the Kinnerton Sandstone Formation are found in well 110/9-1, where a 39 m-thick, red-brown, upward-fining, micaceous silty sandstone occurs directly beneath the Sherwood Sandstone Group. In Formby No. 1 (Wray and Cope, 1948; Kent, 1948), 30 m of dark red-brown, gypsiferous and marly sandstone with millet-seed quartz grains of probable aeolian origin occurs beneath well-cemented, sparsely pebbly, Triassic sandstone of the St Bees Sandstone Formation.

LOWER AND UPPER PERMIAN MARGINAL SEQUENCES

Dating and internal correlation of the clastic sequences is particularly tentative outside the evaporite basins and towards the margins of the depositional basins. Successions consist of two types: firstly breccias, termed brockrams in Cumbria, that occur near fault-bounded massifs and contain mostly Lower Palaeozoic and/or Dinantian clasts, and secondly sandstones with minor shales (Kinnerton Sandstone Formation) in less rugged topography which were derived from a Namurian/Westphalian hinterland.

In south-west Scotland (Figure 35), the highest Permian is not preserved, and no argillaceous uppermost Permian rocks are seen. Smith et al. (1974) suggested that some of the water-laid breccias which overlie aeolian sandstones may be of Late Permian age, and argued that these debris-flow deposits represent erosion and deposition in a wetter climate than that prevailing during the Early Permian. Part of the Doweel Breccia Formation (Brookfield, 1978) of the Dumfries Basin (Figure 40) may belong to this phase of sedimentation. These water-laid deposits can be viewed as the northern counterparts of the Kinnerton Sandstone Formation.

In the marginal breccias of the Ballyalton borehole (Bazley, 1975) in the Strangford Lough Basin (Figure 35), the change from Lower to Upper Permian sedimentation may be reflected in the change from lower sand-matrix breccias (248 m thick) to 53 m-thick, upper, marl-matrix breccias.

On the Ramsey–Whitehaven Ridge, supposed Upper Permian shales which rest on Lower Permian breccias and sandstones are found on the Isle of Man and in the Silloth 1A borehole (Figure 35). The 63 m-thick shales at Silloth 1A are dark red-brown and micaceous ('Eden Shales'); they contain an 11 m-thick, basal, carbonate-rich unit, and an overlying unit which has a high gamma-ray signal. The units may be equivalent to the BS1 and basal BS2 cycles respectively of the East Irish Sea Basin. At the top, slightly more arenaceous, 11 m-thick shales containing two thin siltstone beds may be equivalent to the BS4 cycle. On the Isle of Man, there is a similar upward-coarsening succession some 30 m thick in which interbedded red sandstones become thicker upwards; a basal carbonate phase was not recorded (Lamplugh, 1903).

Adjacent to the Lake District highlands and the Lake District Boundary Fault, the Lower and Upper Permian rocks (in west and south Cumbria) thicken and pass eastwards into a marginal breccia, the Brockram, which crops out in a belt 8 to 9 km wide (Smith, 1924; Figure 42). Individual alluvial fans and lobes can be identified from variations in the locally derived clast content.

In the Humphrey Head borehole near Grange-over-Sands (Figure 35), the Brockram exceeds 257 m in thickness (Institute of Geological Sciences, 1975; Rose and Dunham, 1977). The coarser-grained conglomeratic phases can possibly be equated with the Basal (Lower Permian) and Fleswick breccias elsewhere, and may have been deposited by flash floods during periods of increased rainfall or tectonic activity. The intervening sandstone-rich units with anhydrite aggregates may correspond to drier phases in the Late Permian during which evaporites were forming in the basin centre. Olivine-dolerite and vesicular-basalt pebbles, probably derived from a former outlier of Early Permian volcanic rocks (now totally removed by erosion), predominate in the basal 100 m. Dolomitised Dinantian limestone clasts are dominant in the highest 103 m, and indicate progressive unroofing of cover rocks in the nearby hinterland.

In the north-west Cheshire Basin (Figure 35), the Manchester Marls pass southwards (Colter and Barr, 1975; Colter, 1978) into a sandstone facies, the upper leaf of the largely aeolian Kinnerton Sandstone Formation (Warrington et al., 1980), which is generally indistinguishable from the underlying Collyhurst Sandstone. The Kinnerton Sandstone Formation (formerly the Lower Mottled Sandstone) thickens west of Warrington towards the Mersey Estuary, and exceeds 262 m at Sealand near the Dee Estuary (Wedd and King, 1924). In the Heswall borehole on the Wirral Peninsula (Wade, 1910), mudstone and evaporite stringers provide faint signs of the cyclic deposition in the basin to the north, and indicate a Late Permian age for the 338 m-thick upper portion of the Kinnerton Sandstone Formation.

If contamination from the overlying drift is eliminated, the unconsolidated (63 m thick) red sand apparently containing Upper Permian palynomorphs in BGS borehole 70/07 (Figure 41) is provisionally regarded here as a facies equivalent belonging to the Kinnerton Sandstone Formation (Wilkinson and Halliwell, 1979).

7 Triassic

Triassic strata are widespread in the report area, where they lie conformably upon, and locally overlap, Late Permian rocks. At a few localities in and around the report area, they are overlain by Lower Jurassic strata, but the Triassic generally subcrops Quaternary sediments. The Triassic sequence is subdivided into the Sherwood Sandstone, Mercia Mudstone and Penarth groups (Figure 44). Seismic evidence indicates a maximum offshore thickness of over 2000 m for the Sherwood Sandstone Group, and 3200 m for the Mercia Mudstone Group. The Penarth Group has not been proved offshore; onshore it ranges merely to 23 m in thickness, but is widely distributed and is likely to underlie Jurassic strata in the Keys Basin and perhaps elsewhere (Figure 45). The Triassic has been proved in shallow BGS boreholes and in wells; the uppermost part of the Sherwood Sandstone Group is the reservoir for the Morecambe Gasfield and most other discoveries in the East Irish Sea Basin. The overlying Mercia Mudstone Group provides the seal.

The major Lower Palaeozoic highs of the Longford–Down Massif, the Southern Uplands Massif, the Lake District Massif, the Isle of Man, the Ramsey–Whitehaven Ridge, and the Welsh Massif, continued to influence sedimentation patterns during the Triassic (Figure 46). Triassic sediments thin towards these massifs, and only crestal areas of the most elevated examples probably remained above the general level of deposition. Sedimentation was probably continuous, albeit reduced, between the Kish Bank and East Irish Sea basins.

During Triassic times, there was a northward drift of the area from a palaeolatitude of 19°N to 30°N (Clegg et al., 1954; Smith and Briden, 1977). In Scythian to early Anisian times, during the deposition of the Sherwood Sandstone Group, the climate was semiarid, and the palaeowind direction was from the east-south-east or east-north-east (Thompson, 1969). Clemmensen (1978) attributed this directional variation to seasonal changes brought about by the migration of the trade-wind belt. During deposition of the Mercia Mudstone Group, from Anisian times until the onset of the late Norian/Rhaetian transgression, the climate was extremely hot. At certain times, evaporation exceeded precipitation, leading to the deposition of considerable thicknesses of evaporites, chiefly halite.

During the Early Permian, independently subsiding grabens and half-grabens in the UK had commonly acted as isolated depocentres. Such was the case for the East Irish Sea and Cheshire basins, but in later Permian and Early Triassic times, the two became linked into a single depositional milieu, initially related to the northerly palaeoslope of the Bakevellia Sea, and later to a single northward-draining river system sourced in the Variscan highlands of Brittany. This drainage system (Figure 46), the 'Budleighensis river' of Wills (1956), carried considerable quantities of coarse-grained sediments into the area, which was also the site of aeolian deposition especially towards the margins. The river system may have expired distally in a series of terminal fans

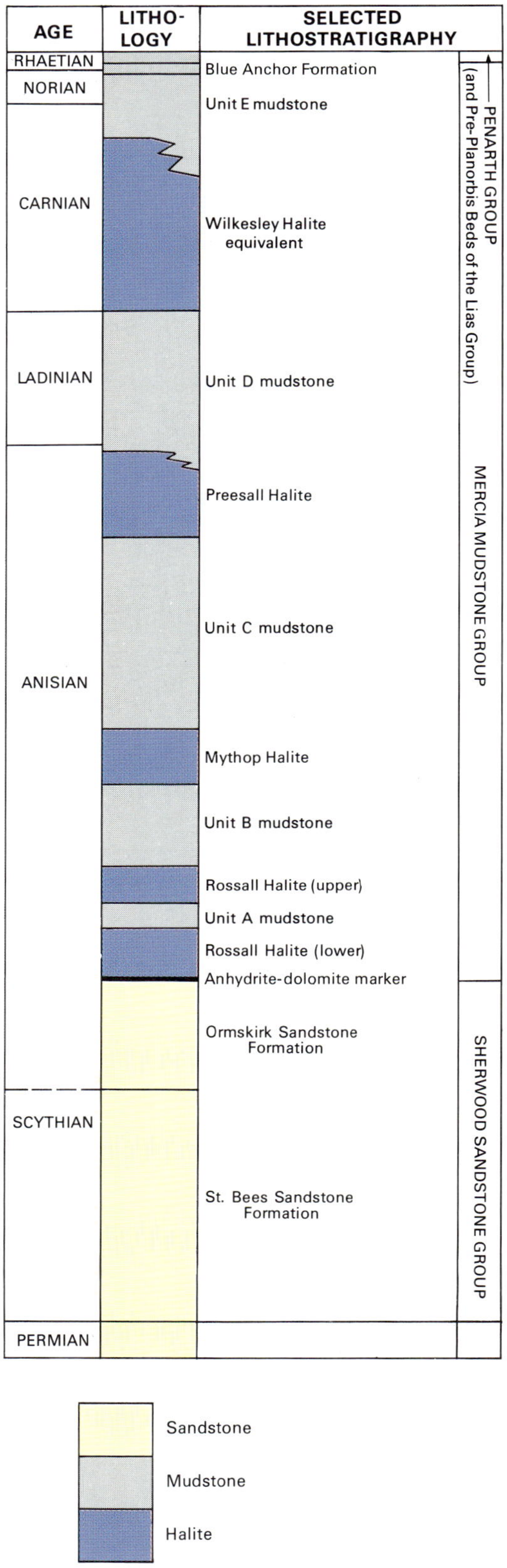

Figure 44 Lithostratigraphy of the Triassic of the Irish Sea region. Not to scale.

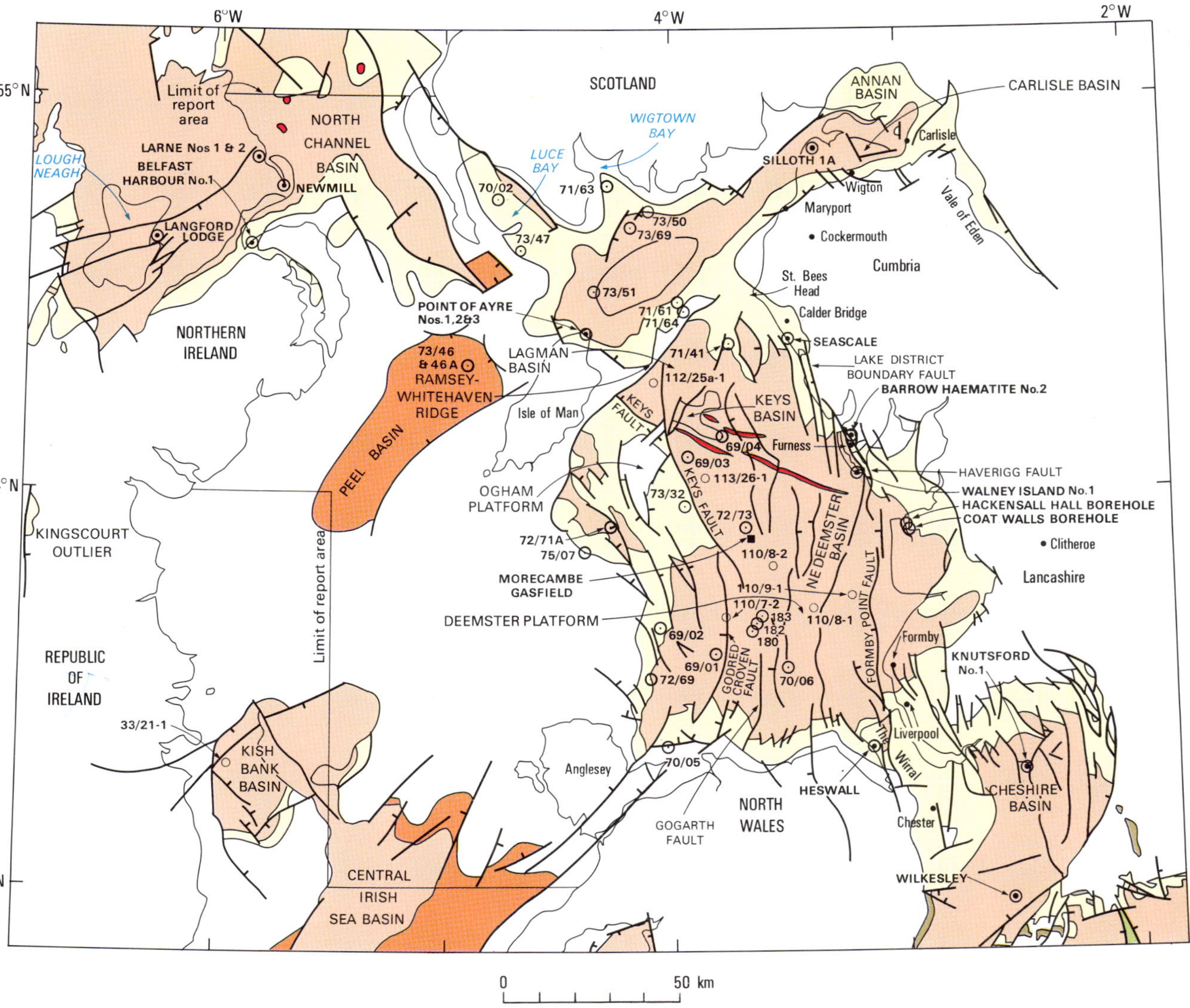

Mercia Mudstone Group and younger strata

Sherwood Sandstone Group

Permo-Triassic, undivided

Sandstone/pebbly conglomerate (Sherwood Sandstone Group)

Tertiary intrusion

○ 113/26-1 Commercial well

⊙ 69/01 BGS boreholes and shallow cores which proved Triassic strata

⦿ Borehole on land

BOREHOLE NUMBER	LITHOLOGY	DEPTH (m)	*AGE
ISLE OF MAN SHEET			
70/02	Sandstone, red-brown	31.0 to 42.5	-
71/63	Sandstone, red-brown	64.0 to 69.5	-
73/46A	Mudstone, red	62	-
73/47	Mudstone, red	37	-
73/50	Sandstone, red-brown	38.0 to 47.0	-
73/51	Siltstone, red-brown with gypsum	42.5 to 45.0	
73/69	Mudstone, red and red-green, with gypsum	26.0 to 46.7	ANISIAN
LAKE DISTRICT SHEET			
69/03	Mudstone, red-brown ,with gypsum	40.4 to 45.0	-
69/04	Mudstone, red and green, with gypsum	22.9 to 25.9	-
71/41	Sandstone, red	34.0 to 45.0	-
71/61	Sandstone, red-brown	18.5 to 31.5	-
71/62	Mudstone, red-brown and grey-green	73.5 to 90.0	ANISIAN
71/64	Sandstone, red	75.0 to 92.0	-
ANGLESEY SHEET			
69/02	Mudstone, red, green, grey mottling	11.0 to 14.0	LATE SCYTHIAN TO ANISIAN
70/05	Sandstone, pale grey	19.5 to 23.5	
72/69	Mudstone, red with grey mottling	23.5 to 30.1	EARLY TO MID TRIASSIC
72/71A	Silty mudstone, red and green with gypsum	21.0 to 24.0	
75/07	Sandstone, red, fine to medium grained	12.65 to 27.1	
LIVERPOOL BAY SHEET			
69/01	Mudstone, red-brown	6.7 to 17.67	
70/06	Mudstone, reddish brown	12.5 to 21.0	-
72/73	Mudstone, red-brown	23.5 to 34.5	-
180 Shallow core	Mudstone, red-brown and green, silty	-	-
182 Shallow core	Mudstone, red-brown, silty	-	-
183 Shallow core	Mudstone, red-brown, silty	-	-
73/32	Sandstone, red, micaceous	16.5 to 58.5	-

* Ages from palaeontological determination (Wilkinson and Halliwell, 1979)

Figure 45 Distribution of the Sherwood Sandstone, Mercia Mudstone and Penarth groups; and the results of BGS boreholes drilled into Triassic strata.

as a result of evaporation and downward filtration (Collinson, 1986), or it may have drained westwards, via Kingscourt in central Ireland, into the shallow-marine environment proposed by Croker and Shannon, (1987), in the Porcupine Bank area farther west.

By early Anisian times, the hinterland had been reduced to one of lower relief, and the 'Budleighensis river' system had lost much of its transporting power. Deposition of fluvial and aeolian sands ceased, and the environment changed to low-lying plains affected by periodic influxes of sea water (Taylor, 1983). Water-laid, laminated, silty mudstones, evaporites and wind-blown dust derived by the lateritic weathering of adjacent terrain and deposited as loess, are the dominant lithologies of the Mercia Mudstone Group (e.g. Arthurton, 1980). Subsidence continued in all the major depocentres throughout the Late Triassic and into Jurassic times; the Scythian to Norian succession of the East Irish Sea Basin is the thickest in or around the UK, and is one of the thickest in north-west Europe (Ziegler, 1990). The late Norian to Rhaetian transgression brought increasingly marine conditions to the area, leading to the deposition of the dark grey mudstones with limestones and sandstones that form the Penarth Group.

SHERWOOD SANDSTONE GROUP

The Sherwood Sandstone Group has been proved in a number of wells and BGS boreholes (Figure 45), and is estimated to be over 2000 m thick in the Keys Basin (Figure 47), but averages 1000 to 1500 m in thickness. In the East Irish Sea, Solway Firth and Kish Bank basins, it is overlain conformably, but abruptly, by the Mercia Mudstone Group. A nonsequence may mark the contact. The sonic velocity for the Sherwood Sandstone Group averages 4420 m/s, generally with a range between 4415 and 4723 m/s, although velocities exceeding 5334 m/s are recorded in the deeply buried, silicified lower sandstones.

The Sherwood Sandstone Group shows a distinctive response on seismic-reflection records, and can be divided into upper and lower packages (Jackson et al., 1987). The lower, thicker package is characterised by lower-amplitude, low-frequency events; it corresponds to the St Bees Sandstone Formation (Figure 48), which is recognised onshore north of Formby (Audley-Charles, 1970a; Warrington et al., 1980), and correlates with the combined Chester Pebble Beds and Wilmslow Sandstone formations of the Cheshire Basin (Figure 48). The upper seismic package corresponds with the Ormskirk Sandstone Formation of the East Irish Sea Basin (Jackson et al., 1987), and contains several high-amplitude, high-continuity events. The Ormskirk Sandstone Formation correlates with the Kirklinton Sandstone Formation of the Solway Firth and Carlisle basins, and with the Helsby Sandstone Formation of the Cheshire Basin (Figure 48). In Northern Ireland, only sediments of the lower package are recognised.

Near the middle of the St Bees Sandstone Formation in the East Irish Sea Basin, there are two, prominent, widely traceable seismic reflectors — the Brown and Yellow reflectors that divide the formation into three (Figure 48; Jackson et al., 1987). The Brown reflector marks the top of a thick, micaceous, silicified sandstone, and the resulting sonic-log break correlates with the Top Silicified Zone of Colter and Barr (1975) and Colter (1978). The reflector is strongest in the north and east of the basin, especially to the east of the Tynwald Fault Complex (Figure 47). The most clear-cut geophysical-log breaks occur in wells 110/8-2 and 113/26-1 (Figures 45 and 49), where the unit is most deeply buried.

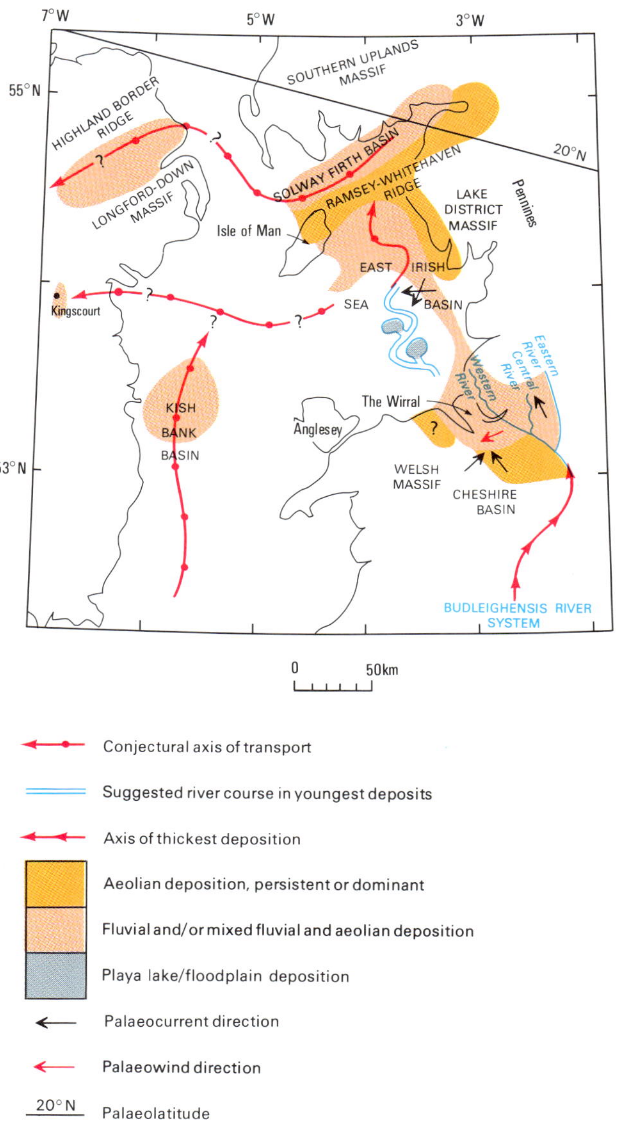

Figure 46 Possible palaeogeography during deposition of the upper Sherwood Sandstone Group (Ormskirk/Helsby Sandstone formations and correlatives). See text for attribution.

The impedance contrasts responsible for the seismic reflections and the sonic-log changes may mark a diagenetic effect resulting, for example, from a prolonged water-table stillstand (Burley, 1984), and they may therefore be diachronous. Nevertheless, the widespread occurrence of the Brown reflector overlying a thick, silicified sandstone unit provides a valuable marker. The Brown and Yellow reflectors are here correlated tentatively with the base and top of the Middle Sandstone Member at Kingscourt in the Republic of Ireland (Visscher, 1971), and on sonic-log character, the Silicified Zone correlates with the Bunter Shale Formation of the southern North Sea (Fisher and Mudge, 1990).

Towards and on the northern margin of the Cheshire Basin where the Chester Pebble Beds Formation becomes less pebbly (Wedd et al., 1923; Wray and Cope, 1947; Taylor et al., 1963), Jackson et al. (1987) proposed that the Top Silicified Zone also corresponds to the junction between the Chester Pebble Beds Formation and the overlying Wilmslow Sandstone Formation of the Cheshire Basin. However, the Chester Pebble Beds Formation is differentiated from the Wilmslow Sandstone Formation on the basis of primary sedimentological characteristics (Warrington et al., 1980); the

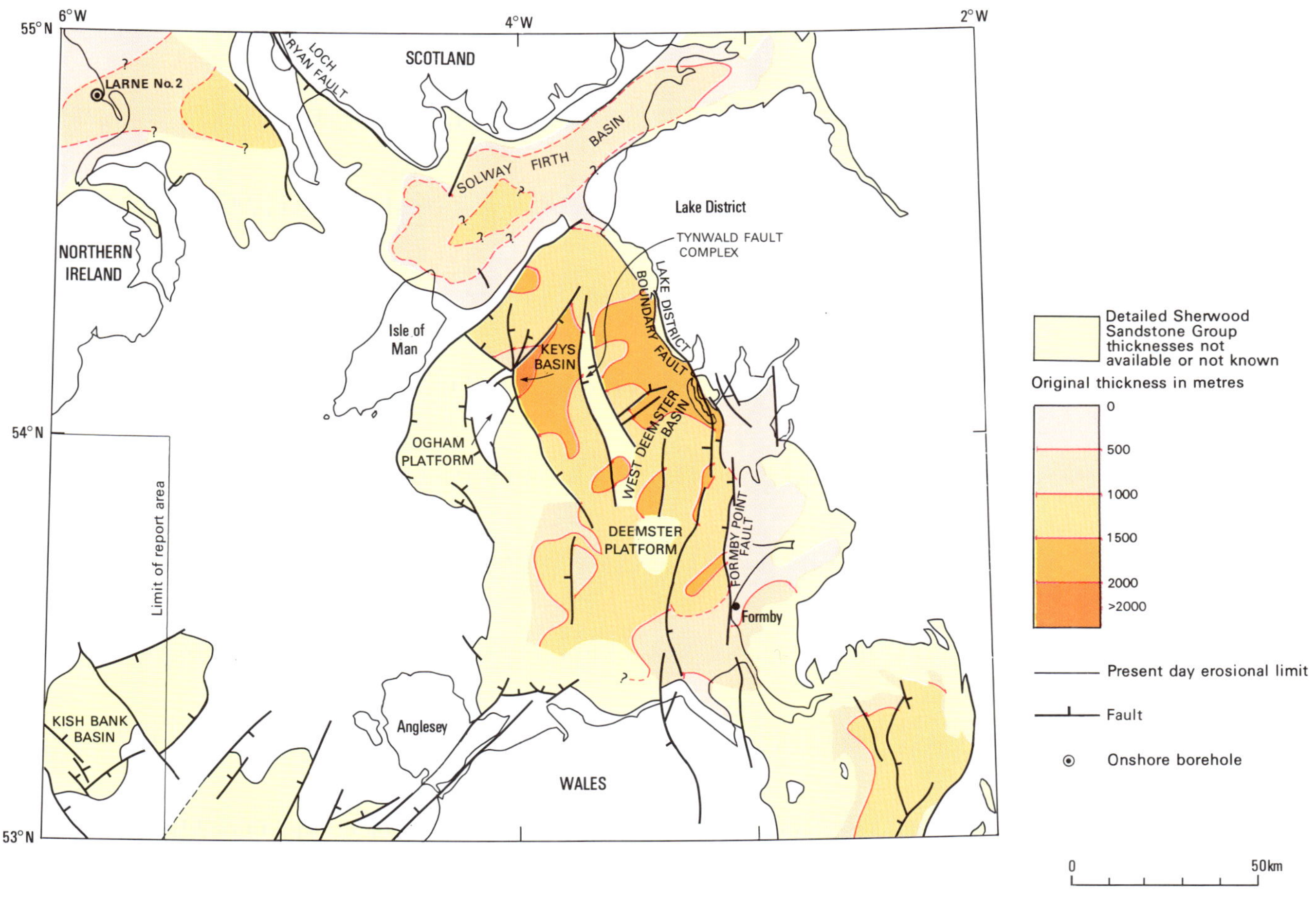

Figure 47 Thickness of the Sherwood Sandstone Group in and around the Irish Sea.

mainly pebble-free beds belong to the Wilmslow Sandstone Formation (Figure 48). It has not been demonstrated that the sedimentological and diagenetic boundaries are coincident, so the Jackson et al. (1987) interpretation must be treated with caution.

The Sherwood Sandstone Group sediments record a complex interplay of fluvial and aeolian deposition. Aggradation was more or less balanced by subsidence, so that a graded river profile was maintained, avoiding choking of the alluvial system. Two major depositional cycles (Thompson, 1970a; b) correspond with the St Bees Sandstone and Ormskirk Sandstone formations and their equivalents (Figure 48). In the East Irish Sea Basin, rapid deposition occurred in a major braided-river system (Turner, 1980). In the Ormskirk Sandstone Formation, channel sandstones alternate with sheet-flood sandstones deposited during the migration of the drainage system (Stuart and Cowan, 1991). The St Bees Sandstone and Ormskirk Sandstone formations are generally pebble-free, but the sediments are less mature than their correlatives in the Cheshire Basin. Litharenites occur in the north, and arkose clasts are found to the south (Burley, 1984), suggesting provenance in the Southern Uplands and Pennines respectively.

Aeolian sediments are scarce in the centre of the East Irish Sea basin, but are more common in the upper part of the sequence and towards the basin margins, for example near Calder Bridge in Cumbria (Figure 45; Trotter et al., 1937). The cleaner, winnowed, aeolian sandstones are coarser grained than the fluvial sandstones, and are poorly cemented or uncemented, although some have dolomite or calcite cement indicative of incipient caliche (Burley, 1984).

During deposition of the Chester Pebble Beds Formation to the south in the Cheshire Basin, adjacent to an area of considerable relief, there was gravel deposition on migrating-bar platforms and mid-channel bars in the confined channels of low-sinuosity rivers (Steel and Thompson, 1983). Thick, pebbly conglomerates occur in the Chester Pebble Beds Formation, and less extensively and in thinner beds within the Helsby Sandstone Formation, chiefly in the Delamere Member (Thompson, 1970a). Scattered pebbles in the St Bees Sandstone Formation persist northwards to a line running approximately through Clitheroe, Formby and well 110/7-2 (Figure 45; Audley-Charles, 1970a; b). The pebbles show an increase in roundness and a decrease in size northwards, and consist largely of protoquartzite (Thompson, 1970a, b; Steel and Thompson, 1983).

Towards the end of Scythian times, the first depositional cycle, as represented by the St Bees Sandstone Formation and its correlatives, was terminated by a tectonic event reflected widely in western Europe as the Hardegsen Disconformity (Geiger and Hopping, 1968; Colter and Barr, 1975). In the British Isles, its effects appear greatest in the English Midlands and in Northern Ireland (Warrington, 1970a; Warrington et al., 1980; Evans et al., 1993). Within the east Irish Sea, deposition was probably virtually continuous, and the tectonic change is reflected only by evidence of renewed uplift of adjacent source areas; uplift and reworking of the Chester Pebble Beds Formation occurred in two or three pulses, and resulted in deposition of the locally pebbly Helsby Sandstone Formation.

In the Helsby Sandstone Formation of the Cheshire Basin (Figure 48), fluvial, pebbly sandstones dominate the Alderley Conglomerate and Delamere members. It has been proposed that three rivers (Figure 46) transported clastic material north-north-westwards through the Cheshire Basin

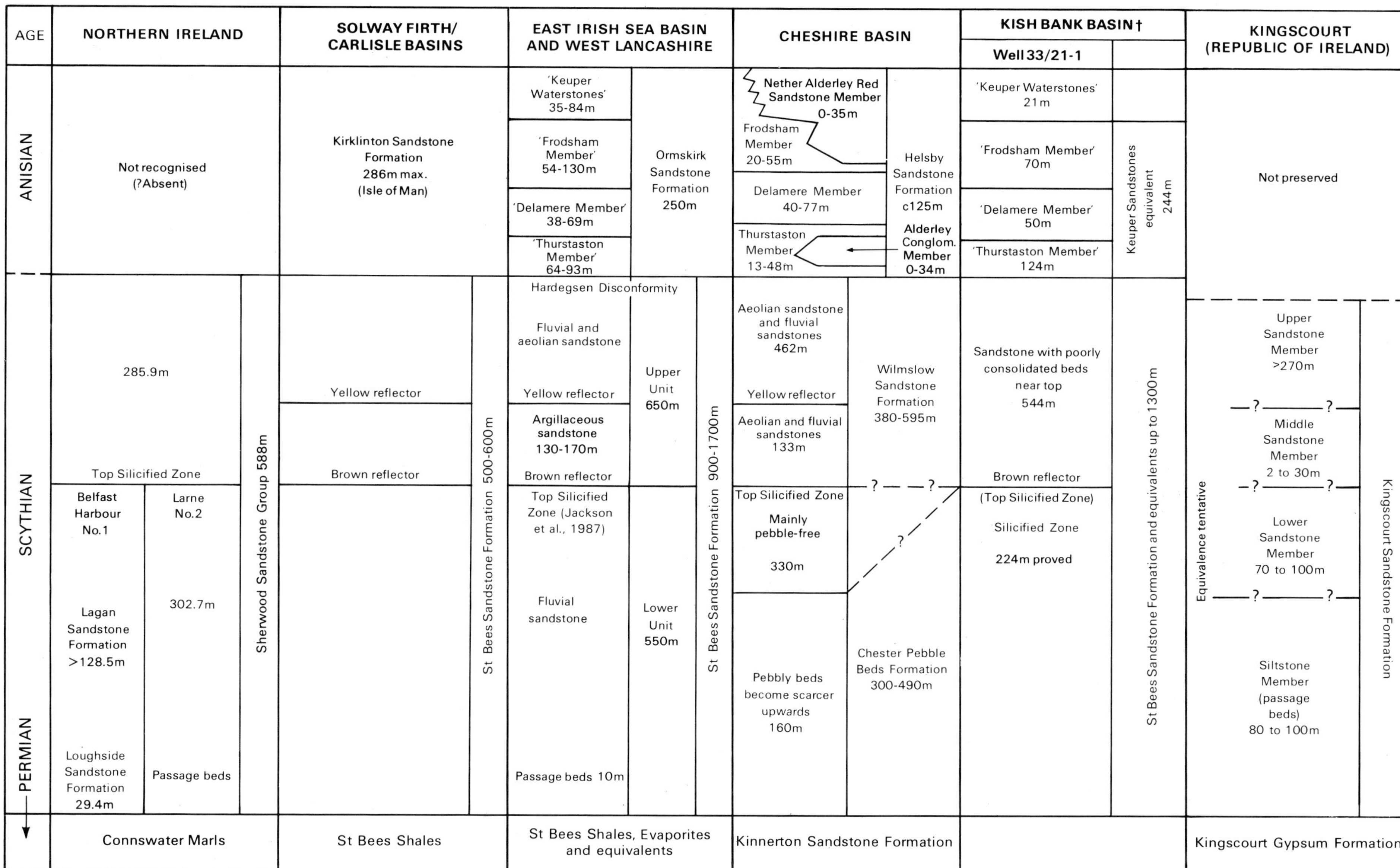

† Published with the permission of Petroleum Affairs Division, Department of Energy, Republic of Ireland.

Figure 48 Correlation of the Sherwood Sandstone Group in the Irish Sea and adjacent areas. Based on sources quoted in text.

(Thompson, 1970b). Aeolian sandstones and reworked aeolian debris occur in the Thurstaston Member, and are dominant in the Frodsham Member (Thompson, 1969). Aeolian influences are especially noticeable in the Wirral Peninsula, which lay astride a more slowly subsiding inter-basinal divide, the Llŷn–Rossendale Ridge of Jackson and Mulholland (1993). Here, the aeolian sands are generally finer grained than the fluvial types, a reversal of the relationship observed in the East Irish Sea Basin.

A gradual lowering of topographic relief during the late Scythian led to a progressive decrease in the transporting power of the rivers. This is reflected in an upward decrease in clast size, which is especially noticeable in the Chester Pebble Beds Formation (Steel and Thompson, 1983). There was a reduction in the supply of pebbles to the more distal areas north-west of the Wirral Peninsula, and deposition of finer-grained, fluvial sandstones occurred over the whole region.

On the southern flanks of the Longford–Down Massif (Figure 46), a brief marine incursion is recorded by palynomorphs in the Middle Sandstone Member of the Kingscourt Sandstone Formation (Visscher, 1971). Kingscourt lies in the direction of the Porcupine Basin, west of Ireland, where a shallow-marine Triassic sandstone facies is recorded (Croker and Shannon, 1987; Tate and Dobson, 1989), indicating a possible seaway across the centre of Ireland at that time. An associated rise in the water table may be marked in the East Irish Sea Basin by a slightly higher proportion of siltstone in the probable equivalent part of the St Bees Sandstone Formation, namely within the Brown-to-Yellow reflector interval (Figure 48; Burley, 1984). A rise in water-table level may also account for the widespread silicification of the St Bees Sandstone Formation below the level of the Brown reflector; this process could have continued with burial and compaction, depending on the composition of circulating pore fluids, and may have proceeded throughout the remainder of the Triassic and into the Early Jurassic (Colter, 1978; Bushell, 1986).

Arenaceous deposition in the region was terminated by a widespread marine transgression, possibly from the south-east, across a landscape of very low relief. This is correlated with the Röt transgression of the southern North Sea (Ireland et al., 1978; Cameron et al., 1992). In marginal areas, seismic-reflection records and geophysical-log correlations (Figure 49) show that the top of the Sherwood Sandstone Group is sharply defined. In the basinal areas, the boundary is more difficult to place because thin playa mudstones interdigitate with the sandstones in the highest beds of the Sherwood Sandstone Group.

The very scarce, largely continental, long-ranging macrofauna and trace fossil assemblages known from the Sherwood Sandstone Group onshore (Pattison et al., 1973; Warrington et al., 1980; Pollard, 1981; 1985) are not suitable for dating and correlation offshore. This is achieved by geophysical-log correlation, and indirectly from regional stratigraphical relationships. Palynological study of the Sherwood Sandstone Group offshore has proved disappointing (Wilkinson and Halliwell, 1979).

The age of the base of the Sherwood Sandstone Group is constrained by the presence of Late Permian miospores in the BS2 cycle and possibly the BS3 cycle in well 110/8-2 (Figures 45 and 49) below the St Bees Sandstone Formation. A comparable microflora occurs in the Upper Mudstone Member of the Kingscourt Gypsum Formation (Figure 43), less than 20 m below the base of the Early Triassic Kingscourt Sandstone Formation (Visscher, 1971). The Siltstone Member and the marine-influenced Middle Sandstone Member of the Kingscourt Sandstone Formation have been dated as early Scythian, and correlated with the Middle Bundsandstein of Germany (Visscher, 1971). However, doubt has been cast on their equivalence with UK sections, for miospores recovered from the basal Helsby Sandstone Formation in Cheshire, that used to be regarded as indicative of a late Scythian age (Warrington, 1970a), are now considered to be of early Anisian age (Warrington in Benton et al., 1994). Miospores indicating Anisian ages have been recorded from the overlying Tarporley Siltstone and Hambleton Mudstone formations at the base of the Mercia Mudstone Group (Figure 44 and see Figure 54; Earp and Taylor, 1986; Wilson and Evans, 1990). The Sherwood Sandstone Group is now assessed as Scythian and earliest Anisian in age.

St Bees Sandstone Formation and equivalent formations

The St Bees Sandstone Formation occurs throughout the Solway Firth, Carlisle and Annan basins, along the Cumbrian and west Lancashire coasts, and in the East Irish Sea Basin (Figure 45). It correlates (Figure 48) with the combined Chester Pebble Beds and Wilmslow Sandstone formations of the Cheshire Basin (Warrington et al., 1980), and with the combined Loughside Sandstone and Lagan Sandstone formations of Northern Ireland (Smith, 1986). An equivalent unit has been proved in the Kish Bank Basin and at Kingscourt (Figures 45 and 48). A mixed depositional environment is envisaged; the formation is dominantly fluvial, with some lacustrine/ephemeral-playa mudstone deposition, but aeolian components increase in importance upwards within the unit.

East Irish Sea Basin

An average thickness of about 1200 m is estimated from seismic profiles for the St Bees Sandstone Formation in the East Irish Sea Basin; in wells it ranges from 937 m in 112/25a-1 to 1188 m in 110/8-2 (Figure 49). The formation shows some thickening towards major growth faults, such as the Keys and Gogarth faults in the west, and the Lake District Boundary and Formby Point faults in the east (Figure 45). The formation reaches 1700 m adjacent to the Lake District Boundary Fault, and thins to about 900 m in the south of the basin.

The formation rests conformably on Upper Permian strata over most of the basin. The base is generally sharp over the Upper Permian marls, and is recognised by a marked downward increase in radioactivity, as in well 113/26-1 in Figure 49, although mudstone interbeds are common in the lowest part of the formation. Along the southern margin, where the underlying marls are absent, the gamma-ray log marker nevertheless persists at the junction with the Kinnerton Sandstone Formation. The top of the formation is marked by a characteristic sonic-velocity decrease, and with a seismic-reflection event that is correlated with the so-called Hardegsen Disconformity (Colter and Barr, 1975), manifestations of which are restricted to minor truncation locally of the highest St Bees Sandstone Formation reflections.

In the East Irish Sea Basin and adjacent onshore areas, the St Bees Sandstone Formation is divided into Lower and Upper units at the level of the Brown reflector (Jackson et al., 1987). Jackson et al. (1987) considered that the Lower unit correlates with the Chester Pebble Beds Formation of the Cheshire Basin, and the Upper unit with the Wilmslow Sandstone Formation (Figure 48), although this proposal is not universally accepted.

The Lower unit has an average thickness of 550 m, and consists of reddish brown, fine-grained, argillaceous, highly micaceous sandstone. There is usually a tight, siliceous cement, although this is locally dolomitic or anhydritic. Numerous, thin, red or greyish green mudstone beds are scattered throughout, but are more common in the lower part, as in well 112/25a-1 (Figure 49). There is some evidence of small-scale, upward-fining cycles, as seen in the lowest beds onshore at St Bees Head (Figure 45) where they are bounded by prominent erosional surfaces (Turner, 1980; Burley, 1984). The cycles comprise a basal-lag conglomerate passing up into cross-bedded sandstones interpreted as the deposits of linguoid bars and migrating dunes in a low-sinuosity sandy braided river, with a thin, persistent, micaceous, overbank mudstone to complete each cycle of deposition.

The Upper unit, with a typical thickness of 650 m, comprises fine- to medium-grained sandstones with subangular and rounded grains, but includes some thicker mudstone-rich units especially at the top. The beds are less micaceous than the Lower unit. The sandstones contain an abundant detrital-clay matrix, together with a late-stage authigenic platy illite which reduces the extent of later quartz and carbonate cementation (Colter and Ebbern, 1978; 1979). Although commonly hard and well cemented, thin zones of soft, uncemented sandstone with millet seed or frosted grains are present. These are probably aeolian in origin, and tend to increase in number and thickness upwards, concomitant with a slight upward decrease in both mica content and radioactivity, as in well 113/26-1 (Figure 49).

Sonic velocities appear to be largely dependent on the type of cementation: velocities are low in uncemented sandstone, moderately high where there is a carbonate or dolomitic cement, and very high where the cement is silica. The seismic records and well logs show a greater lateral and vertical facies variation than in the Lower unit. There are both upward-fining and upward-coarsening cycles, which are generally thicker and more clearly defined than in the Lower unit.

An argillaceous sandstone section, 130 to 170 m thick, is recognised at the base of the Upper unit in the centre of the East Irish Sea Basin; it may correspond with the Brown-to-Yellow seismic interval. This interval is 104 m thick at Seascale (Gregory, 1915) on the basin margin (Figure 45), where mudstones make up a smaller proportion of the sequence (Jackson et al., 1987). A second mudstone-rich interval occurs in the uppermost 180 m of the Upper unit offshore, where interbedded mudstones are again thicker and more numerous.

Five deep boreholes in the Formby area of west Lancashire (Figure 52; Kent, 1948; Falcon and Kent, 1960; Colter and Barr, 1975) are critical in correlating the successions of the East Irish Sea and Cheshire basins. At Formby, the beds display marked variations in thickness, colour (partly due to the reducing influence of hydrocarbons), cementation, and in the ratio of aeolian to fluvial sediments. In several Formby boreholes, the lowest beds contain scattered pebbles of quartz and quartzite, but in a boring at Well Lane, Bootle, the pebbly sandstone passes up into a cemented pebble-free sandstone (Wray and Cope, 1948) whose top is believed to coincide with the Top Silicified Zone; the entire unit is correlated by Jackson et al. (1987) with the Chester Pebble Beds Formation, and hence with the Lower unit of the St Bees Sandstone Formation. In west Lancashire, the thickness of the Lower unit varies from 200 to 464 m, and the generally well-cemented sediments are fluvial in origin. At Formby, the overlying beds equivalent to the Wilmslow Sandstone Formation of the Cheshire Basin consist of 183 to 347 m of poorly cemented, mottled sandstone with alternating well- and poorly consolidated beds, and interbedded marls. Formerly the Upper Mottled Sandstone (Warrington et al., 1980), these beds are correlated with the Upper unit of the St Bees Sandstone Formation, and a mixed fluvial-aeolian environment is envisaged.

Southwards on the Wirral Peninsula, the well-cemented Chester Pebble Beds Formation is reduced to 158 m thickness at Heswall (Figure 45; Wade, 1910). It rests sharply but conformably on the softer Late Permian Kinnerton Sandstone Formation, and a two-fold division is again recognised; the lower 70 m contain quartzitic pebbles, and the upper 88 m has intraformational mud-pellet conglomerates.

Cheshire Basin

In the type area around Chester, the Chester Pebble Beds Formation is up to 305 m thick and contains scattered quartzitic pebbles in a medium- to coarse-grained, well-cemented, sandstone matrix (Earp and Taylor, 1986). Interbedded, friable, nonmicaceous sandstones with millet-seed grains probably indicate some winnowing of wind-blown sand onto gravel bars. Locally, a very pebbly 50 m-thick unit occurs near the base; this may be the proximal correlative of the only record of pebbles in the East Irish Sea Basin in well 110/7-2 (Figure 45).

In the Knutsford No. 1 borehole, the Chester Pebble Beds Formation is, according to Jackson et al. (1987), represented by 490 m of tightly cemented, fluvial sandstone (Figure 49). The sandstone shows an invariant sonic-log trace and a higher average gamma-ray response than the underlying, more-porous, aeolian sandstone (Kinnerton Sandstone Formation) of Permian age, indicating a higher proportion of mica and mudstone. The lower part contains pebbles, whereas the upper, largely pebble-free, but nevertheless well-cemented, division contains some beds of more-porous, aeolian sandstone which are identified by their lower sonic velocity (Steel and Thompson, 1983). A pronounced gamma-ray and sonic-log break at the top of the pebble-free division marks the top of the Silicified Zone, although Colter and Barr (1975) and Evans et al. (1993) placed the pebble-free unit within the Wilmslow Sandstone Formation.

The Wilmslow Sandstone Formation correlates with the Upper unit of the St Bees Sandstone Formation (Figure 48). It consists of uniform, loosely cemented, reddish brown, cross-bedded sandstones of aeolian origin with some interbedded, laminated, shaly sandstones that are probably of fluvial origin. The proportion of aeolian sandstone increases towards the margin of the basin in the Chester area, where the base of the formation is transitional (Steel and Thompson, 1983).

In the Knutsford No. 1 borehole, the Wilmslow Sandstone Formation is 595 m thick according to Jackson et al. (1987). In the lowest 133 m, porous, aeolian sandstone, distinguished by a lower sonic velocity, alternates with more-tightly cemented sandstone in a unit that may correlate with the Brown-to-Yellow reflector interval of the East Irish Sea Basin (Figure 49; Jackson et al., 1987). This is overlain by slightly more-argillaceous, fluvial sandstone that passes up into well-cemented sandstone.

Solway Firth Basin

The Solway Firth Basin is separated from the Carlisle Basin by a low interbasinal ridge, and from the East Irish Sea Basin by the Ramsey–Whitehaven Ridge, from which the higher beds of the Sherwood Sandstone Group have been eroded (Figure 45). The beds of the lower Sherwood Sandstone Group represent a finer-grained equivalent of the St Bees Sandstone Formation, in which the Brown and Yellow reflectors are both strongly developed on seismic reflection records.

In the Silloth 1A borehole on the north coast of Cumbria (Figure 49), the St Bees Sandstone Formation is 395 m thick. The micaceous sandstone of the Lower unit contains abun-

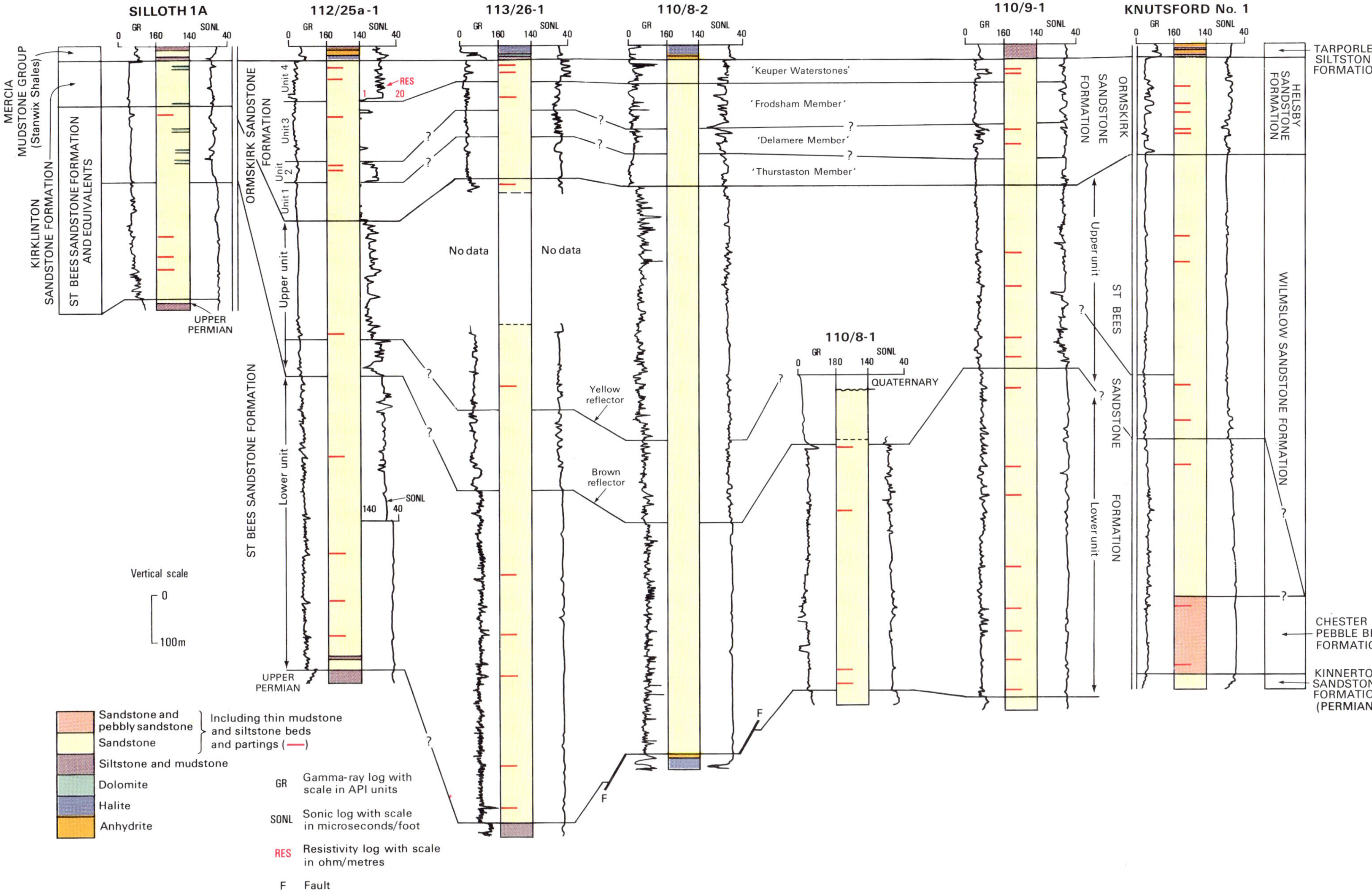

Figure 49 Correlation of well logs in the Sherwood Sandstone Group. See Figure 45 for locations.

dant mudstone interbeds at the base, but becomes cleaner and less micaceous upwards. Individual cycles up to 10 m thick recognised on the gamma-ray log, which apparently coarsen upwards, may be related to fluctuations in K-mica content. In the Upper unit, sandstone from the Brown-to-Yellow interval is cleaner, and contains thinner mudstone beds than that of the Lower unit; a clay matrix is reflected in lower sonic velocities. The clean sandstone shows the lowest and most variable sonic-log velocities, and contains only two mudstone beds near the top. Two thin, poorly cemented layers in the sandstone may indicate aeolian influence, suggesting a mixed fluvial-aeolian origin in common with the Wilmslow Sandstone Formation.

On the Isle of Man, the lowest beds in the St Bees Sandstone Formation comprise 200 m of red, fine-grained, micaceous sandstones with some interbedded grey mudstones (Lamplugh, 1903). The highest beds (330 m) are more-uniform red sandstones with a few grey bands (Gregory, 1920). In the Stranraer Basin, which is contiguous with the Solway Firth Basin, the maximum depth to Lower Palaeozoic basement is estimated at 1370 m. The basin is thought to be filled with sandstone of Permo-Triassic age (Mansfield and Kennett, 1963); reddish brown sandstone was proved in BGS borehole 70/02 (Figure 45) and is believed to be from the St Bees Sandstone Formation, rather than the Lower Permian as previously thought.

North Channel Basin

The sequence in the North Channel Basin has not been drilled, and its nature can only be inferred from onshore boreholes. The Larne No. 2 borehole (Figure 47; Downing et al., 1982; Penn et al., 1983) proved a 588 m-thick Sherwood Sandstone Group succession of Triassic age in which the Top Silicified Zone of the East Irish Sea Basin was identified 302 m above the base. Below this marker, the sandstone is very fine grained, and in the lowest 126 m is subordinate to siltstone and mudstone. Above the Top Silicified Zone, the sandstone is fine to medium grained, with a variable anhydrite/dolomite cement and rare mudstone beds. In the Belfast Harbour No. 1 borehole (Smith, 1986), small pebbles occur at the base of individual sandstone beds in the Loughside Sandstone Formation (Figure 48). The position of the Hardegsen Disconformity is taken at the top of the Sherwood Sandstone Group (Penn et al., 1983); thus, the equivalent of the Ormskirk Sandstone Formation has not been recognised or is absent. This may be due to increased erosion on the outer margins of the basin, where tectonic events would have had a greater influence, or to non-deposition.

Kish Bank Basin

An equivalent of the St Bees Sandstone Formation proved in well 33/21-1 consists of fine- to medium-grained sandstone (Figure 48). The lowest 224 m comprise a tightly cemented sandstone that probably equates with the Silicified Zone and thus with the Lower unit of the St Bees Sandstone Formation in the East Irish Sea Basin.

Ormskirk Sandstone Formation and equivalent formations

A log break, believed to correspond to the position of the Hardegsen Disconformity (Colter and Barr, 1975; Colter 1978), generally separates the Ormskirk Sandstone Formation and equivalents, with their cleaner, more-thickly bedded sandstones and lower sonic velocity, from the underlying, argillaceous, finer-grained sandstone of the upper part of the St Bees Sandstone Formation. The Ormskirk Sandstone Formation (Warrington et al., 1980) has a similar distribution to the St Bees Sandstone Formation, but has been eroded from some intrabasinal horsts. It equates with the Kirklinton Sandstone Formation of the Solway Firth Basin and the Helsby Sandstone Formation of the Cheshire Basin. An equivalent has not been identified in Northern Ireland. Seismic records show the Ormskirk Sandstone Formation and its equivalents as a discrete package with four or five, persistent, equally-spaced, internal reflections. Thickening towards growth faults is slight.

East Irish Sea Basin

In the East Irish Sea Basin, the Ormskirk Sandstone Formation (Jackson et al., 1987), 250 m thick on average, comprises most of the sandstones forming the reservoir of the Morecambe Gasfield (see Figure 79). Nomenclature from the Cheshire Basin has been applied to the four subdivisions of the reservoir sandstones (Colter and Barr, 1975; Colter, 1978; Bushell, 1986); the so-called 'Thurstaston', 'Delamere' and 'Frodsham' members are overlain by the 'Keuper Waterstones' (Figures 48 and 49). These four subdivisions are based on variation in sandstone:mudstone ratio and sedimentological characteristics (Colter and Ebbern, 1978; Bushell, 1986), and have been correlated across the Morecambe Gasfield without major lithological or thickness variation. However, Ebbern (1981) and Stuart and Cowan (1991) have expressed difficulties concerning subdivision of the Ormskirk Sandstone Formation in the North Morecambe Gasfield. The four lithostratigraphical subdivisions have been retained in this account, but placed in single quotation marks. However, it should be noted that the Helsby Sandstone Formation (Warrington et al., 1980) of the Cheshire Basin, from which the names are largely derived is divided into only three members, each with precisely defined sedimentological characteristics (Figure 48; Thompson, 1970a; b). Thus it is most unlikely that, for example, the Delamere Member of the Cheshire Basin is equivalent to the 'Delamere Member' of the Morecambe Gasfield. In this account, the 'Keuper Waterstones' of the Morecambe Gasfield are regarded as a local facies equivalent of the upper Frodsham Member, and thus analogous to the Nether Alderley Red Sandstone Member (Thompson, 1970b) of the Cheshire Basin. This interpretation is supported by seismic evidence, and by well-log characteristics at Formby (Colter, 1978). Furthermore, the Thurstaston Member is excluded from the Helsby Sandstone Formation of the Cheshire Basin by Earp and Taylor (1986) and Evans et al. (1993).

The sandstones of the Ormskirk Sandstone Formation (Figure 48) are generally pale grey or pale brown, with red-brown and orange mottling. They appear to be red only where early carbonate cement was precipitated. In a few hard bands, contemporaneous reduction of red sandstone may have been aided by perched water tables above local mudstone aquicludes, especially in the 'Delamere Member' and 'Keuper Waterstones' (Burley, 1984). The sandstones are generally fine to medium grained, with poor to moderate sorting and subangular to subrounded quartz grains, although some thin bands are uncemented, coarse grained and well sorted. The latter contain frosted quartz grains indicating either aeolian deposition on temporarily emergent fluvial bars, or penecontemporaneous redeposition of aeolian sand. The sandstones have a variable detrital-clay matrix, now mainly of illite which may have replaced primary smectite, and are variously cemented by quartz, calcite and dolomite. Thin, grey-green and red-brown, noncalcareous mudstones up to 1 m thick, and exceptionally up to 5 m (Bushell, 1986), occur throughout, and are more common in the 'Delamere Member' and 'Keuper Waterstones'.

The 'Thurstaston Member' (Figure 48) consists of cross-bedded sandstones with sporadic shale partings, and is interpreted as the deposit of a fluvial system of low-sinuosity channel type in which little of the interchannel argillaceous sediment has been preserved (Ebbern, 1981). This member shows the highest porosity (15.3 per cent) and lowest permeability (1.24 mD) values of the formation (Ebbern, 1981).

The sandstones of the overlying 'Delamere Member' are cross-bedded, with mica-rich partings, mud-pellet lag conglomerates, and a matrix of detrital clay and authigenic platy illite. Interbedded shales make up as much as 35 per cent of the sequence, and individually are up to 3.3 m thick in well 110/9-1 (Figure 49). Porosity and permeability values are low, and the member is interpreted as a floodplain sediment, possibly deposited in ephemeral channels (Ebbern, 1981).

The 'Frodsham Member' is lithologically similar to the 'Thurstaston Member', but contains a relatively higher proportion of interbedded shale which becomes commoner upwards and towards the centre of the basin. Porosity is moderate to good at an average of 14.1 per cent, and permeability is good at 1.02 mD. These beds are interpreted as channel sands deposited from low-sinuosity streams (Ebbern, 1981).

In the 'Keuper Waterstones' at the top of the formation (Ebbern, 1981), high-porosity, cross-bedded sandstones and low-porosity, argillaceous sandstones occur in upward-fining cycles up to 15 m thick. The permeability is good. Two mudstone beds up to 6 m thick can be correlated widely; the lower is interpreted as a playa-lake deposit, and the upper as a floodplain mudstone. The member is interpreted as a distal, floodplain deposit.

On the coast of south-west Cumbria, mottled, grey and pale brown, oil-impregnated sandstone exceeding 28 m thick in the Barrow Haematite No. 2 borehole (Figure 45; Rose and Dunham, 1977) is overlain conformably by the Hambleton Mudstone Formation of the Mercia Mudstone Group. The sandstone can be assigned to the uppermost beds of the Ormskirk Sandstone Formation. Seismic evidence offshore suggests that the Ormskirk Sandstone Formation also occurs in the Seascale borehole (Figure 45; Jackson et al., 1987), where it is over 180 m thick (Gregory, 1915; Trotter et al., 1937).

Around Formby (Figure 45), the 200 m-thick Ormskirk Sandstone Formation is transitional between the locally pebbly Helsby Sandstone Formation of the Cheshire Basin and the finer grained Ormskirk Sandstone Formation of the Morecambe Gasfield. Interbedded mudstones may be up to 5 m thick, but are rare; this scarcity suggests that aeolian influences increase eastwards away from the centre of the basin. In the Formby boreholes (Kent, 1948), the beds are pale grey to brown due to reduction by hydrocarbons, but at outcrop, brownish yellow and reddish colours prevail. The beds vary from soft, pebbly or coarse-grained sandstone to fine-grained sandstone with marl partings (Wray and Cope, 1948). The sandstones are better cemented than the underlying Wilmslow Sandstone Formation.

Cheshire Basin

The Helsby Sandstone Formation is 125 m thick on average and regarded here as the direct equivalent of the Ormskirk Sandstone Formation (Warrington et al., 1980); five constituent members were recognised by Thompson (1970b), although only three are of basinwide extent. The Thurstaston Member consists of interbedded, fluvial and aeolian sandstones, and is best developed on the margins of the Cheshire Basin and on the Wirral Peninsula (Thompson, 1970a; b). Where the Thurstaston Member is absent, it is difficult to separate the Wilmslow and Helsby Sandstone formations; indeed, Earp and Taylor (1986) and Evans et al. (1993) consider that the Thurstaston Member is a part of the Wilmslow Sandstone Formation. The member passes laterally into the flood conglomerates of the Alderley Conglomerate Member, and upwards into the distinctive, fluvial, pebbly sandstones of the Delamere Member which, unlike their claimed analogue in the East Irish Sea Basin, contain exotic clasts. The Delamere Member is overlain by aeolian sandstones of the Frodsham Member (Thompson, 1969). The Nether Alderley Red Sandstone Member is an argillaceous sandstone of fluvial origin which, like the 'Keuper Waterstones' of the East Irish Sea Basin with which it may be correlated, was deposited only in the most actively subsiding areas and floodplains (Thompson, 1970b); it passes laterally into the middle and upper portions of the Frodsham Member.

Solway Firth Basin

In this basin, the Kirklinton Sandstone Formation, the lateral equivalent of the Ormskirk Sandstone Formation, reaches a proven maximum thickness of 286 m on the Isle of Man (Gregory, 1920; Colter, 1978). It consists of soft, bright red, coarse-grained, cross-bedded, nonmicaceous sandstone, with some uncemented beds and scattered millet-seed grains (Gregory, 1920). The fourfold subdivision of the Morecambe Gasfield cannot be recognised, and interbedded mudstones are absent. In the Silloth 1A borehole (Figure 49) the 94 m-thick sandstone is bimodal, consisting of either fine-grained and silty, or medium- to coarse-grained beds. Lower sonic velocities compared to the underlying St Bees Sandstone Formation reflect higher porosity and permeability.

Similar clean sandstones occur in the Maryport (Eastwood, 1930), Carlisle (Dixon et al., 1926), and Cockermouth districts (Eastwood et al., 1968), (Figure 45). This distribution probably indicates a greater aeolian influence in the basin-margin areas flanking the braided channels of the east Irish Sea depocentre (Bushell, 1986). At Wigton (Figure 45), a 12 m succession may reflect local overstep of the formation by the Mercia Mudstone Group (Eastwood et al., 1968).

Kish Bank Basin

Beds equivalent to the Ormskirk Sandstone Formation are thought to have been proved in well 33/21-1, where they have been subdivided into four members, named as in the Morecambe Gasfield (Figure 48). The 265 m-thick succession consists of unconsolidated sandstone of aeolian origin, alternating with finer grained, argillaceous sandstone. At the top, the 'Keuper Waterstones' consist of interbedded siltstone, mudstone and sandstone, capped by 5 m of unconsolidated sand.

MERCIA MUDSTONE GROUP

The Mercia Mudstone Group in the report area is up to 3200 m thick (Figure 50; Jackson et al., 1987), and comprises reddish brown silty mudstones with some greenish grey mudstones. In the areas of most rapid subsidence, thick interbedded halites are common. The group is of Anisian to early Rhaetian age (Figure 44), and is overlain, probably nonsequentially, by the Penarth Group. Like the Sherwood Sandstone Group, the Mercia Mudstone Group was deposited over most of the East Irish Sea Basin, but erosion from Mid-Jurassic times onwards has removed part or all of the sequence from many localities, so that the full thickness of Triassic is rarely preserved. The highest deposits of the Mercia Mudstone Group, and Penarth Group if present off-

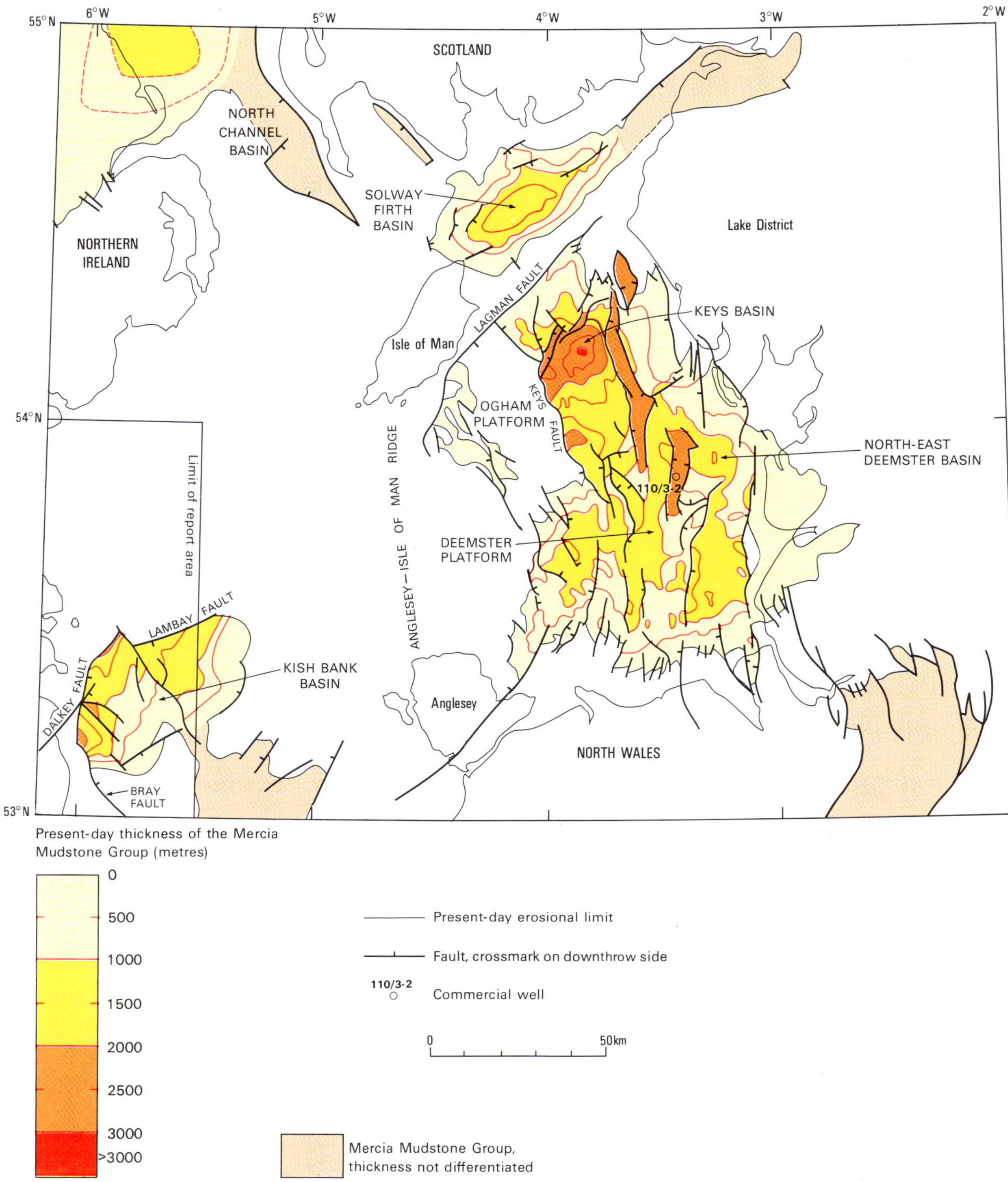

Figure 50 Thickness of the Mercia Mudstone Group.

shore, are restricted to a few sites where they have been preserved under a cover of Lias Group.

On seismic sections, the Mercia Mudstone Group is distinguished from the Sherwood Sandstone Group in displaying more reflections, and by the lateral variability of its seismic character. Minor faulting, glide planes and halokinetic features such as salt flow locally are seen offshore where the base of the lowest halite acts as a *décollement* between the competent Sherwood Sandstone Group and the incompetent Mercia Mudstone Group.

The general absence of Penarth Group and Jurassic rocks offshore prevents detailed assessment of the original overall geometry of basin fill of the Mercia Mudstone Group, but an indication of the original thickness is shown in Figure 51. At present, the thickest sequence, over 3000 m, is restricted to the Keys Basin (Figure 50), with over 2000 m deposited in the Kish Bank Basin and over 1500 m preserved in the Solway Firth and North-East Deemster basins. In both the Keys and North-East Deemster basins, thickening towards the basin centre suggests true thermal-relaxation subsidence, but elsewhere there is substantial thickening towards major syndepositional or growth faults. The Mercia Mudstone Group also shows a pronounced regional northward increase in thickness away from the North Wales coast towards the Lagman Fault at the Ramsey–Whitehaven Ridge.

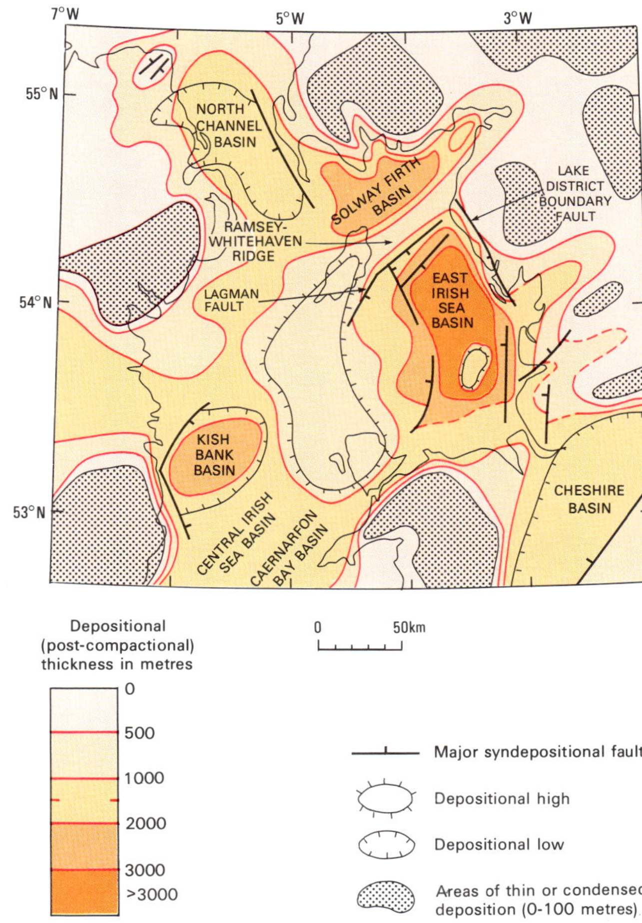

Figure 51 Estimated maximum original depositional thickness (post-compaction) of the Mercia Mudstone Group.

The Mercia Mudstone Group is well known from land studies, and has been proved in several BGS boreholes drilled in the Irish Sea (Figures 45 and 52; Wright et al., 1971; Fletcher and Ransome, 1978; Wilkinson and Halliwell, 1979). The sediments have also been proved in a number of offshore wells; the maximum drilled thickness of 2030 m (with a faulted basal contact) is recorded in well 110/3-2 (Figure 50 and see Figure 57).

The base of the Mercia Mudstone Group is everywhere sharp, generally with a marked upward lithological change from a sandstone- to a mudstone-dominated sequence. Seismic-reflection records in the East Irish Sea and Solway Firth basins demonstrate that there is no angular unconformity with the Sherwood Sandstone Group. The boundary is reported to be diachronous, younging from north to south (Warrington, 1970b; Warrington and Ivimey-Cook, 1992).

A coastal depositional environment for the group is indicated by a mix of land and marine biota. Very rare *Equisetalean* (horse-tail fern) remains, and miospores comprise the continental flora, whereas sparse microplankton (acritarchs) represent the marine biota. The climate and conditions of deposition have been likened to the present-day semiarid region of the Ranns of Kutch in western India (Glennie and Evans, 1976), which is sited in a high-pressure zone north of the trade-wind belt.

Palaeocurrent data from Cheshire suggest that the north-westerly directed flow seen in the Sherwood Sandstone Group continued during deposition of the lower part of the Mercia Mudstone Group (Ireland et al., 1978). Palaeocurrent data are few elsewhere, but higher in the sequence within the Thornton Mudstones around Blackpool (Wilson, 1990), siltstone bands become less numerous towards the north-west, again suggesting derivation from the south-east. A local source is envisaged for the clastic material.

At the end of the Early Triassic, peneplanation of the landscape had occurred over a wide area. During the Anisian, depositional areas expanded gradually, thereby re-establishing continuity with the southern North Sea Basin (Warrington and Ivimey-Cook, 1992), and bringing a change in depositional environment from fluvial and aeolian desert conditions to intertidal mudflats. If the marine sandstones recorded in the Porcupine Basin west of Ireland (Croker and Shannon, 1987; Tate and Dobson, 1989) are of Early Triassic age, it is also possible that the Röt marine transgression may have approached the Irish Sea area from the west through the Irish Midlands and the Kish Bank Basin.

At the base of the Mercia Mudstone Group, a local arenaceous facies dominates in the south of the Cheshire Basin (Figure 53; Earp and Taylor, 1986). This passes laterally north-westwards into thinly interbedded sandstones and siltstones, a facies that characterises the Tarporley Siltstone Formation of the Cheshire Basin and the Lagavarra Formation of Northern Ireland (Figure 54); it also occurs as unnamed units higher in the sequence (Warrington et al., 1980; Earp and Taylor, 1986). The Tarporley Siltstones comprise thinly interbedded, fine-grained sandstones with interlaminated siltstones and mudstones in units about 3 m thick (Ireland et al., 1978). Subaerial exposure is indicated by mudcracks, clay galls and pseudomorphs after halite or anhydrite. The sequence was deposited in an intertidal environment ranging from intertidal sand-flat to high-intertidal mudflat with sand bars in tidal channels. A trace-fossil assemblage found in the tidal-channel and intertidal-sand-flat deposits indicates communities mostly of suspension feeders in moderate-energy environments, but with some feeding burrows and arthropod tracks thought to be characteristic of moist, low-energy sites, possibly sluggish streams or the shorelines of ephemeral lakes (Ireland et al., 1978; Frey and Pemberton, 1987).

The Tarporley Siltstones pass northwards and north-westwards into grey, interlaminated mudstones and siltstones of the Hambleton Mudstones of west Lancashire and Merseyside (Figures 53 and 54; Wilson and Evans, 1990). In the latter area, beds contain giant pseudomorphs of calcite after halite thought to have been formed in a near-coastal sabkha (Wilson, 1990). The mudstones pass basinwards into a dolomite zone, and then into an anhydrite zone; this is the anhydrite-dolomite marker of the East Irish Sea Basin (Figure 54; Colter, 1978). The marker beds fail towards the basin margin, such as in well 110/9-1 (Figure 52), and onshore in west Lancashire and Walney Island. Two thin anhydrite beds occur near the base of the Lagavarra Formation in Northern Ireland (Manning and Wilson, 1975).

A pattern of mudstone and halite deposition (Figure 55) was widely established in early Anisian times, and persisted throughout much of the remainder of the Triassic. Onshore at Blackpool, Walney Island, and in Cheshire, the mudstone consists of two main lithological types: blocky, unbedded mudstones, and interlaminated siltstones and mudstones (Arthurton, 1980).

The blocky mudstones are red, and contain gypsum nodules and veins; these are likely to have originated largely as wind-blown dust that settled onto dry or nearly dry ground (Taylor et al., 1963; Taylor, 1983). The absence of bedding may be in part due to penecontemporaneous mixing and resettling of sediment before final consolidation, a process similar to the zardeh or 'ploughed ground' in Iran

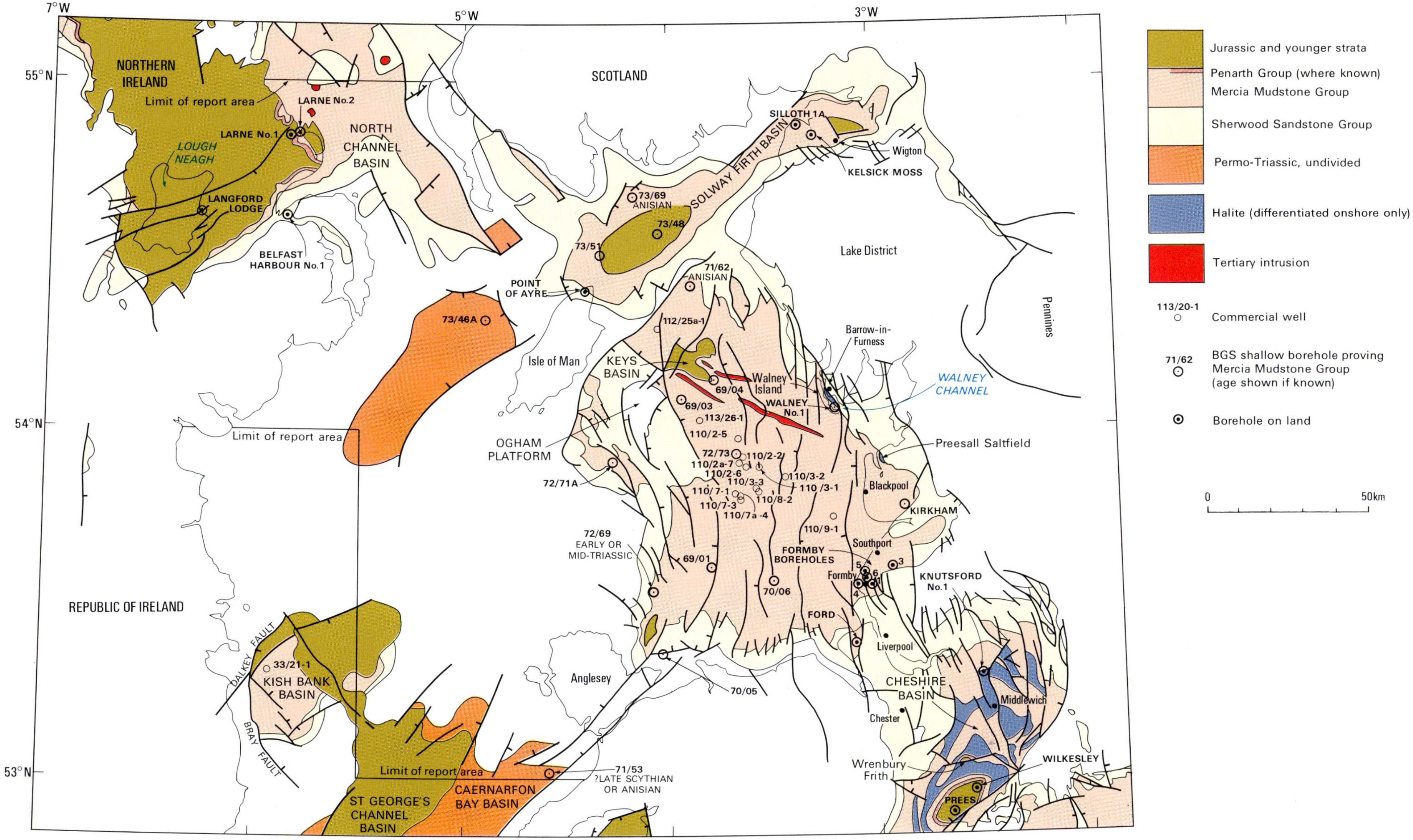

Figure 52 Distribution of the Mercia Mudstone Group.

(Arthurton, 1980). Such mudstones are the dominant facies in the Singleton and Breckells mudstones of west Lancashire.

The interlaminated siltstones and mudstones occur in the Thornton Mudstones of west Lancashire (Figure 54). They are virtually devoid of macrofossils, but contain scarce acritarchs indicative of a connection to the sea. Desiccation cracks are abundant in the Blackpool district, indicating periodic drying out. Current ripple-lamination, small-scale cross-bedding, and minor cut-and-fill structures are common, and are thought to be indicative of deposition in shallow water. The nearest modern depositional analogue is probably the Ranns of Kutch, a low-lying tidal to supratidal sabkha environment just above normal high-tide level, in which the monsoon causes seasonal flooding up to 2 m deep for up to 150 km across the desert. Evaporation of the sea water results in the precipitation of halite, and the growth of halite and gypsum crystals in the surface sediments. Conditions are generally inimical to life; the few foraminiferal species recorded are thought to be replenished by the annual flooding (Glennie and Evans, 1976).

An alternation of the blocky and laminated facies in the Coat Walls Mudstones of Blackpool and in equivalent beds in Cheshire (Wilson, 1990; Arthurton, 1980), points to an alternation of aeolian and water-lain deposits.

Halite comprises 35 to 55 per cent of the basinal succession, and occurs at five main levels (Figure 54). The depositional area appears to have increased with time, for the Rossall Halites and equivalents are more restricted than the later Preesall Halite and its equivalents (Figure 56). The thicker halites are massive, and consist of pale brown or

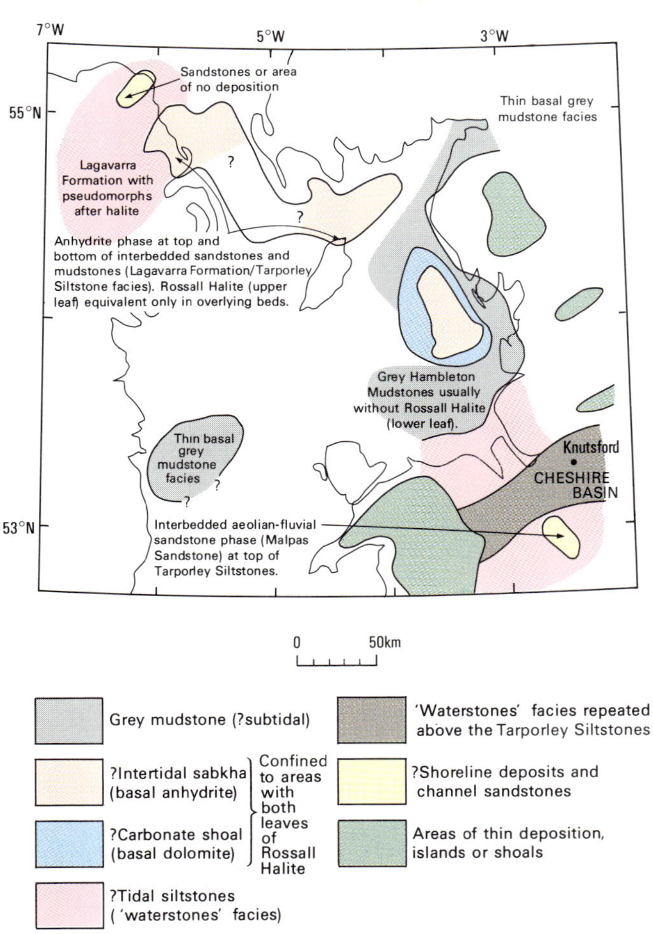

Figure 53 Facies relationships in the lower part of the Mercia Mudstone Group. See text for discussion.

colourless, primary halite, and some recrystallised, pink halite. The halites contain some impurities, including gypsum, and calcium or magnesium carbonate, usually as dolomite. Clay and silt occur throughout, both as inclusions in the salt and as units of mudstone up to 30 m thick.

The halites are interpreted as the deposits of intertidal and supratidal coastal-marine sabkhas, but with strong continental influences. The deposits of the Irish Sea lack the strong zonation patterns and fringing, concentric, facies belts of halites formed in continental playas (Hsü, 1972), where there is a sharper division between halite, sulphate, carbonate, and clastic rocks. Halites of coastal-marine sabkhas, both intertidal and supratidal, are generally characterised by displacive halite and chaotically mixed halite and mudstone or haselgebirge (Hsü, 1972), a feature common to the Irish Sea halites (Wilson, 1990). Within the Triassic of the Irish Sea, the areas of halite deposition may have formed a string of linked brine flats, or salt pans, separated by mudflats formed from sediment carried by rivers from low-lying hills, and supplemented by wind-blown dust.

The top of the Mercia Mudstone Group is marked by the characteristically hard, grey-green mudstone of the Blue Anchor Formation (formerly Tea Green Marl) and its equivalents (Figures 44 and 54; Warrington et al., 1980). It is locally disconformable upon the underlying red mudstone, and indicates less-oxidising conditions established during the transgression of late Norian/Rhaetian times, which flooded the low-lying sabkha and mudflat plains of the earlier Triassic. The Norian and Rhaetian appear to have been times of more uniform sedimentation without the differential subsidence characteristic of earlier Triassic times.

East Irish Sea Basin and west Lancashire

The East Irish Sea Basin possessed one of the highest subsidence rates in north-west Europe during Mid- to Late Triassic times (Ziegler, 1990), and is the thickest depocentre of the Mercia Mudstone Group in and around the British Isles. The offshore successions of the Irish Sea are twice, and locally three times as thick as their onshore counterparts in west Lancashire and Furness. Due to erosion, and drilling on relative highs, only the lower and middle parts of the sequence have been penetrated in wells; the maximum thickness proved is 2030 m in well 110/3-2 (Figure 57). However, seismic records indicate a preserved maximum of 3200 m in the Keys Basin, and stratigraphical reconstructions (Figure 51) suggest a former maximum of possibly 4000 m near the Lake District Boundary Fault.

Correlation with the thinner onshore sequence is complicated by halokinesis and low-angle faulting. The halites provide the most reliable means of correlation, but distinctive mudstones can be matched at certain stratigraphical intervals. Wilson (1990) has used the characteristic alternations of colour-banded mudstone in the Thornton Mudstones onshore (Figure 54) to effect a correlation with well 110/8-2 offshore (Figure 52); this may also be possible for other offshore wells. Caution is needed in using merely the sonic velocity of mudstone for correlation, as this is influenced by the maximum depth of burial (Colter, 1978).

Offshore, the halites have been defined arbitrarily where the salt content of major rock units exceeds 50 per cent; four major halites have been recognised, following the work of Wilson and Evans' (1990) in the Blackpool district (Figure 54). These are the Rossall (lower and upper leaves), Mythop, and Preesall halites (Jackson et al., 1987). A higher, fifth halite, which would correlate with the Wilkesley Halite of the Cheshire Basin (Wilson and Evans, 1990) is recorded in

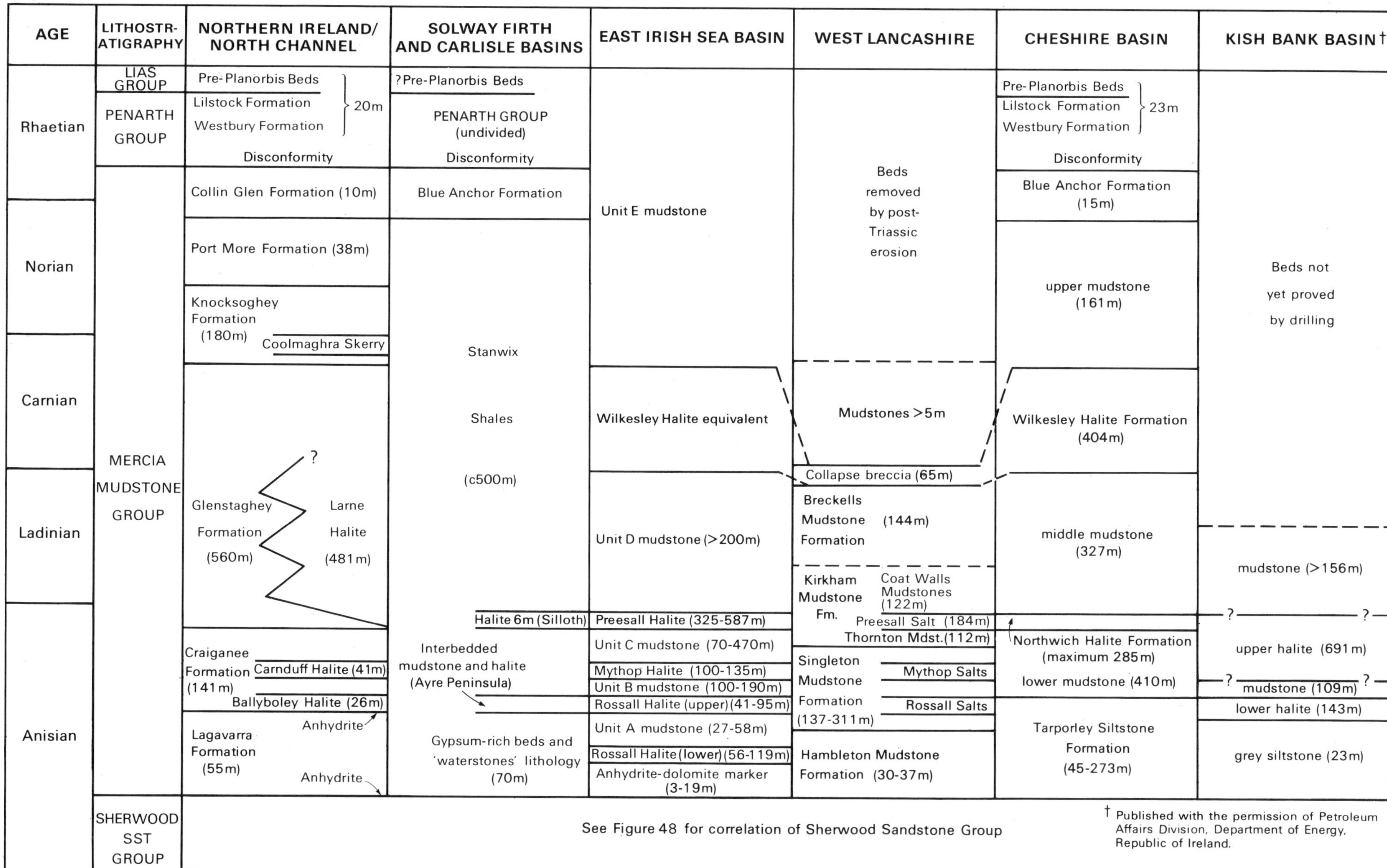

Figure 54 Correlation of the Mercia Mudstone Group in and around the Irish Sea. Based on sources quoted in text.

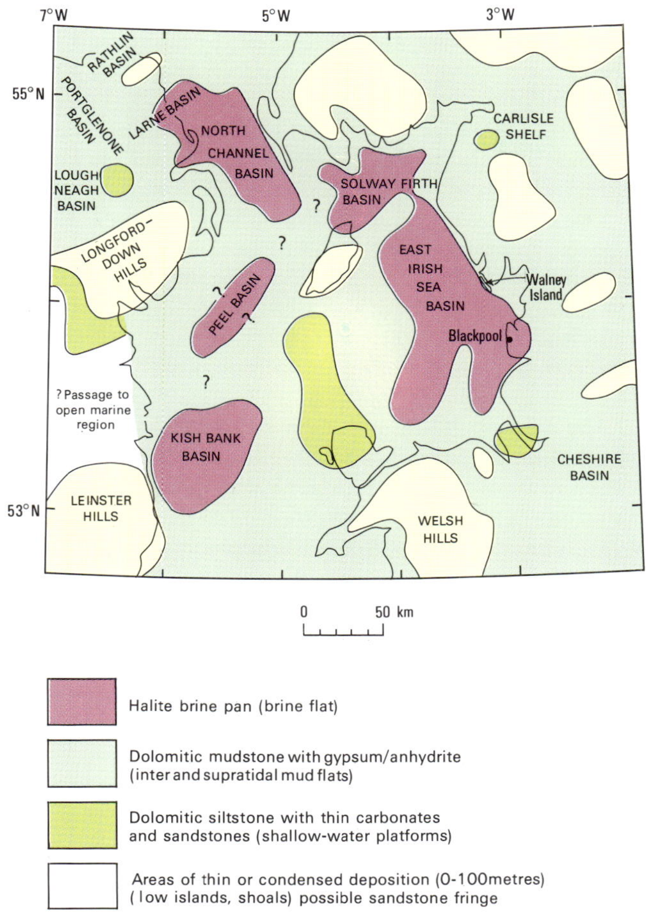

Figure 55 Suggested palaeogeography and facies during deposition of the Rossall/Mythop halites and equivalents (Anisian). Substantially modified from Audley-Charles (1970b), Warrington (1970a, 1974); Bennett (1983), and Wilson (1990). See also Warrington and Ivimey-Cook (1992).

unreleased commercial wells. The major interbedded mudstones, units A to D, all contain stringers and beds of halite up to 10 m thick (Jackson et al., 1987).

The basal beds offshore comprise the anhydrite-dolomite marker bed (Colter, 1978), which is present in many unfaulted sequences in the northern and central parts of the basin, except those in wells 110/9-1 (Figure 58) and 112/25a-1 (Figure 52). Anhydrite is confined to the central parts of the East Irish Sea Basin, with dolomite in the north-east, and fine-grained clastic rocks to the south-west (Figure 53). The marker has not been reported onshore in England, but correlates with the transgressive Lower Röt Mudstone Member of the southern North Sea (Cameron et al., 1992). The dolomite succession, 3 to 19 m thick, typically comprises three to four dolomite beds, each up to about 3 m thick, interbedded with silty mudstones. The anhydrite succession is generally thinner with anhydrite developed as a single bed, 4 to 6 m thick; it rests on, and passes up into, silty mudstones.

The Rossall Halite, overlying the basal marker, is divided into lower and upper leaves by the Unit A mudstone (Jackson et al., 1987). The Rossall Halite (lower leaf) is 56 to 119 m thick, and has no onshore correlative in the Blackpool district, Walney Island, or in Northern Ireland. It consists of three main halites separated by two persistent mudstones; the halites show a decrease in thickness, and contain more numerous mudstone interbeds, towards the top. There is a slight but characteristic upward convergence of gamma-ray and sonic-log traces (Figure 58). The apparent doubling in thickness of the Rossall Halite (lower leaf) seen in well 113/26-1 arises from halokinesis and a salt swell and the regional northward increase in thickness.

The Unit A mudstone is 27 to 58 m thick, and typically gives high gamma-ray readings. In places it contains thin siltstones and silty sandstones towards the base, as in wells 110/2-2, 110/3-1 and 110/7a-4 (Figure 52). A median halite with a thin overlying stringer are commonly present, although halites are absent in the thinner successions of wells 110/2-2 and 110/3-3.

Onshore, the Hambleton Mudstones (30 to 37 m), and the lowest 10 m of the Singleton Mudstones which underlie the Rossall Salts, are jointly equivalent to the combined anhydrite-dolomite marker, the Rossall Halite (lower leaf) and the Unit A mudstone. The Hambleton Mudstones are a shallow-water sequence of grey, sandy mudstones and siltstones bearing large pseudomorphs, up to 10 m across, of calcite after halite (Wilson and Evans, 1990). They have yielded macrofossils which include the crustacean *Euestheria minuta*, sporadic trails, and finely comminuted plant debris. Within the Hambleton Mudstones, a breccia with soft-sediment deformation has been interpreted as a seismite, reflecting an earthquake shock at the time of the Röt trangression (Wilson and Evans, 1990).

The upper leaf of the Rossall Halite (Figure 54) is rather variable in thickness offshore, ranging from 41 to 95 m; it commonly comprises four individual halites separated by three persistent mudstone beds. It equates with the Rossall

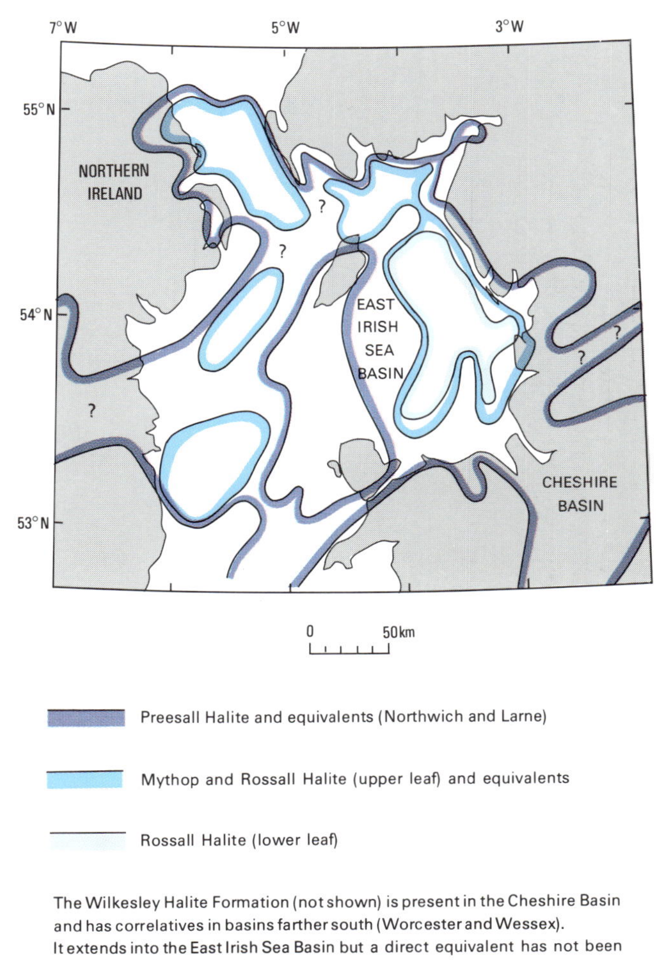

Figure 56 Approximate depositional limits of individual halites in the Mercia Mudstone Group.

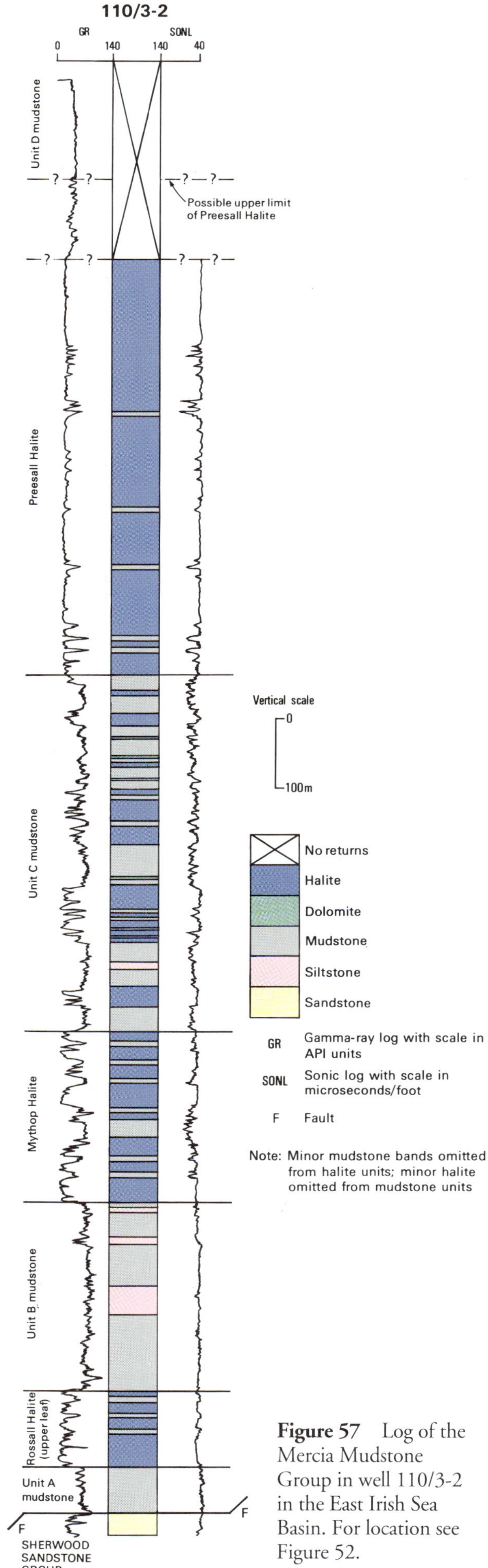

Figure 57 Log of the Mercia Mudstone Group in well 110/3-2 in the East Irish Sea Basin. For location see Figure 52.

Salts onshore. The halites become increasingly cleaner and thicker upwards, and thus in contrast to the Rossall Halite (lower leaf) a slight upward divergence of gamma-ray and sonic-log traces is observed (e.g. wells 110/2-5 (Figure 58) and 110/3-3). The result is an 'hour-glass' or 'waisted' gamma-ray and sonic-log profile for the Rossall Halites and Unit A mudstone; this forms a correlatable feature.

The distinctive Unit B mudstone is distinguished by the highest gamma-ray values recorded in the Mercia Mudstone Group, and is 100 to 190 m thick (Figure 54). Halites are commonly absent, (well 110/3-2, Figure 57), and thus the unit forms a characteristic marker in thick successions. However, up to seven halite stringers occur in the upper half of the unit in well 110/2-6 (Figure 58).

The Mythop Halite is 100 to 135 m thick and contains more mudstone interbeds than the Rossall Halite, with a tendency to develop two more-persistent halite beds. Lateral variability appears to be more pronounced than in underlying beds, for it displays a more variable geophysical-log profile, both vertically and laterally. Wells 110/2-5 and 110/2-6 demonstrate the characteristically waisted gamma-ray/sonic-log profile where the halite contains a thick median mudstone (Figure 58).

Onshore on Walney Island and in west Lancashire (Rose and Dunham, 1977; Wilson and Evans, 1990), the 137 to 311 m-thick Singleton Mudstones include the Rossall Salts at about 10 m above the base, and the Mythop Salts near the top. The mudstones correlate offshore with the topmost part of Unit A mudstone, the Rossall Halite (upper leaf), Unit B mudstone, the Mythop Halite and the lower part of Unit C mudstone (Figure 54). The Singleton Mudstones are an aeolian and shallow-water sequence of reddish brown, structureless or poorly bedded mudstones cut by numerous veins of gypsum. The Rossall Salts are represented by thin halite beds and penecontemporaneous solution breccias. The Mythop Salts are mainly of the haselgebirge type, representing halite crystal growth within mudstone; the division is cleaner and thicker in the north-west, and passes south-eastwards into penecontemporaneous solution breccias (Wilson, 1990; Wilson and Evans, 1990).

Recent work in the Blackpool district (Warrington in Wilson and Evans, 1990) indicates that the Hambleton Mudstones to lowermost Coat Walls Mudstones succession is largely, if not entirely, Anisian in age (Figure 54). This district also provides evidence for the position of the only stage boundary to be recognised within the Triassic in and around the report area. In the Coat Walls borehole (Figure 45), the Anisian pollen *Stellapollenites thiergartii* Mädler has been recorded up to about 17 m above the base of the Coat Walls Mudstone. *Retisulcites perforatus* (Mädler) and *?Ovalipollis pseudoalatus* (Thiergart), indicative of a Ladinian age, occur at 20 and 28 m above the base respectively.

Unit C mudstones vary in thickness from about 70 to 470 m, and contain a higher proportion of thin sandstones, siltstones and interbedded halites than the earlier mudstone units. They correlate with the uppermost Singleton Mudstones and the distinctive Thornton Mudstone Member of the Kirkham Mudstone Formation onshore (Figure 54; Wilson and Evans, 1990), and can thus be assigned an Anisian age. The Thornton Mudstones comprise eleven 'colour-paired' cycles which are remarkably persistent and allow precise correlation between the onshore boreholes of Walney Island and Blackpool (Wilson, 1990). Each 'colour-paired' cycle consists of approximately equal proportions of alternating greenish grey and reddish brown mudstones with thin siltstones and dolomitic siltstones, imparting a banded appearance to the entire succession. Sedimentary structures

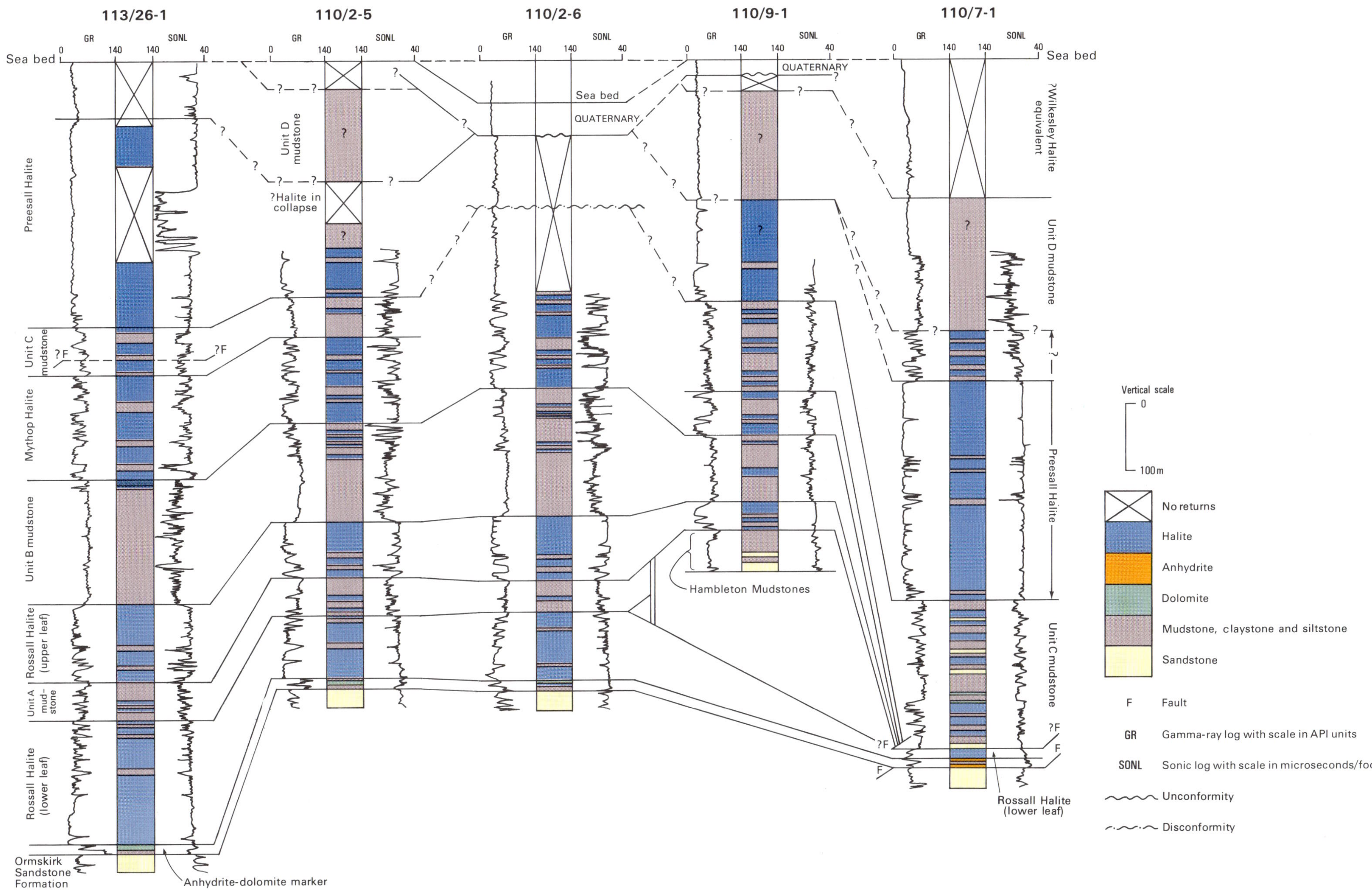

Figure 58 Logs of the Mercia Mudstone Group in selected boreholes in the East Irish Sea Basin. See Figure 52 for locations.

include desiccation cracks, current-ripple lamination, cut-and-fill structures, and convolute bedding. This lithology has been identified offshore in well 110/8-2 (Figure 52; Wilson, 1990), and is probably widespread.

Offshore, the Preesall Halite is a thick, clean halite that contains thin, red, mudstone partings which can be correlated individually over a distance of at least 15 km between the 587 m-thick development of halite in well 110/3-2 (Figure 57) on the downthrow side of the Crosh Vusta Fault Complex, and the 325 m thick unit in well 110/7-1 (Figure 58). It is likely that the halite was formerly more widespread, but erosion in post-Triassic times has restricted present-day occurrence. The full thickness is preserved only in the major basins and deeper grabens. In addition, near contemporary land surfaces, solution of the halite will have taken place continuously since the climate of the British Isles became temperate during the mid-Tertiary. A residual collapse breccia is likely to be found over extremely wide areas offshore, where the Preesall Halite is at or near the present-day level of erosion. The collapse breccia may be up to 335 m thick onshore (Wilson and Evans, 1990).

In west Lancashire, the Preesall Salt contains more and thicker mudstones than offshore and thins southwards and towards the basin margin (Wilson and Evans, 1990). The complete halite sequence, 100 to 130 m thick on average, is preserved only at depth, beyond the reach of groundwater circulation. At surface, the halite has been dissolved away by circulating groundwater, and the position of the 'outcrop' is marked by subsidence hollows. The residual collapse breccia consists of mudstone and gypsum, and averages about 70 m in thickness (Evans, 1970). On Walney Island the collapse breccia is at least 99 m thick; the Walney Channel (Figure 52) is situated close to the core of a north-north-westerly trending syncline (Rose and Dunham, 1977; Evans and Wilson, 1975), and may be a solution-subsidence feature coincident with a former brine run. In the Kirkham borehole (Figure 52), halite is completely absent down to 320 m, suggesting ingress of fresh water and dissolution of halite (Wilson and Evans, 1990).

The thickest halite in the offshore sequence was initially correlated by Jackson et al. (1987) with the Wilkesley Halite Formation (Figure 54) of the Cheshire Basin, on the basis of sparse, yellow, Carnian spores recovered from the halite in well 110/3-2. However, these spores are now regarded as contaminants, and Wilson (1990) has correlated the thick halites in wells 110/3-2 and 110/7-1 (Figures 57 and 58) with the Preesall Salt, and hence with the Northwich Halite Formation of the Cheshire Basin. This suggests that an offshore equivalent of the Wilkesley Halite Formation occurs within the topmost 1600 m of the Mercia Mudstone Group.

Offshore, the Unit D mudstones overlie the Preesall Halite, but clear sections are currently recorded only in wells 110/7-2 (Figure 45) and 110/7-1 (Figure 58) where only 200 m above the Preesall Halite have been logged. Elsewhere, thick mudstones overlying the Preesall Halite may comprise, in parts, foundered mudstones and collapse breccias, for example in wells 110/2-5, 110/2a-7, 110/7-3, 110/9-1 and 113/26-1 (Figure 52), but have also been included in Unit D mudstones. Miospores recovered from well 110/7-1 suggest the presence of Ladinian or younger Triassic deposits above the Preesall Halite; marine microplankton show a slight concentration in the mudstones spanning the Preesall Halite, indicating continued marine influence.

Onshore in the Blackpool district (Wilson and Evans, 1990), the sequence above the Preesall Salt is subdivided into the Coat Walls Mudstones and the Breckells Mudstones. The Coat Walls Mudstones consist of alternations of blocky, reddish brown mudstone and laminated, greenish grey or reddish brown mudstone. About 200 m of the Breckells Mudstones are preserved, but the original thickness is not known. The formation consists dominantly of red, blocky mudstones with gypsum nodules and scarce, grey-green bands. In the Hackensall Hall borehole (Figure 45; Wilson, 1990; Wilson and Evans, 1990), a collapse breccia near the top of the Breckells Mudstones was inferred to include the residue of a totally dissolved halite, probably equivalent to the Wilkesley Halite Formation of the Cheshire Basin (Figure 54).

Solway Firth and Carlisle basins

The Mercia Mudstone Group has been recovered onshore in boreholes on the Isle of Man at the south-western margin of the Solway Firth Basin (Gregory, 1920), and in a number of boreholes north of the Lake District. At the Point of Ayre in the Isle of Man (Figure 52), the lowest 70 m consists of an interbedded mudstone and sandstone 'waterstones' lithology, with gypsum-rich units at the base and top. This unit is tentatively correlated with the anhydrite-dolomite marker, Rossall Halite (lower leaf) and Unit A mudstone of the East Irish Sea Basin (Figure 54). It passes upwards into mudstone with interbedded halite interpreted as the equivalent of the Rossall Halite (upper leaf).

Offshore, it is possible that a full sequence of the Mercia Mudstone Group, about 800 m thick, is present in the central part of the Solway Firth Basin, where it may be overlain by Jurassic strata. The group was proved in BGS boreholes 73/51 and 73/69, (Figures 45 and 52).

In the Carlisle Basin, the Stanwix Shales are chiefly known from boreholes which suggest a total thickness of about 500 m (Figure 54). The succession consists of red-brown, blocky mudstones alternating with grey-green, silty mudstones which are more common near the base and include indications of evaporitic conditions of deposition (Dixon et al., 1926). In the Kelsick Moss borehole (Figure 52), a sandy, gypsum-bearing succession at the base of the Stanwix Shales is probably the lateral equivalent of the Hambleton Mudstones of west Lancashire (Figures 53 and 54). Similarly, in the Silloth 1A borehole, the lowest 40 m is distinguished by a more-calcareous base and a relatively high radioactive trace on the gamma-ray log (Figure 59). The only marker within the Silloth succession is a 6 m-thick halite some 200 m above the base of the Stanwix Shales. This can be correlated with a thin halite within gypsum-rich mudstones in the Kelsick Moss borehole, and may be the lateral equivalent of the Preesall Halite (Figure 54).

Cheshire Basin and the Formby district

The entire Mercia Mudstone Group, 1340 m thick, is preserved beneath the Lias Group at Prees in the south of the Cheshire Basin (Figure 52) where two major halite units, the Northwich Halite Formation and the overlying Wilkesley Halite Formation, occur (Poole and Whiteman, 1966; Evans et al., 1968). In the Formby and Southport areas, only the sequence below the Preesall Halite is preserved (Figure 54; Wilson, 1990).

The lowest unit of the Mercia Mudstone Group, the Tarporley Siltstone Formation, consists of interlaminated mudstones and siltstones with thin beds of sandstone — the 'waterstones' lithology. It attains a thickness of 273 m near Knutsford (Figure 53; Taylor et al., 1963). Locally, the upper part of the Tarporley Siltstone Formation passes laterally south-eastwards into a sandstone of aeolian origin (Figure 53). In the Formby district, the Tarporley Siltstone Formation is 45 to 58 m thick. As in Cheshire, its base is marked by an upward increase in gamma radiation due to the presence of micaceous mudstones, which are silty in places, with some interbedded fine-grained sandstones (Rowley in Wilson, 1990).

In Cheshire and at Formby, the lower mudstone division of the Mercia Mudstone Group consists of poorly laminated, reddish brown mudstone, with local repetition of the 'waterstones' lithology around Knutsford (Earp and Taylor, 1986). Strata directly beneath the Northwich Halite Formation are well-bedded, reddish brown and greenish grey interlaminated mudstones and siltstones, rather like the Thornton Mudstones of the Blackpool district (Wilson, 1990).

The Northwich Halite Formation consists of a number of beds of halite with partings of reddish brown mudstone; it attains a maximum thickness of 285 m north of Middlewich (Figure 52), and thins towards the basin edge (Evans et al., 1968). It is equivalent to the Preesall Salt, and to the Muschelkalk Halite of the southern North Sea (Warrington et al., 1980). The 327 m-thick middle mudstone division of the Mercia Mudstone Group directly overlies the Northwich Halite; it consists of distinctive alternations of structureless, reddish brown mudstones, with greenish grey and reddish brown, interlaminated mudstones and siltstones, with gypsum nodules near the top (Arthurton, 1980).

The 404 m-thick Wilkesley Halite Formation is preserved fully towards the southern edge of the basin, where it comprises halite with a number of beds of reddish brown mudstone (Poole and Whiteman, 1966). The 161 m-thick upper mudstone division consists of reddish brown, structureless mudstone with nodules and bands of anhydrite.

The 15 m-thick Blue Anchor Formation at the top of the group consists of hard, grey-green mudstones and siltstones with some gritty bands (Poole and Whiteman, 1966); it shows evidence of sun cracks and minor erosion surfaces. Fossils include *Euestheria minuta* and fish remains, indicating a change from the oxidising conditions which prevailed throughout most of the Triassic. A minor erosion surface occurs at the base of the overlying Penarth Group.

North Channel Basin and Northern Ireland

Offshore, the Permo-Triassic of the North Channel Basin has not been penetrated by commercial wells, but it is likely that the Mercia Mudstone Group occurs extensively. A complete sequence of the Mercia Mudstone Group, 967 m thick, has been drilled in the Larne No. 1 borehole (Manning and Wilson, 1975), at the coast of Northern Ireland (Figure 52), and a large part of the sequence (958 m) was proved in Larne No. 2 (Figure 59; Downing et al., 1982; Penn et al., 1983). In Northern Ireland, there is a rapid lateral transition from halite-bearing to halite-free successions. Significant halites occur in the North Channel Basin, whereas halite-free successions have been proved in the Langford Lodge borehole to the south-west (Figure 52; Warrington, in Manning et al., 1970), and in the Port More and Magilligan boreholes in the north (Wilson and Manning, 1978). Only in the Langford Lodge borehole have all the Triassic stages been recognised (Warrington in Manning et al., 1970).

Six formations were erected (Figure 54) in both the halite-bearing and the halite-free successions (Manning

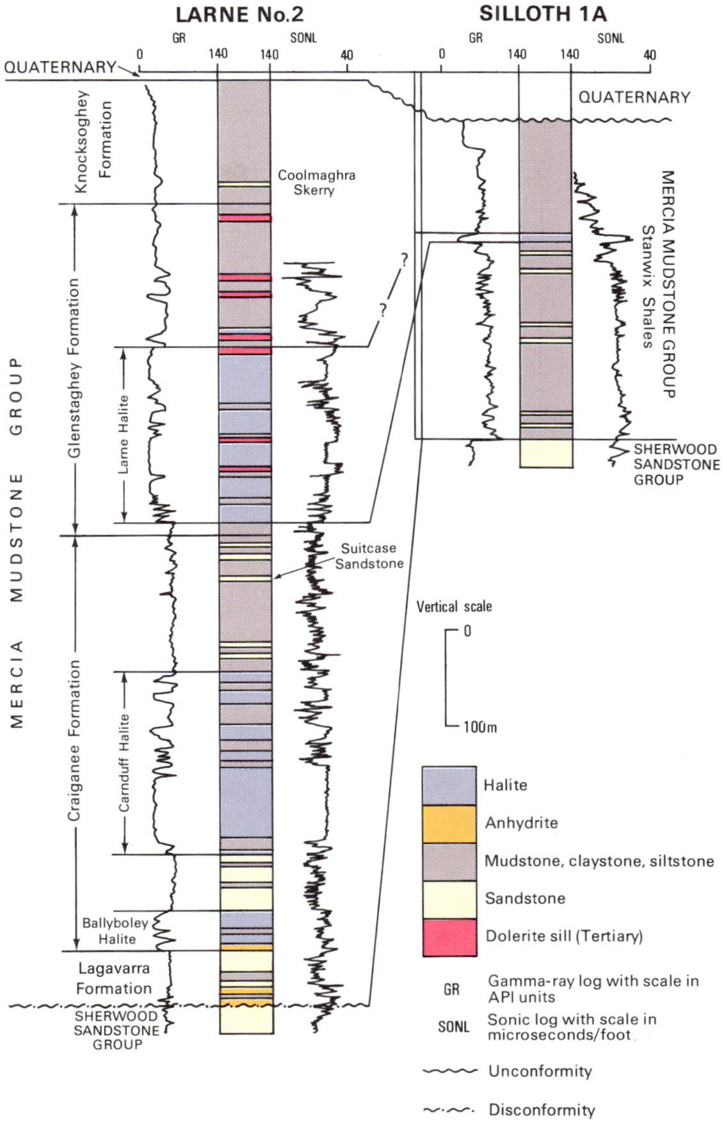

Figure 59 Logs of the Mercia Mudstone Group in the Larne No. 2 and Silloth 1A boreholes. For locations see Figure 52.

and Wilson, 1975). The basal Lagavarra Formation is up to 55 m thick at Larne and consists of striped beds similar to the 'waterstones' lithology, with interbedded anhydrite at the base. This intertidal sabkha sequence is similar to beds in the East Irish Sea Basin and on the Isle of Man (Figure 53).

The clastic rocks of the overlying Craiganee Formation consist mostly of interlaminated siltstones and mudstones, with two thin sandstones near the top. Two units of halite, the Ballyboley and Carnduff halites (Figures 54 and 59), appear to thicken eastwards at Larne. Anhydrite nodules occur in the mudstones in Larne No. 1 beneath the halites, and miospores from beds between the halites (Warrington and Harland, 1975) are now assessed to be Anisian in age (G Warrington, written communication, 1992).

The Glenstaghey Formation is 560 m thick in Larne No. 1, in which it consists of reddish brown mudstone dominated by the 481 m-thick Larne Halite (Manning and Wilson, 1975). This massive halite contains scattered pockets and interbeds of red and green mudstone incorporating halite crystals. Thickness variations suggest that part of the halite and underlying mudstone is cut out by faulting in Larne No. 2, since the halite fails to follow the trend of easterly thickening shown by the lower halite beds.

The Knocksoghey Formation is up to 180 m thick and consists of massive, reddish brown mudstone with patches and bands of gypsum and anhydrite. It is similar to the upper mudstone of the Mercia

Mudstone Group of Cheshire, and the Triton Anhydrite Formation of the southern North Sea (Cameron et al., 1992) in which sulphate development reaches a maximum (Warrington, 1974). The Coolmaghra Skerry (Figures 54 and 59), up to 10 m thick, occurs about 53 m above the base, and is prominent on gamma-ray logs; it is the most persistent sandstone skerry in Northern Ireland and can be regarded as a time marker approximating to the top of the Carnian (Geiger and Hopping, 1968; Warrington et al., 1980). It is correlated with the Arden Sandstone Member of the English Midlands (Manning and Wilson, 1975), and the Schilfsandstein of the southern North Sea (Warrington, 1970a).

The Port More Formation is defined in the Port More borehole to the north of the report area; it is up to 38 m thick and lithologically similar to the Knocksoghey Formation, although sulphate minerals are absent. The Norian to Rhaetian, Collin Glen Formation (formerly the Tea Green Marl) at the top of the group comprises greyish green calcareous mudstones up to 10 m thick, and is correlated with the Blue Anchor Formation in England (Warrington et al., 1980).

Kish Bank, Caernarfon Bay and St George's Channel basins

In the Kish Bank Basin, the Mercia Mudstone Group, or 'Keuper' of Jenner (1981), is locally fully preserved and conformably overlain by the Lias Group (Dobson and Whittington, 1979). The sequence thickens towards the bounding faults, the Bray Fault to the south-west, and the Dalkey and Lambay faults to the north-west, and is estimated to total approximately 2250 m (Figure 50).

On shallow-seismic profiles, the sequence is divisible into a lower and upper unit (Dobson and Whittington, 1979). The lower unit consists of dipping, thickly bedded reflections interpreted as halite, alternating with thinly bedded, high-amplitude reflections considered to be mudstone. The upper unit shows only widely spaced, gently flexed reflectors. Deep seismic-reflection records display similar reflector characteristics to those in the East Irish Sea and Solway Firth basins, with evidence of salt swells and pillows, and synclinal collapse zones possibly caused by halite solution at depth (cf. Jenyon and Cresswell, 1987). A basal *décollement* with the Sherwood Sandstone Group suggests the presence of the Rossall Halite equivalent, and steeply dipping, high-amplitude reflections adjacent to the Dalkey Fault (Jenner, 1981) imply minor slippage on younger halite layers.

The lower part of the sequence has been proved in well 33/21-1 (Figures 52 and 54). The lowest beds consist of dark grey, silicified siltstone with traces of anhydrite near the base. These pass up into a 143 m-thick lower halite division which contains interbedded mudstone towards both the base and top. The halite is overlain by 109 m of red-brown mudstone with interbedded siltstone and sandstone; both the halite and the mudstone are claimed to be late Scythian in age. The 691 m-thick upper halite consists of a thicker lower unit of alternating halite and mudstone which has yielded both late Scythian and Anisian miospores, and an upper unit of thick halites with a few thin mudstone beds. The overlying red-brown mudstone has yielded Anisian miospores. The dating is not necessarily at variance with the East Irish Sea and Cheshire basins, bearing in mind that it was undertaken before the re-positioning of the Scythian–Anisian boundary by Benton et al. (1994).

In the Caernarfon Bay Basin, sandstone and siltstone from BGS borehole 71/53 (Figure 52) contain late Scythian or Anisian palynomorphs (Wilkinson and Halliwell, 1979). They are tentatively equated with the Tarporley Siltstones, although similar forms are known from the Helsby Sandstone Formation of the Sherwood Sandstone Group (Figure 48; Warrington, 1970a; Benton et al., 1994). In the St George's Channel Basin, the lower part of the Triassic sequence is considered to subcrop beneath Quaternary, but has not been proved.

PENARTH GROUP

The Penarth Group, of Rhaetian age (Figures 44 and 54), was deposited chiefly under lagoonal and shallow-marine conditions during a marine transgression. It rests disconformably on the Blue Anchor Formation, and passes up conformably into the fully marine Lias Group. The Penarth Group has not been proved offshore, but despite only being up to 23 m thick, its widespread occurrence beneath the Lias Group in Northern Ireland, and in the Carlisle (Ivimey-Cook et al., in press) and Cheshire basins (Figure 52), suggests that it is likely to occur under those sediments in the Keys and Solway Firth basins.

The Penarth Group in Northern Ireland rests disconformably on the Collin Glen Formation (Figure 54; Griffith and Wilson, 1982). It is subdivided into two formations. The lower, Westbury Formation, is the thicker and consists of dark grey to black, fissile mudstones that are commonly pyritic and contain a few thin limestone, siltstone, and fine-grained sandstone beds. It contains *Rhaetavicula contorta* (Portlock) and *Protocardia rhaetica* (Merian), with some conglomeratic fish-bearing layers near the base. The overlying Lilstock Formation consists of a lower division of drab-grey, locally reddish grey, calcareous mudstones, and an upper division of calcareous mudstones, with shales containing thin limestones. This is overlain conformably by soft grey mudstones of the Lias Group, the lowest beds of which — the Pre-Planorbis Beds — lack ammonites, and are of Rhaetian age (Ivimey-Cook, 1975; Cope et al., 1980).

In the Cheshire Basin, an erosion surface marks the junction between the dark grey to black, fossiliferous mudstones of the Westbury Formation and the overlying, less fossiliferous Lilstock Formation (Poole and Whiteman, 1966). The Pre-Planorbis Beds are the youngest Triassic rocks, and represent the inception of fully marine conditions.

8 Jurassic and Cretaceous

The marine Rhaetian transgression that spread across southern Britain towards the end of the Triassic Period advanced over an area of widespread low relief, causing uniform depositional conditions to prevail during latest Triassic and Early Jurassic times. In the report area, sediments of Mid-Jurassic to mid-Cretaceous age are believed to be absent, perhaps in part due to later erosion, but marine deposition recurred in Late Cretaceous times following a eustatic rise in sea level (Cope, 1984; Haq et al., 1987).

On the periphery of the offshore area, Lower Jurassic rocks are preserved in Northern Ireland beneath the protective capping of early Paleogene lavas, and small outliers of Lower Jurassic sediments are preserved in the Carlisle and Cheshire basins (Figures 2 and 60). Offshore in the East Irish Sea Basin, proven rocks of Early Jurassic age are restricted to an erosional outlier in the Keys Basin (Jackson et al., 1987). Other occurrences are inferred in the Solway Firth and Berw basins, largely on the basis of seismic interpretation and regional considerations. Jurassic sediments have been proved in the Kish Bank Basin to the west of the report area, and are possibly present in the Central Irish Sea and Caernarfon Bay basins. No Cretaceous rocks are known offshore other than small outcrops emerging from beneath the lavas of Northern Ireland (Figure 60).

Evidence from Northern Ireland and the St George's Channel suggests that widespread earth movements took place at the end of the Early Jurassic and during the Early Cretaceous (Fletcher, 1977; Barr et al., 1981). Jackson et al. (1987, p.202) postulated a period of pre-end Cretaceous block faulting 'probably as part of the Late Cimmerian movements', with a later period of uplift, possible compression and inversion, in latest Cretaceous times. Traditionally, the main phase of inversion, uplift and postdepositional faulting of the Irish Sea basins has been assumed to have been during the Tertiary (Moseley, 1972). Employing the calculations of Francis (1982), a minimum of 500 m uplift and erosion has occurred in the Keys Basin above the subcrop of the Fleetwood Dyke Group. Nevertheless, the continuity of the Fleetwood Dyke Group in crosscutting the Tynwald Fault Complex (Figures 2, 17 and 52) indicates that the main period of post-Triassic faulting preceded dyke intrusion during the early Paleocene (Figure 2; see also Arter and Fagin, 1993; Jackson and Mulholland, 1993).

JURASSIC

The base of the Jurassic in Britain is placed just above the base of the Lias Group, at the first appearance of the ammonite *Psiloceras* (Cope et al., 1980; Warrington et al., 1980). The Lias Group was typically deposited in fully marine conditions, and consists of rhythmic alternations of monotonous, blue-grey, calcareous mudstone, thin beds of calcareous siltstone, and bands or concretions of argillaceous

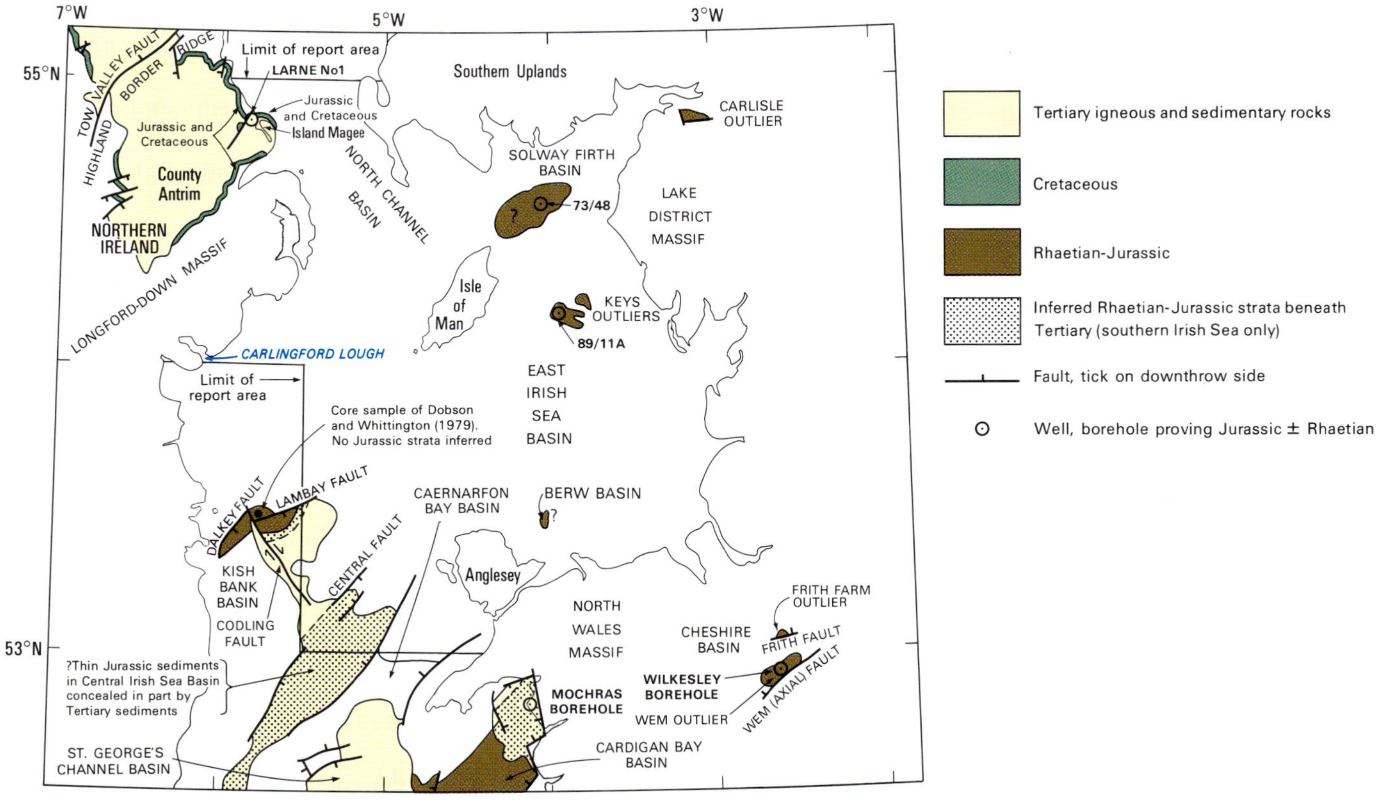

Figure 60 Distribution of Jurassic and Cretaceous strata. Compiled from BGS map sheets, Fletcher (1977), Jenner (1981), and Dobson and Whittington (1979).

limestone. The group shows few lateral facies changes, and contains a rich and diverse marine fauna dominated by bivalves and ammonites; the latter provide the basis for a detailed zonation.

The Lias Group is exposed along the County Antrim coast of Northern Ireland, with a narrow outcrop extending offshore east of Island Magee (Figure 60; BGS Isle of Man Solid Geology sheet). In the southern part of Antrim, only the Lower Lias is represented; Hettangian and Sinemurian beds (Figure 61) rest conformably on Rhaetian deposits, and are overlain unconformably by Cretaceous rocks (Wilson, 1972). The Larne No. 1 borehole (Figure 60) proved 52 m of dark grey mudstone of Hettangian age (Manning and Wilson, 1975), but a thickness of some 100 m, with a wider stratigraphical range, has been estimated from coastal exposures (Ivimey-Cook, 1975). Beneath the base Cretaceous unconformity, the Lias thins irregularly north-westwards towards the Highland Border Ridge, where the Cretaceous sediments rest on Triassic or older rocks (Fletcher, 1977). Younger Liassic rocks occur in the north of Antrim, but no Middle or Upper Jurassic strata have been proved anywhere in Northern Ireland; their absence may be due either to non-deposition or to their removal during the Early to mid-Cretaceous interval.

Lias Group limestones and shales containing Hettangian ammonites form a small outlier near Carlisle (Figure 60), where they give rise to a plateau above the less-resistant Triassic rocks (Dixon et al., 1926; Cope et al., 1980). Boreholes in this outlier proved Lias Group resting on Triassic Penarth Group (Ivimey-Cook et al., in press). Outliers of the Lias Group also occur to the south-east of the report area (Figures 60 and 61), near the axis of the Cheshire Basin (Poole and Whiteman, 1966; Colter and Barr, 1975).

The Keys Basin contains the thickest post-Permian succession in the East Irish Sea Basin; the top of this succession displays a set of high-continuity seismic reflectors up to 600 m thick (Figure 61), which rest disconformably on reflectors taken to be near, or at the top of, the Mercia Mudstone Group (Jackson et al., 1987). BGS borehole 89/11A, drilled near the centre of the outlier, proved 9.15 m of dark grey, finely laminated, fissile siltstone and sandy limestone with abundant bivalves and a few bands of thin interbedded limestone. The presence of the ammonites *Arnioceras* and *Caenisites*? is indicative of a Sinemurian age (Semicostatum to Obtusum zones, probably Turneri Zone — H C Ivimey-Cook written communication, 1990). The high-amplitude reflectors at the base of the seismic package are equated tentatively with the limestones of the Penarth Group or basal Lias Group. Smaller Lias Group outliers may be preserved in fault-bounded slivers within the Tynwald and Crosh Vusta fault complexes (Figure 17), and Lower Jurassic strata may form much of the c.200 m of youngest sediment postulated, from seismic evidence, in an outlier in the Berw Basin (Figure 60; Jackson and Mulholland, 1993).

In the axial part of the Solway Firth Basin, a 650 m-thick sequence is observed on seismic sections to be similar to that in the Keys Basin. The sequence rests disconformably upon the Mercia Mudstone Group in an elongate centroclinal fold along strike from the Carlisle Outlier. In contrast to both the Carlisle and Keys outliers, the bounding faults are not seen. BGS borehole 73/48 (Parkin and Crosby, 1982; Wilkinson and Halliwell, 1979), proved 4.5 m of barren, red and greenish grey, slightly calcareous siltstone in these sediments. On lithological criteria, the beds would be assigned to the Mercia Mudstone Group, but this would imply that the angular discordance lies within the Mercia Mudstone Group. A substantial intra-Mercia Mudstone Group disconformity is not recorded elsewhere in the British Isles at this level. This incongruity cannot be accounted for at present; a possible explanation is that the siltstone may equate with red siltstones of Mid- or Late Jurassic age that rest unconformably on Lower Jurassic strata in the Cardigan Bay Basin (Penn and Evans, 1976; Dobson and Whittington, 1987). The lower and middle parts of the 650 m-thick package may thus comprise the Lias Group.

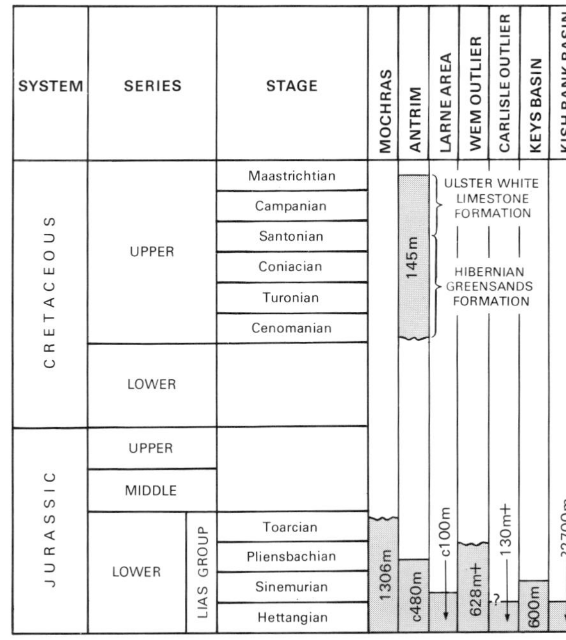

Figure 61 Age range of Jurassic and Cretaceous rocks in and around the report area. See text for sources.

In the Kish Bank Basin (Figure 60), two core samples of calcareous mudstone and limestone containing an Early Jurassic microfauna were recovered on the upthrow side of the Lambay Fault (Dobson and Whittington, 1979). Seismic evidence suggests the presence of up to 300 m (Jenner, 1981), or possibly 2700 m (Naylor et al., 1993), of strongly reflective Lias Group, apparently conformably overlying Permo-Triassic strata. Within the report area to the east of the Kish Bank Basin, Jurassic rocks may be present locally beneath Tertiary sediments in the Central Irish Sea Basin (Dobson and Whittington, 1979; BGS Anglesey Solid Geology sheet). Their occurrence in the Caernarfon Bay Basin also remains a possibility (Dobson and Whittington, 1987; BGS Cardigan Bay Solid Geology sheet).

To the south of the report area at the northern end of the Cardigan Bay Basin (Figure 60), seismic data indicate a thickness of 1800 m for the closely spaced, persistent reflectors of the Lias Group (Dobson and Whittington, 1987). The Mochras borehole drilled 1306 m of downfaulted, Hettangian to Toarcian, dark silty mudstone at the northern end of this basin (Ivimey-Cook, 1971).

These preserved erosional remnants of the probable former continuous cover of marine Lower Jurassic rocks over and around the Irish Sea relate to two tectonic settings. Firstly the thicker sequences (at Mochras and in the Keys, Kish Bank, Cheshire and possibly Solway Firth basins), that were deposited in axial or fault-related settings. Secondly, thinner successions deposited in more marginal or platform areas (Northern Ireland, Carlisle, and Berw Basin). In all cases, the Jurassic sediment thicknesses are proportional to the respective Upper Triassic thicknesses, indicating that a similar tectonic setting obtained during the Early Jurassic. Their preservation in synclinal cores or hanging-wall outliers indicates continued post-Early Jurassic warping.

CRETACEOUS

Adjacent to the report area, Cretaceous deposits are found only in Northern Ireland, where they reach an aggregate maximum thickness of 145 m (Wilson, 1972; Fletcher, 1977). Only Late Cretaceous sediments are found in Northern Ireland, where the marine succession is divided into two parts (Figure 61). The lower sequence, the Hibernian Greensands Formation, comprises up to 25 m of glauconitic sandstones that were deposited during a transgressive phase. The sands are Cenomanian to mid-Santonian in age, and include one minor and one major nonsequence (Wilson, 1981). The upper part of the Cretaceous sequence, the mid-Santonian to Maastrichtian Ulster White Limestone Formation (Fletcher, 1977), consists of pelagic limestone with accessory flint bands. A basal conglomerate deposited after a further transgression disconformably overlies the greensand (Fletcher, 1977). The chalk is a clean, coccolith-foraminiferal micrite with regular and persistent bands of flint nodules. However, unlike its English counterparts, the Ulster White Limestone Formation is a hard, nonporous, stylolitic limestone whose durability is thought to be due to hydrothermal alteration and compaction during eruption of Tertiary lavas. The Campanian and Maastrichtian succession along the coast of Northern Ireland, although thin, is one of the most complete and best exposed in Western Europe.

Although terrestrially-derived detrital input to the Chalk Sea was negligible, the Highland Border Ridge to the north of the area, and to a less visible extent the Longford–Down/Southern Uplands Massif (Figure 60), exerted a persistent influence on Cretaceous erosion and sedimentation. The Cretaceous sequence shows progressive onlap across these highs, which were submerged only in the late Campanian (Fletcher, 1977).

It is likely that thin Upper Cretaceous limestone and chalk were deposited extensively over much of Ireland and the Irish Sea area (Wilson, 1972, 1981; Cope, 1984). However, Senonian chalk preserved in an exhumed Dinantian limestone swallow hole in County Kerry (Walsh, 1966) is the only other preserved evidence, although chalk has been recovered on the continental shelf to the west of Ireland (Croker and Shannon, 1987) and in the North Celtic Sea Basin to the south (Naylor and Shannon, 1982).

Proven offshore occurrence of Cretaceous strata in the report area is restricted to a small outcrop on the west side of the North Channel Basin (Figure 60); this is a continuation of the Northern Ireland outcrop (BGS Isle of Man Solid Geology sheet). On the basis of the distribution of flint erratics found on the eastern coast of Anglesey, Greenly (1919) suggested that a Cretaceous outlier might occur in the East Irish Sea Basin. It remains possible that a chalk veneer was removed by Pleistocene erosion from the Lias Group outlier in the centre of the Keys Basin or elsewhere, but the absence of Antrim Lava Group erratics in Anglesey makes it unlikely that the flint erratics were derived from Northern Ireland. Another possible source could lie nearby in the Berw Basin (Figure 60), an independently subsiding half-graben formed in the hanging-wall of the Berw Fault, that may also contain Jurassic strata as part of a thin post-Triassic succession similar to that in Northern Ireland (Jackson and Mulholland, 1993).

9 Paleogene and Neogene

Following widespread inundation by the Chalk Sea in the Late Cretaceous, the beginning of the Tertiary was a time of considerable change in the vicinity of the British Isles. Rifting and increased thermal activity associated with the opening of the North-East Atlantic in the late Paleocene (56 Ma — Dewey and Windley, 1988) resulted in uplift and extensive volcanic activity on the continental margins around the incipient ocean, including northern and western Britain. To the south of Britain, the northward movement of Africa towards Europe produced the Alpine orogeny, which reached its peak during the mid-Tertiary. Compression related to the Alpine mountain building resulted in uplift or inversion in parts of Britain. Throughout these events, continued northward continental drift contributed to climatic change; the Paleogene climate was subtropical with seasonal rainfall (Wilkinson et al., 1980), but cool, temperate conditions prevailed by the end of the Neogene.

Tertiary rocks are extensively preserved in Northern Ireland (Figures 2, 60 and 62), mainly as early Paleogene igneous rocks, but including the Lough Neagh Clays of late Oligocene age. In north-east Wales, probable Tertiary clays and lignites are preserved in solution subsidence cavities in Carboniferous Limestone (Walsh and Brown, 1971), but more significant are the terrestrial Tertiary sediments that largely occur offshore in the Cardigan Bay Basin, and have been drilled at Mochras (Woodland, 1971) and elsewhere (Tappin et al., 1994). Offshore in the report area, a limited population of small intrusive igneous bodies have a wide but scattered distribution; the basalts of Northern Ireland extend only a short distance eastwards into the North Channel. Tertiary sediments occur in the Central Irish Sea and Kish Bank basins.

Evidence points to the removal of a considerable thickness of Permian to Early Triassic sediment, both before and during end-Cretaceous compression and inversion, and also during late Tertiary uplift (Jackson et al., 1987; Jackson and Mulholland, 1993). The marine Jurassic and Cretaceous sequences which are likely to have covered much of the area now exist only under the protective cover of basalts in Antrim, and as small outliers elsewhere. From studies of sonic velocities and vitrinite reflectance of the Mercia Mudstone Group and Carboniferous respectively, Colter (1978) estimated uplift for the Mercia Mudstone Group in well 110/3-2 in the East Irish Sea Basin (Figures 52 and 57) to be in excess of 2000 m. In north-west England, 3000 m of Tertiary erosion has been suggested from the results of fission-track analysis (Lewis et al., 1992), though this value has been rejected by Holliday (1993) and McCulloch (1994).

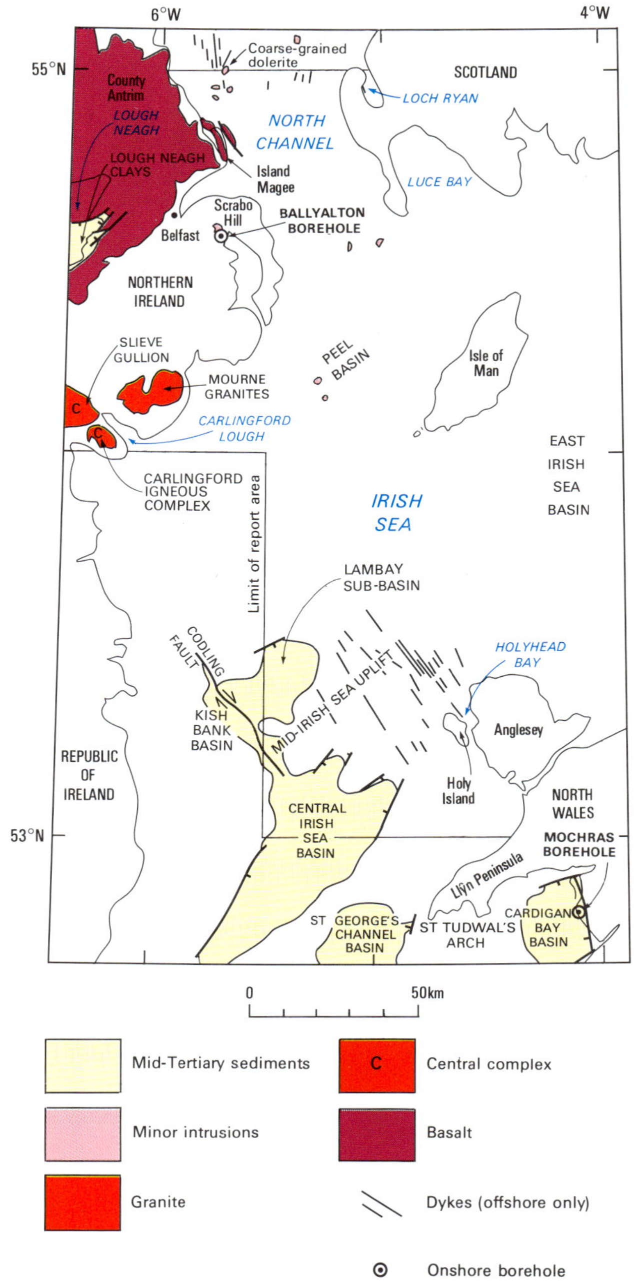

Figure 62 The Tertiary geology of part of the Irish Sea and adjacent areas. Land dykes are not shown; offshore dykes are generalised after Wright et al. (1971), Kirton and Donato (1985), Caston (1975), and BGS maps.

IGNEOUS ROCKS

The coastal cliffs of eastern Antrim (Figure 62) form the recessional basalt scarp of the Antrim Lava Group, which rest on the eroded surface of the Upper Cretaceous Ulster White Limestone Formation (Fletcher, 1977; Griffith and Wilson, 1982). Inland, the lavas can be up to 800 m thick, and have been divided into three main units, separated by two major lateritic beds that represent prolonged episodes of subtropical weathering (Wilson, 1972; Preston, 1981). The lavas are dominantly aa-type massive flows with blocky tops and bases, and were derived from fissure eruptions. Individual flows are lenticular in form, and the dominant composition is alkaline olivine basalt. The lavas are thought to span the interval 61 to 58 Ma (Mussett et al., 1988), and thus predate the opening of the North-East Atlantic (Dewey and Windley, 1988). At the coast bordering the report area, only the 250 m-thick Lower Basalt Formation is present. Associated with the lavas are a series of volcanic plugs and

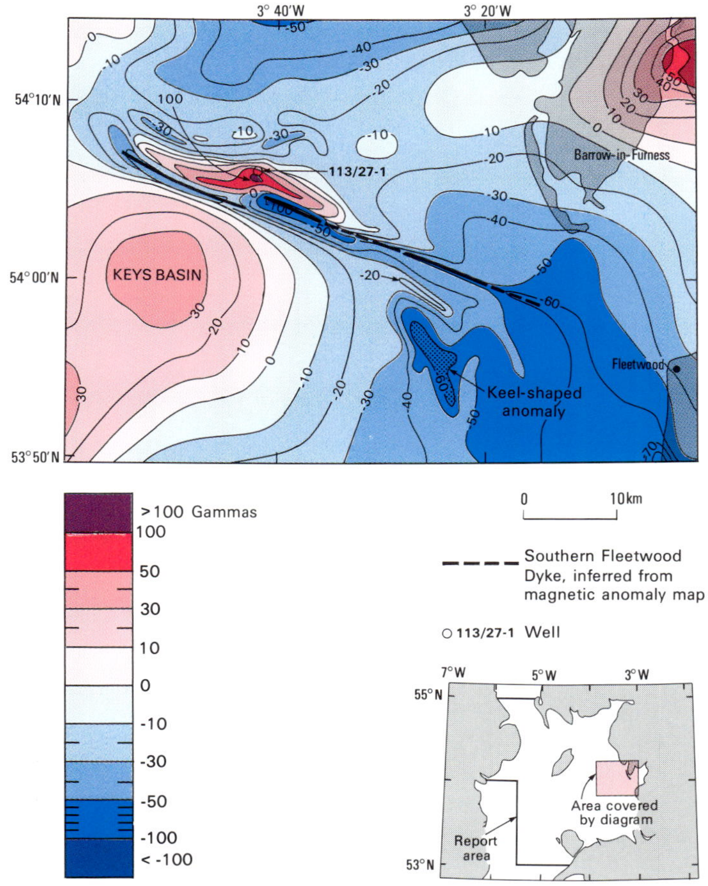

Figure 63 The Fleetwood Dyke Group magnetic anomaly. Taken from BGS maps; dyke locations after Kirton and Donato (1985).

doleritic dykes which may have acted as conduits for the lava flows (Preston, 1981).

Three major intrusive complexes occur near the Irish Sea coast of Northern Ireland, although their outcrops do not extend to the shore (Figure 62). The Slieve Gullion shield volcano may have formed at 58 to 56 Ma (Meighan et al., 1988), towards the end of, or after, the fissure eruptions. Activity then spread south-eastwards to the Carlingford complex. Both these igneous complexes have an evolutionary history of multiple injection culminating in ring dykes and cone sheets. The Mourne Granites are a late phase of acid intrusive rocks emplaced as five, separate, asymmetrical lobes dated as 56 to 51 Ma (Gibson et al., 1987; Thompson et al., 1987).

The earliest dykes were emplaced during the eruption of the lavas. A dyke from the Kingscourt Swarm to the southwest of Carlingford, dated at 51 Ma (Preston, 1981), may, together with the latest Mourne Granite, be the youngest intrusion. Crustal extension to accommodate the dykes locally exceeded 4 per cent in the Belfast area (Manning et al., 1970). Numerous Tertiary sills, generally of olivine dolerite, occur in Northern Ireland; a thick sill which caps Scrabo Hill, east of Belfast (Figure 62), is 159 m thick in the Ballyalton borehole (Bazley, 1975).

Offshore in the North Channel, extrusive igneous rocks of the Lower Basalt Formation are largely confined to a small, faulted outlier north-east of Island Magee (Figure 62; BGS Isle of Man Solid Geology sheet); their original extent is unknown. Dolerites intruded into Permo-Triassic sediments have been identified in the North Channel, both as rock islets and in the subsurface. As the intrusions are harder than the host rock, they tend to form ridges and knolls at the sea bed (Caston, 1975; 1976), and produce hyperbolic reflectors on seismic records. A pinnacle of coarse-grained dolerite at the northern limit of the report area (Figure 62; Eden et al., 1973) is seen on sparker profiles as an isolated, steep-sided stock (Deegan, 1978). The stock lies at the southern end of a group of dykes with a dominant north–south trend (Caston, 1975). Two similar outcrops, which have been interpreted as Tertiary volcanic necks (Wright et al., 1971), are mapped at the south-eastern entrance to the North Channel (Figure 62; Caston, 1976). Two minor intrusions have been mapped in the Peel Basin west of the Isle of Man, intruded into rocks of probable Permo-Triassic age (Figure 62; BGS Isle of Man Solid Geology and Aeromagnetic Anomaly sheets) and more intrusions are likely to be present (Bacon and McQuillin, 1972; Wright et al., 1971). In view of the composition of the plugs and dykes on land, these offshore intrusions are probably composed of dolerite (Charlesworth, 1937). If the rocks of the Peel Basin are Permo-Triassic in age (Figure 45), at least some of the intrusions are likely to be of Tertiary age. If the dykes or volcanic rocks were feeders for lava, then the Antrim Lava Group may have extended farther south than at present.

Dykes of probable Tertiary age have also been recorded on the south-west coast of Scotland, in Loch Ryan, and on the Isle of Man where Lamplugh (1903) noted a significant decrease in numbers above 30 m OD, and their absence above 213 m OD. A number of isolated dykes of probable Tertiary age occur in north-west England (e.g. Earp et al., 1961; Poole and Whiteman, 1966), including the extensive Cleveland Dyke Echelon (Kirton and Donato, 1985).

Major, isolated, *en-échelon*, west-north-westerly trending dykes (Young, 1965; Wright et al., 1971; Jackson et al., 1987) have been traced over a distance of some 50 km in the Keys Basin, where they intrude Mercia Mudstone Group sediments (Figure 63). The dykes, termed the Fleetwood Dyke Group by Kirton and Donato (1985), generate characteristic high-frequency magnetic signatures (Figure 23; BGS Liverpool Bay and Lake District Aeromagnetic Anomaly sheet), and on seismic records can be identified by diffraction patterns extending from the sea bed (Figure 64). The dykes generally dip very steeply in a northerly direction (Arter and Fagin, 1993), and locally assume a sill-like character. Widths of 100 m in the west, and 600 m in the east, have been estimated by Kirton and Donato (1985) for the southern dyke.

Near to the anomalies, well 113/27-1 (Figure 63) drilled dolerite from one of the Fleetwood dykes; K-Ar whole-rock dating carried out for BGS gave ages of 61.4 ± 0.8 Ma and 61.5 ± 0.8 Ma (K Hitchen and J D Ritchie, oral communication, 1990), whereas another dating by the same method yielded an age of 65.5 ± 1.0 Ma (Arter and Fagin, 1993). These ages imply that the dyke is older than most of the igneous rocks of Northern Ireland but more similar to ages derived from dykes in North Wales (Evans et al., 1993).

The width of dykes in Northern Ireland does not normally exceed 30 m (Wilson, 1972), and those in the North Channel measure less than 13 m in width (Caston, 1975). The anomalous length and width of the Fleetwood dykes is matched most closely adjacent to the report area by the Doraville Dyke, that is up to 100 m wide in County Fermanagh in Northern Ireland; Preston (1967) believed it to be related to a buried igneous complex in South Tyrone. The north-westerly trending keel-shaped magnetic anomaly immediately south-east of the Fleetwood dykes (Figure 63) may be a deeply buried igneous body or plug not easily resolved on seismic records. The identification elsewhere in the Irish Sea, such as at the entrance to Luce Bay (Figure 62), of seismic patterns similar to those in Figure 64 suggests that minor

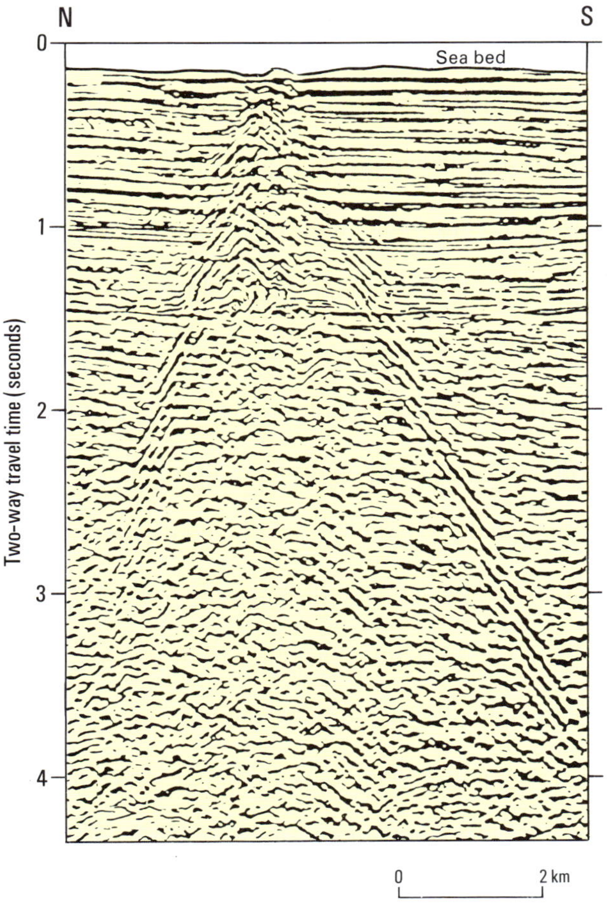

Figure 64 Seismic-reflection section across the southern Fleetwood dyke, showing diffraction patterns. After Kirton and Donato (1985).

intrusions may be more common than shown on current maps (e.g. Jackson et al., 1987).

To the west and north-west of Anglesey (Figure 62), a group of north-westerly trending negative magnetic anomalies (Wright et al., 1971) is believed to represent a swarm of Tertiary dykes intruded into Precambrian rocks (Figure 23; Kirton and Donato, 1985). The dykes are traceable through Holy Island to southern Anglesey, where they are up to 30 m wide (Greenly, 1919). The largest anomaly in Holyhead Bay has been ascribed to a body 70 m wide that consists of a group of dykes (Al-Shaikh, 1969).

SEDIMENTARY ROCKS

Following the volcanicity in north-east Ireland and western Scotland during the early Paleogene, there is probably a gap of over 20 Ma in the geological record before sediments of late Oligocene age were deposited (Curry et al., 1978). During this interval, the basalts of Northern Ireland were faulted and warped, notably to form a gentle synclinal sag in which the lacustrine Lough Neagh Clays are preserved to a maximum known thickness of 363 m (Figure 62; Wilson, 1972; Wilkinson et al., 1980). Weathered basalt 20 m thick preserved directly beneath the Lough Neagh Clays in the Washing Bay borehole (Fowler and Robbie, 1961), at the south-western bank of Lough Neagh, indicates a stable landscape without significant erosion earlier in the Oligocene.

To the south of the report area in the St George's Channel Basin, the Tertiary succession exceeds 1500 m in thickness and is Eocene to Oligocene in age (Tappin et al., 1994). In the Mochras borehole, drilled in the Cardigan Bay Basin (Figure 62), a mid-Oligocene to early Miocene age is indicated for the 524 m of Tertiary sediments (Herbert-Smith, 1979). Many other small basins of terrestrial sediment of largely Oligocene age have now been identified in western Britain, ranging from Devon in the south to the Little Minch off north-west Scotland in the north (Wilkinson et al., 1980; Evans et al., 1991).

The only Tertiary sediments identified offshore within the report area are in the Central Irish Sea and Kish Bank basins (Figure 62; BGS Anglesey Solid Geology sheet). These have not been sampled, and their age is reasoned, from regional considerations, to be Oligocene (Dobson and Whittington, 1979). Shallow-seismic records show these strata to have weak but persistent reflectors, usually flat-lying, but locally folded and disrupted by possible salt movement, especially in the Central Irish Sea Basin where the succession both dips and thickens to the south-west, and is up to approximately 450 m thick (Dobson and Whittington, 1979). The deposits are broadly synclinally disposed, infilling two small grabens that are separated by a south-westerly plunging Lower Palaeozoic anticline (BGS Anglesey Solid Geology sheet).

The Tertiary deposits of the Central Irish Sea Basin are linked, over the Mid-Irish Sea Uplift, with the Kish Bank Basin to the north-west along the line of the dextral strike-slip Codling Fault. The deposits overlie the Mesozoic unconformably, and are thickest near the Codling Fault (Dobson and Whittington, 1979). By analogy with onshore sequences in the Lough Neagh Basin and those in the Cardigan Bay Basin (Dobson and Whittington, 1987), the sediments are perhaps most likely to be of fluvio-lacustrine origin.

The Oligocene lake in which the clays and lignites of Lough Neagh accumulated formerly drained southwards into the Irish Sea (Charlesworth, 1963). Drainage reversal to the north followed Miocene uplift (Wilkinson et al., 1980). The deposits of the Kish Bank and Central Irish Sea basins may formerly have been connected with the Lough Neagh Clays along this original southerly drainage channel through Carlingford Lough (Figure 60). Any connection with the Cardigan Bay Basin was probably ended by uplift of the Llŷn Peninsula and the St Tudwal's Arch (Dobson and Whittington, 1987).

10 Pleistocene and Holocene

The Neogene and early Pleistocene were times of erosion and uplift across the Irish Sea region (Dobson and Whittington, 1987), following which, extensive deposition occurred in the mid- to late Pleistocene. Pleistocene and Holocene sediments are recognised on seismic profiles both by their unconformable relationship with pre-Quaternary strata and by their commonly subhorizontal stratification. Borehole evidence indicates that the Quaternary deposits are unlithified, ranging from soft to very stiff or hard.

The thickness of Quaternary sediments ranges up to some 300 m (Figure 65), but even the thickest sequences include major erosion surfaces, and nowhere has deposition been continuous. The thicker (greater than 100 m) deposits are concentrated in a 25 to 80 km-wide trough roughly coextensive with the deepest waters of the Western Trough (Figure 66), a bathymetric feature which runs through the western Irish Sea. The Western Trough ranges in depth from 60 to 315 m, and forms the central part of the Celtic Trough, which extends 500 km from the Malin Shelf to the Celtic Deep (Figure 67). The Western Trough is flanked to the west and east respectively by the Irish and Eastern platforms; these are low-relief shelves with water depths only locally greater than 60 m (Figure 66). In the south of the report area, the Eastern Platform is referred to as the Welsh Platform, which includes Caernarfon Bay. Quaternary deposits are generally less than 50 m thick on the platforms, and are absent from areas off Anglesey and Ireland (Figures 65 and 68).

Important features in the area, particularly in the Western Trough, are major incisions that cut down over 100 m to form enclosed depressions both on the platforms and in the trough. Many incisions on the platforms are filled with deposits up to 200 m thick in localised, enclosed depressions that are less than 5 km wide and up to 40 km long (see Figure 70). Some incisions, both on the platforms and in the trough, are incompletely filled and are marked by elongate, enclosed bathymetric deeps (Figure 66). These major incisions are thought to have a glacigenic origin (Wingfield, 1990); three generations of incision are recognised in the report area, forming important stratigraphical reference points (Figure 69) that may relate to a crude event stratigraphy for much of the offshore UK area (Wingfield, 1989).

All BGS boreholes, except one, have been drilled on the platforms in the Irish Sea, although some borehole results from farther south in the Celtic Trough relate to the Quaternary sequence in the Western Trough (Tappin et al., 1994). In general, the recognition of the units within the sequence is based on interpretation of seismic profiles, qualified by borehole data. Many of these units either occur only in the Western Trough, or are attenuated and difficult to correlate across the platforms. Four types of seismic facies are recognised:

1. Tabular stratified. These stratiform units are tens of metres thick and are laterally extensive for tens of kilometres. They are internally distinguished by closely spaced, horizontal reflectors.

2. Tabular unstratified. With an overall geometry similar to the tabular stratified, these lack internal continuous reflectors, although there may be discontinuous, dipping reflectors.

3. Lenticular infills. Channel or incision infill deposits from a few, to many tens of metres, in thickness. Acoustic signatures range from chaotic with high-angle and discontinuous reflectors, to draped and cross-stratified with horizontal reflectors. They may also be transparent or have only very faint reflectors.

4. Lenticular upstanding. These deposits form banks or sediment waves, and are largely confined to sea-bed features.

The Quaternary deposits of the Irish Sea report area have been divided into six formations (Hession, 1988; BGS Anglesey Quaternary Geology sheet). These are, in order of decreasing age: the Bardsey Loom, Caernarfon Bay, St George's Channel, Cardigan Bay, Western Irish Sea and Surface Sands formations (Figure 69). These formations are seen on seismic profiles to overlie one another in the Western Trough, although nowhere are all the formations present. Lateral transition, including interdigitation, occurs from the Cardigan Bay Formation into the Western Irish Sea Formation. Informal subdivision can be made of each formation into members and/or facies, although this has not been attempted for the Bardsey Loom and St George's Channel formations (BGS Quaternary Geology sheets).

In view of inconsistencies in correlation between Irish and British stages for the Quaternary (Bowen et al., 1986), north-west European names are used here, following BGS Quaternary Geology sheets for the area. Indeed, only limited correlations can be made with the complex and controversial successions to be found on land surrounding the report area (Garrard, 1977; Huddart et al., 1977; Warren, 1985; Thomas, 1985; Bowen et al., 1986; Eyles and McCabe, 1989).

BARDSEY LOOM FORMATION

The oldest recognised Quaternary strata are confined to the Western Trough in St George's Channel, where they are deeply buried (Figures 68 and 69). The deposits consist of tabular-stratified beds up to 50 m thick which occupy shallow basins some 3 to 10 km wide in rockhead. No samples have been obtained from these deposits, but the acoustic signature of discontinuous, concave-up, cuspate reflectors with onlap on to the basin margins may be tentatively interpreted as rudimentary bedding within shallow channels, possibly indicating a fluviatile or shallow-marine environment.

In BGS borehole 89/10 to the south of the report area in the Celtic Trough (Figure 67), the Bardsey Loom Formation comprises beds of clay, sand, pebbly sand and gravel with layers of peat (Tappin et al., 1994). The deposits have a sparse microbiota indicative of a cold environment (BGS Biostratigraphy Group), and given their position at the base of the Quaternary sequence below the first incision level, they are tentatively ascribed a pre-Elsterian age (Figure 69).

CAERNARFON BAY FORMATION

Within the report area, this formation consists of the Lower Unstratified and Incision Infill members, neither of which

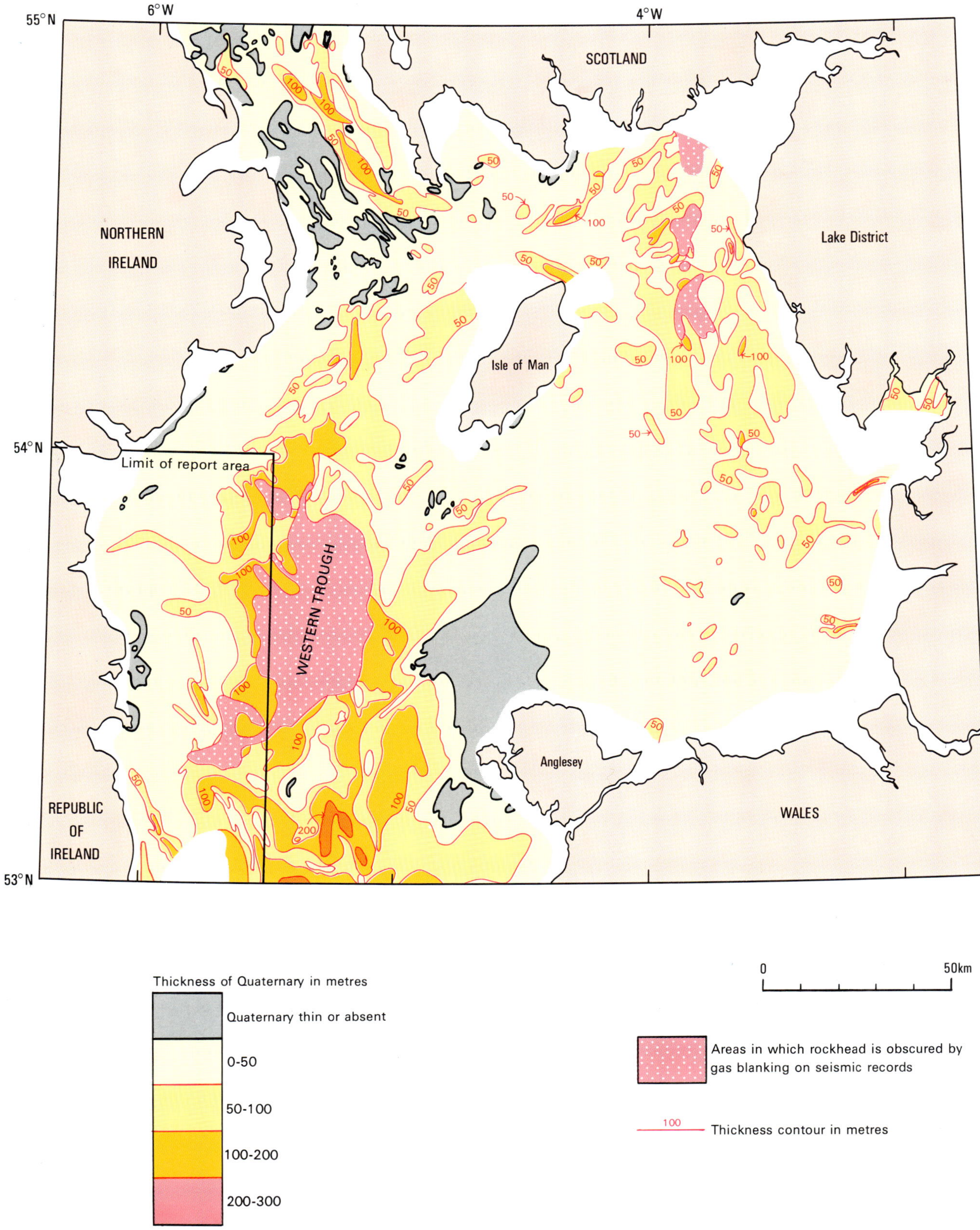

Figure 65 Thickness of Quaternary sediments in the Irish Sea.

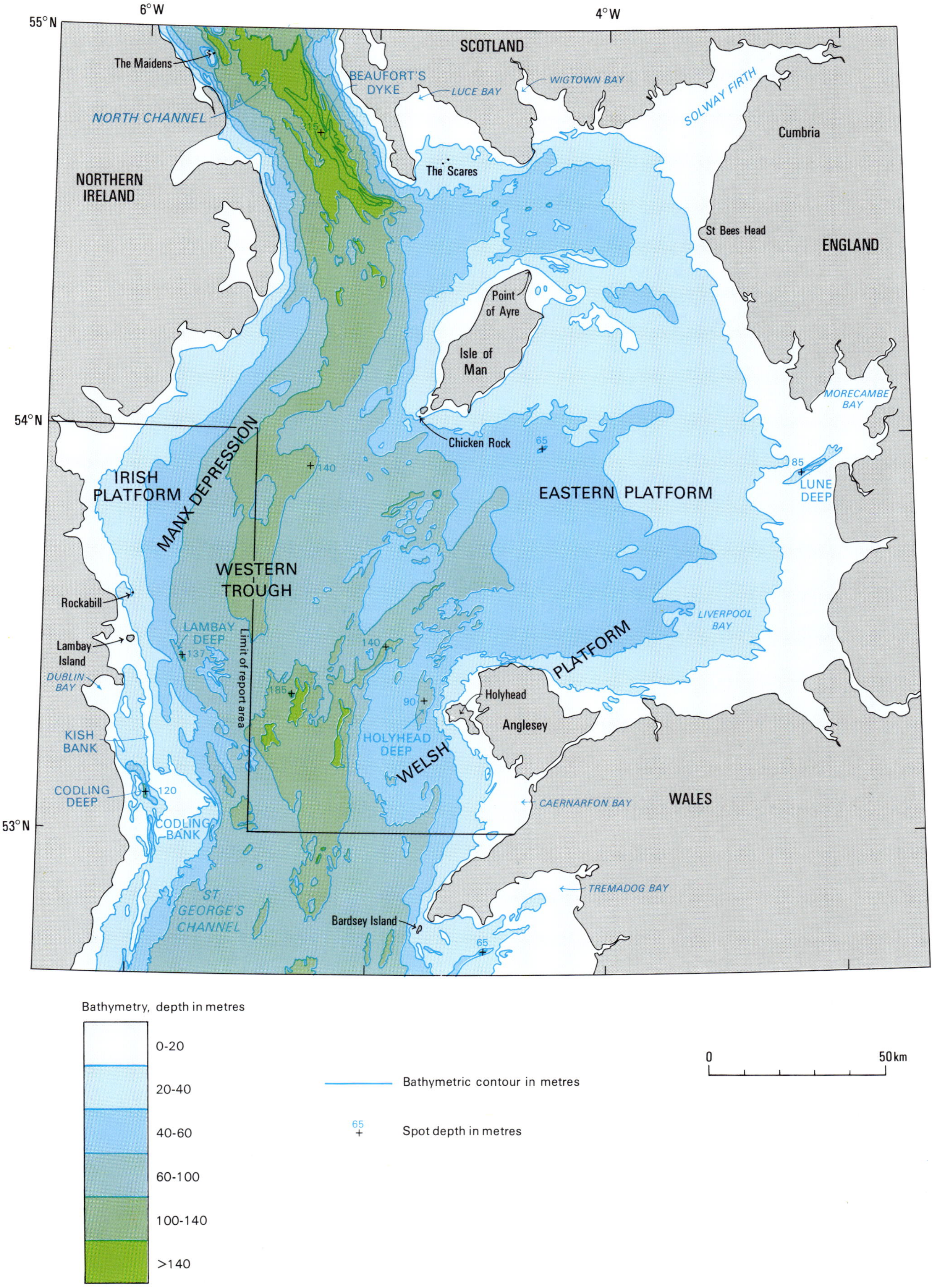

Figure 66 Bathymetry of the report area.

crop out. The members are separated by a major erosion surface marking the first generation of major incisions (Figure 69).

Lower Unstratified member

This unit is formed of tabular-unstratified deposits up to 70 m thick, and occurs largely in the Western Trough, but also on the platform margins in Caernarfon Bay, north-west of Holyhead, and around the Lambay Deep. On seismic profiles, the member generally lacks internal structure, but may exhibit discontinuous reflectors and hyperbolic point sources. The member either rests upon the Bardsey Loom Formation with apparent conformity, or overlies a subhorizontal unconformity above pre-Quaternary rocks.

The acoustic signature of this member suggests that it consists of unsorted deposits laid down in subglacial or ice-proximal conditions. South of the report area, this member is proved in boreholes and cores as an overconsolidated diamicton, interpreted as a subglacial lodgement till (Tappin et al., 1994). Since the incision surface above this member is considered to relate to late Elsterian glaciation (Wingfield, 1989), these sediments are thought to be of earlier Elsterian age.

Incision Infill member

This member is formed of lenticular infills more than 200 m thick that postdate, or were coeval with, the erosion of the boat-shaped depressions cut down through both older Quaternary strata and up to 150 m of pre-Quaternary rocks (Wingfield, 1990). The Incision Infill member completely fills the eroded depressions, though the succeeding tabular unit, the St George's Channel Formation, shows a tendency to thicken above the incisions (Figure 68). These intra-Caernarfon Bay Formation incisions (Figure 70) are up to 25 km long and 5 km wide, and are apparently confined to the Western Trough, unless some of the infilled incisions on the Eastern Platform are of this generation.

Boreholes in the eastern Irish Sea have penetrated up to 77 m of deposits forming incision infill, which cannot be assigned unequivocably to this member (Figure 71). These infill deposits are, in ascending order: diamictons of stiff clay with stones, mud with clasts from pebble to boulder size, sands, and muds and clays. Such deposits are typical of glacigenic incision infills around the UK (Holmes, 1977; Long and Stoker, 1986; Wingfield, 1990). The sediments are considered to have been laid down in the incisions during the late Elsterian deglaciation, possibly with final infill of kettle-holes during the Holsteinian (Figure 69).

ST GEORGE'S CHANNEL FORMATION

This unit consists of tabular-stratified deposits generally 30 to 65 m thick. It occurs in the southern part of the Western Trough (Figure 68), and extends on to the adjacent platforms where it crops out locally and wedges out in Caernarfon Bay and near the Lambay Deep. Its acoustic signature consists of closely spaced, subhorizontal reflectors which are laterally continuous over many kilometres. Minor intraformational disconformities are indicated locally.

The formation has not been sampled in the report area, but to the south, three boreholes through the St George's Channel Formation (Tappin et al., 1994) proved muds with minor shell debris and sporadic pebbles. Although these deposits have been previously described as interglacial marine deposits (Jasin, 1976; Garrard, 1977; Whittington, 1980; Hession, 1988), indications in the microbiota are overwhelmingly of boreal, cold waters that suggest arctic-like, glaciomarine, depositional conditions (Jasin, 1976; D M Gregory and R Harland, written communications, 1985; 1987). Only a minor interval, represented by deposits less than 1.5 m thick in the mid-parts of sections from two boreholes in Cardigan Bay, shows slight signs of cool, temperate conditions. In view of this information, and its position within the Quaternary sequence, an early Saalian age is proposed for the St George's Channel Formation.

CARDIGAN BAY FORMATION

The Cardigan Bay Formation has Upper and Lower Till members formed of tabular-unstratified deposits, which are dominantly made up of till. In between lies the Bedded and Infill member that largely comprises the lenticular infill of intra-Cardigan Bay Formation incisions (Figures 68 and 69).

Lower Till member

This tabular-unstratified member ranges up to 90 m in thickness in the northern part of St George's Channel, where it crops out only locally at the eastern margin of the Western Trough (BGS Anglesey Quaternary Geology sheet). Its base is an erosion surface with gentle topographic variation up to 15 m in amplitude. Internally, the deposits characteristically exhibit discontinuous, high-angle reflectors that produce a

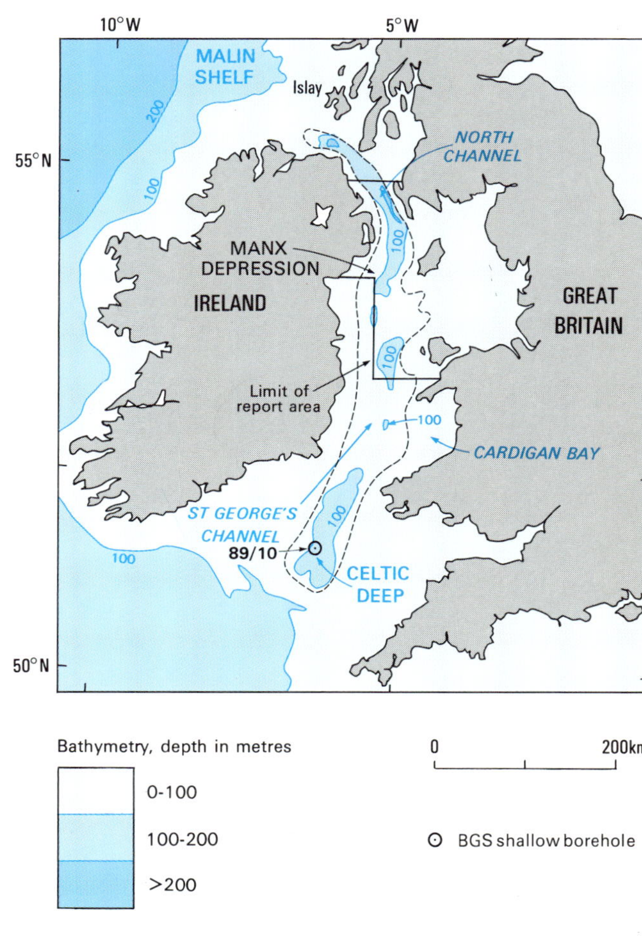

Figure 67 The Celtic Trough, extending both north and south of the report area.

chaotic acoustic character which contrasts strongly with the close, continuous and parallel reflectors of both the underlying St George's Channel Formation and the overlying Bedded and Infill member of the Cardigan Bay Formation. Where the Bedded and Infill member is absent and the Upper Till member directly overlies the Lower Till member, the two till members are difficult to differentiate seismically.

Only one BGS offshore borehole in the report area has proved this member. Borehole 71/54, in Caernarfon Bay (Figure 71), sampled 7 m of very stiff clay with abundant pebbles. This is interpreted as a subglacial lodgement till, and is presumed to have been deposited by Saalian ice prior to the cutting of the next generation of incisions into which the Bedded and Infill member was deposited.

Bedded and Infill member

Two distinct seismic facies are present in this member, although they do not neccessarily occur together. The lower is of the lenticular-infill facies, and the upper is tabular stratified.

The lower facies is at least 85 m thick in the eastern Irish Sea, and occupies incisions of various sizes. The largest incisions, in excess of 75 m deep, are shown on Figure 70. Some of these incisions may however be of an earlier generation, in which cases their infills belong to the Caernarfon Bay Formation. The seismic profiles across these lenticular infills show a range of acoustic characters comparable with those seen in the Incision Infill member of the Caernarfon Bay Formation, with a generally upward-fining trend. They are similarly interpreted to represent a deglaciation episode, in this case the Saalian deglaciation, perhaps including partial infill during part of the Eemian Interglacial (Figure 69).

The tabular-stratified facies is extensively recognised only across northern St George's Channel and in Caernarfon Bay; it is up to 45 m thick on the Welsh Platform (Hession, 1988). It also occurs patchily in the eastern Irish Sea, where it was proved in several boreholes beneath the Upper Till member (Figure 71). The Ayre Marine Silts proved in boreholes on the northern Isle of Man (Figure 71; Lamplugh, 1903; Smith, 1930) and offshore in Luce Bay (Figure 66) probably form the uppermost part of the Bedded and Infill member; these are silts with a cold or boreal marine macrofauna overlying barren, gravelly sands (BGS Isle of Man Sea Bed Sediments and Quaternary Geology sheet). The cold climate in which this facies was deposited may indicate a very late Saalian or early Eemian origin, as the climate warmed prior to the onset of the Eemian Interglacial.

Upper Till member

The Upper Till member crops out extensively at the sea bed in the Irish Sea, especially north of Anglesey (BGS Anglesey Quaternary Geology sheet). It is proved in numerous sea-bed cores as a till comprising stiff or hard clay with clasts ranging in size from sand grains to cobbles and boulders up to 1 m in diameter. In profiles, it forms a tabular-unstratified unit across most of the platforms and in the northern and southern parts of the Western Trough. It is however largely absent over the central part of the trough in the Manx Depression (BGS Anglesey Sea Bed Sediments sheet), except as patches on upstanding features (Figure 68). The Upper Till member ranges from 5 to 35 m in thickness in many offshore boreholes (Figure 71).

To the north-west of Holyhead, thin deposits of this till member are locally interpreted to overlie the Western Irish Sea Formation B (Figure 69). The Upper Till member is considered to be the product of Weichselian glacial processes, and it may be matched with the basal till of the 'Irish Sea Drift' on land (Eyles and McCabe, 1989). If this is so, it lends support to the view that the basal portion of the 'Irish Sea Drift' is subglacial in origin. It also correlates with the similarly deposited basal Devensian deposits on the Isle of Man (Eyles and McCabe, 1989). Nearshore, both in Dublin Bay (Naylor, 1964) and Morecambe Bay (Knight, 1977), the Upper Till member forms the basal Quaternary deposit in numerous boreholes, where it ranges from 2 to 56 m in thickness.

WESTERN IRISH SEA FORMATION

This formation crops out over much of the area (Figure 72), and its sediments are similar in seismic character and geometry to deposits of three older formations: the Caernarfon Bay, St George's Channel and Cardigan Bay formations. The formation comprises both localised incision-infill deposits up to 200 m thick, and more widespread, tabular-stratified deposits. The formation is mainly to be found in extensive east and west belts, each of which in its north-central part is distinguished by extensive muds and muddy sands at the sea bed (see Figure 75). In the finest-grained parts of both belts, widespread gas blanking impedes interpretation of acoustic profiles. Outside the belts, the formation is locally thick in incision infills such as Beaufort's Dyke, but is otherwise only patchily developed (Figures 68 and 72).

In the east belt, the northern St George's Channel and the North Channel, the Western Irish Sea Formation lies unconformably upon, and entirely postdates, the Upper Till member of the Cardigan Bay Formation. The deposits comprise a lower, incision-infill member, and an upper member of tabular-stratified sediments; they are considered to be no older than late Weichselian, as they overlie till thought to be of late Weichselian age (Figure 69).

In the west belt, similar incision infill and tabular-stratified facies are present, but they occur two-fold. There is a lower unit, the Western Irish Sea Formation B, comprising incision infills up to 100 m thick, with younger, tabular-stratified deposits up to 80 m thick. At the top of this lower unit there is a widespread erosion surface above which lies the Western Irish Sea Formation A. This upper unit is seismically similar to the lower, and it is commonly not possible to separate them. Furthermore, their relationship is widely obscured on seismic profiles by gas blanking in the sediments.

A number of observations suggest that the Western Irish Sea Formation B is the lateral equivalent of the Bedded and Infill and the Upper Till members of the Cardigan Bay Formation (Figure 69). Its basal deposits rest on an erosion surface that locally forms incisions at the base of the Western Irish Sea Formation; this surface can be matched with the intra-Cardigan Bay Formation erosion surface and its incisions. Also, the facies and interrelationships of the incision-infill and tabular-stratified portions of the Western Irish Sea Formation B are similar to those observed in the Bedded and Infill member of the Cardigan Bay Formation. However the basal Western Irish Sea Formation incisions are not so large, and their tabular-stratified parts are thicker and more extensive than are those of the Cardigan Bay Formation. The incisions at the base of the Western Irish Sea Formation B cut through the Lower Till member of the Cardigan Bay Formation and older deposits. On a very few profiles, the Upper Till member of the Cardigan Bay Formation appears to pass laterally into the Western Irish Sea Formation B, or overlies it.

SECTION 1

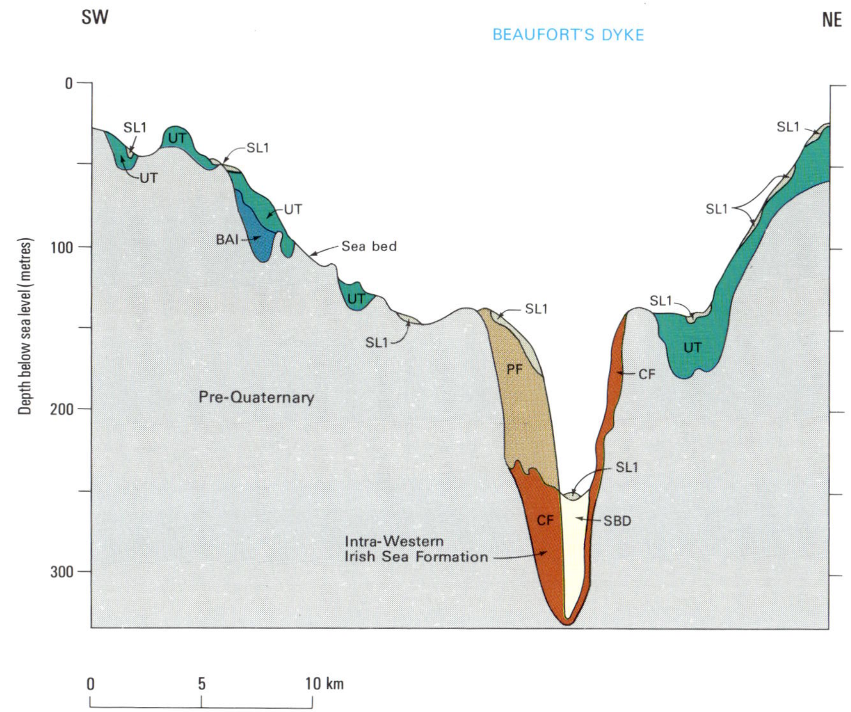

SECTION 2

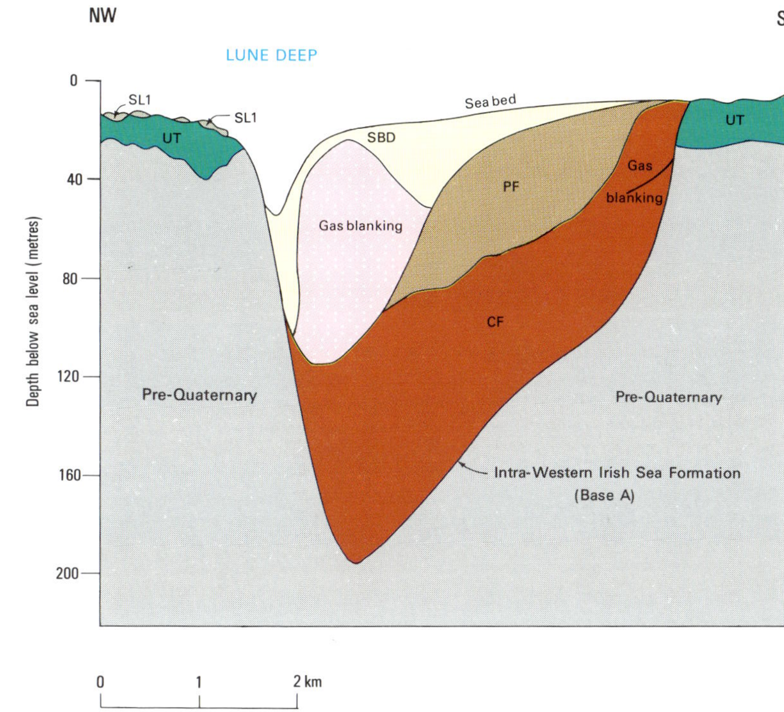

SECTION 3

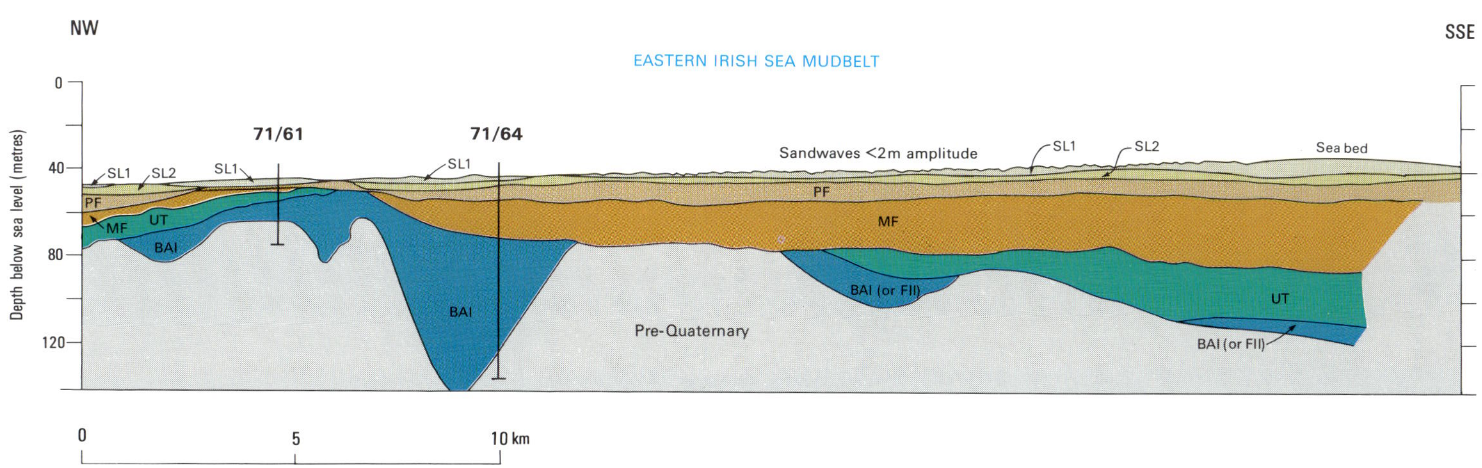

LOCATIONS OF SECTIONS

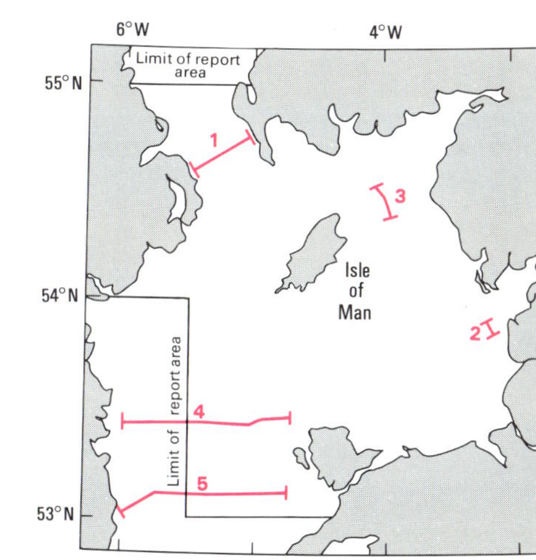

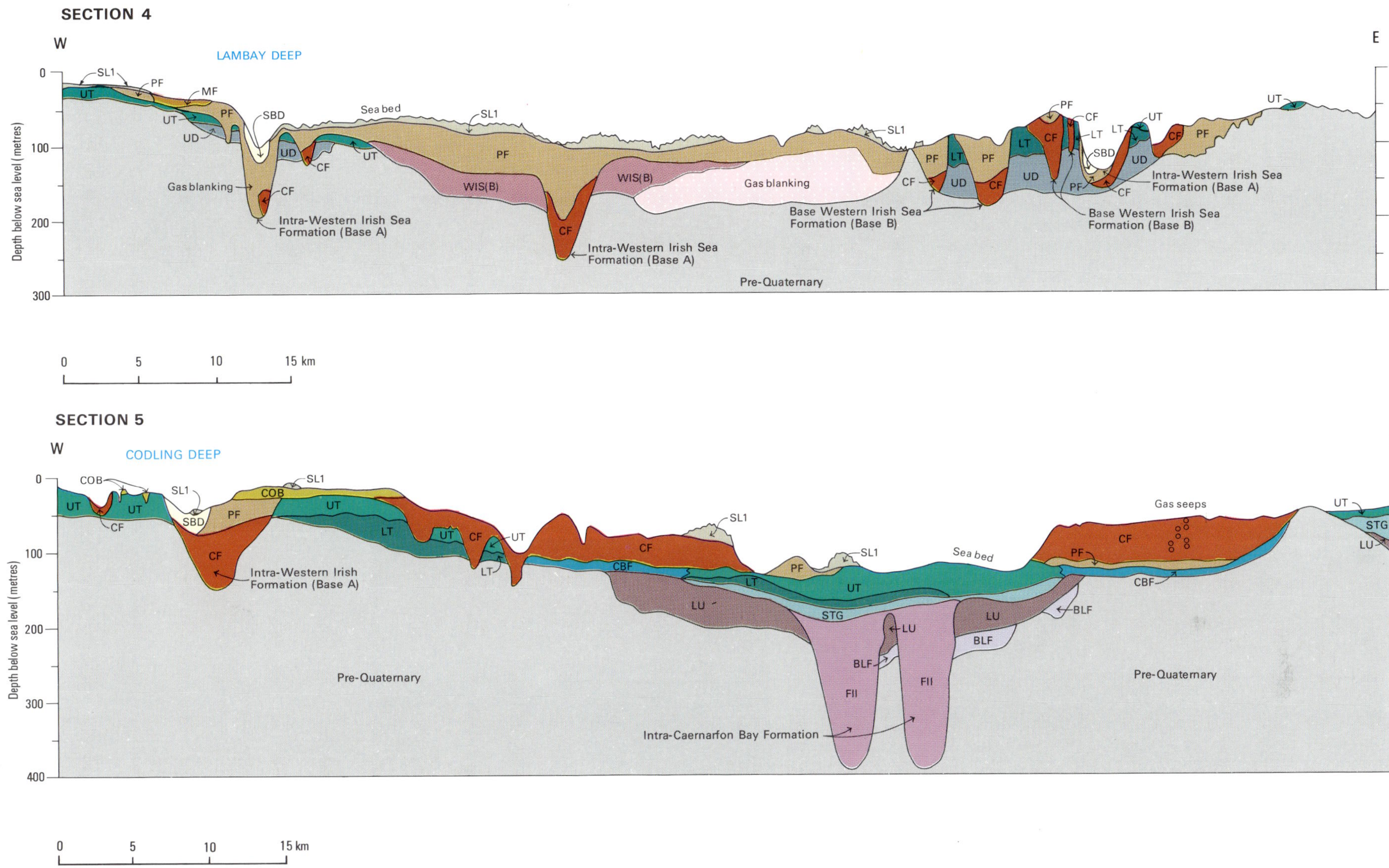

Figure 68 Cross-sections through Quaternary sequences in the Irish Sea. See Figure 69 for key to abbreviations.

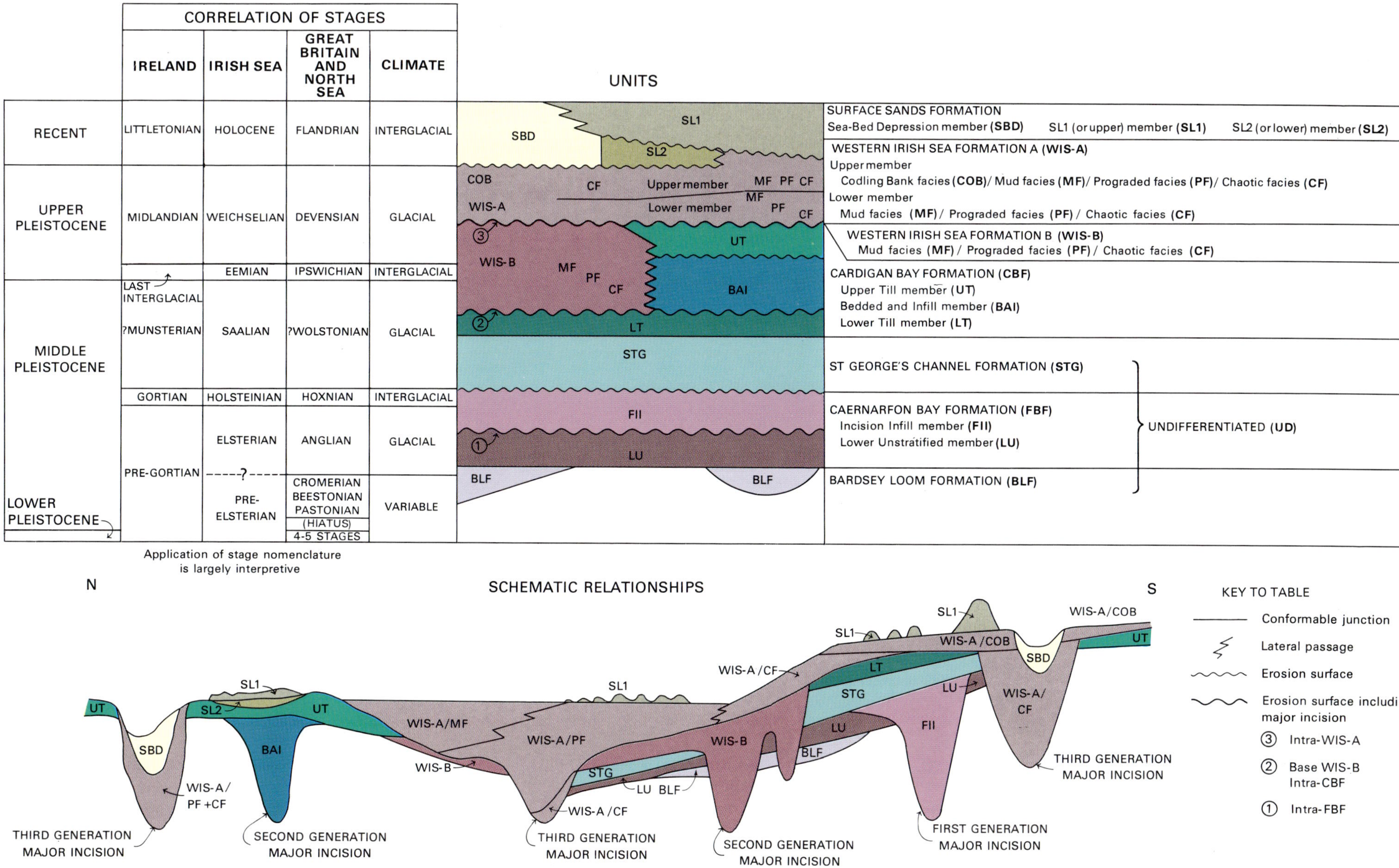

Figure 69 Quaternary stratigraphy in the report area, with schematic section showing the interrelationships of formations. Stage correlation after West (1977); stratigraphy based on BGS Quaternary Geology sheets.

The Western Irish Sea Formation B is overlain unconformably by the Western Irish Sea Formation A on all profiles not made obscure by gas blanking. This unconformity, the intra-Western Irish Sea Formation erosion surface, locally cuts through the Western Irish Sea Formation B into older sediments to produce major incisions (Figures 68 and 69). On the platforms, this erosion surface underlies all the Western Irish Sea Formation sediments, and overlies the Upper Till member of the Cardigan Bay Formation.

The Western Irish Sea Formation B is not divided into members, but the Western Irish Sea Formation A can be informally divided into a lower, incision-infill member, and an upper, tabular-stratified member. All the deposits do however display marked facies changes, both upwards and laterally. These changes are attributed to variations in sedimentation with time, or from proximal to distal settings in relation to sediment supply. Four facies recognised are the chaotic, prograded, mud, and Codling Bank facies; the latter only occurs in the upper member of Western Irish Sea Formation A. In nearshore areas, Naylor (1964) described deposits in Dublin Bay which are equivalents of the upper member of the Western Irish Sea Formation A, and the facies may be closely matched with those described by Eyles and McCabe (1989) from the 'Irish Sea Drifts' of the Screen Hills of south-east Ireland. Equivalents are also identified in Morecambe Bay (Knight, 1977). The Western Irish Sea Formation A ranges in age from late Weichselian to very early Holocene.

Chaotic facies

On seismic profiles, this facies has an amorphous signature that lacks reflectors, or consists of high-angle, impersistent and irregular reflectors. It occurs in the basal parts of major incision infills (Figure 68), as well as in the basal parts of the tabular-stratified deposits; it is from a few metres to 25 m thick. The principal development of the chaotic facies is in the southern part of the west belt (Figure 72), where it forms the entire thickness of the upper member of the Western Irish Sea Formation A (Figures 68 and 69).

Boreholes in both the west and east belts have proved chaotic facies sediments up to 24 m thick; they are dominantly gravels with muds, sands, cobbles and boulders. They contain very sparse foraminifera and dinoflagellate cysts indicative of arctic-like conditions (R Harland and D M Gregory, written communications, 1984–1986). This facies probably formed during glaciation in glaciolacustrine and glaciomarine, ice-proximal conditions.

Prograded facies

Prograding wedges of fine- to medium-grained sands form the bulk of the major incision infills of both the Western Irish Sea Formation B and the lower member of the Western Irish Sea Formation A (Figures 68 and 69). Seismically, these are tabular-stratified deposits that exhibit prograding reflectors. The prograded facies of the upper member is notably developed towards the margins of both belts (Figure 72), with progradation into the centres of the belts. Two exceptions to this configuration are a lobe of prograded facies overlying the mud facies in the northern part of the east belt, and a 35 km-wide zone of prograded facies between the chaotic facies to the south and the mud facies to the north in the southern half of the west belt (Figure 72). The progradation in the latter zone dips northwards or north-westwards into the Western Irish Sea Mudbelt (BGS Anglesey Quaternary Geology sheet).

Where they come to the sea bed, the prograded facies deposits are reworked into bedforms, mostly as sand-wave

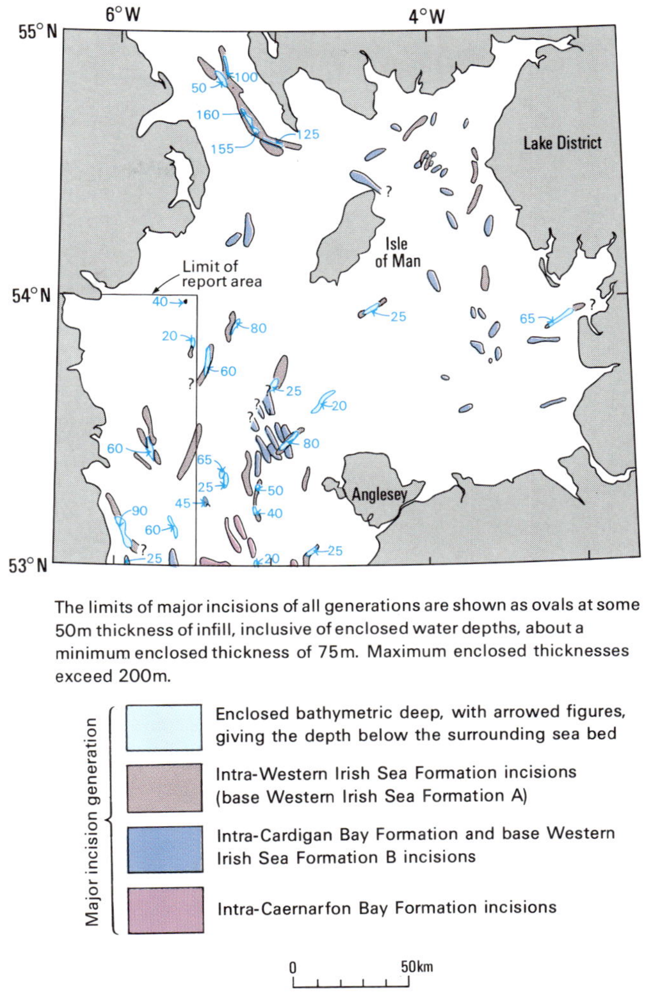

Figure 70 Distribution of major incisions and enclosed deeps.

fields (Whittington, 1980; Hession, 1988). Sea-bed cores and boreholes confirm that the facies consists of sands, up to 43 m thick, with subordinate muddy and pebbly parts (Figure 71). Microfossils are absent or sparse (R Harland and D M Gregory, written communications, 1984–1986), which is not incompatible with a cold-water environment of deposition. The facies is interpretated as being prodeltaic and glaciomarine, representing passage from the ice-proximal chaotic facies to a distal mud facies as ice retreated.

Mud facies

On seismic profiles, this facies has a near-transparent acoustic character with parallel, subhorizontal reflectors forming tabular-stratified units. In the Western Irish Sea Formation A, it is extensively developed in both the west and east belts, where it is over 80 m thick (Figure 72). Deposits with similar acoustic character are identified at depth, and to the same thickness, in the Western Irish Sea Formation B (Figure 68). The facies in both units B and A show dense concentrations of gas, producing acoustic impedence on seismic records. Such gas concentrations are commonly terminated upwards by otherwise indistinct reflectors thought to represent internal stratification in the mud facies. Lateral margins of the gas blanking may be shown by abrupt changes of level of the upper termination. Also present are sporadic hyperbolic reflections interpreted to be caused by single cobbles or boulders; these reflections are commonly concentrated in layers.

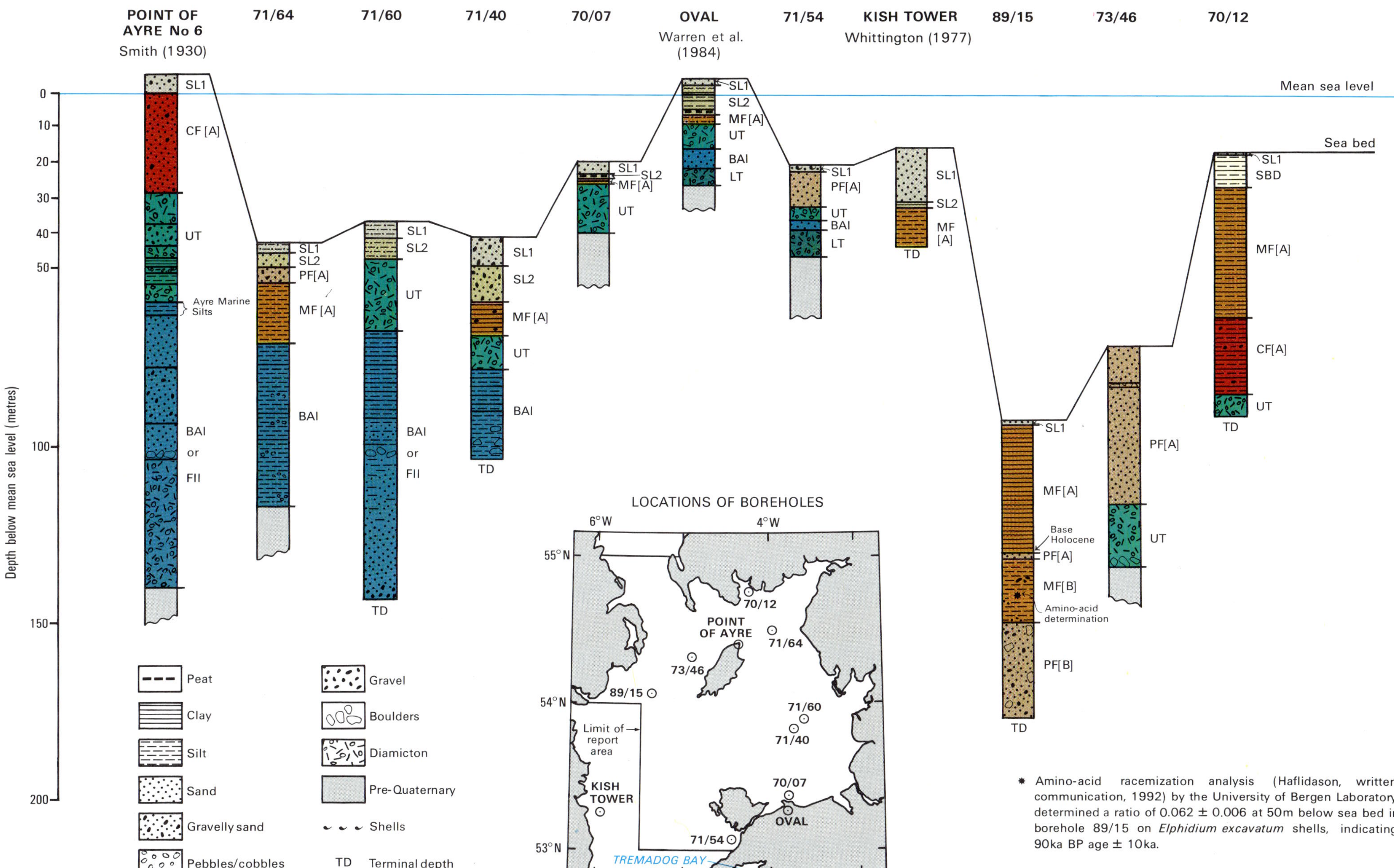

Figure 71 Summary logs of selected boreholes.

* Amino-acid racemization analysis (Haflidason, written communication, 1992) by the University of Bergen Laboratory determined a ratio of 0.062 ± 0.006 at 50m below sea bed in borehole 89/15 on *Elphidium excavatum* shells, indicating 90ka BP age ± 10ka.

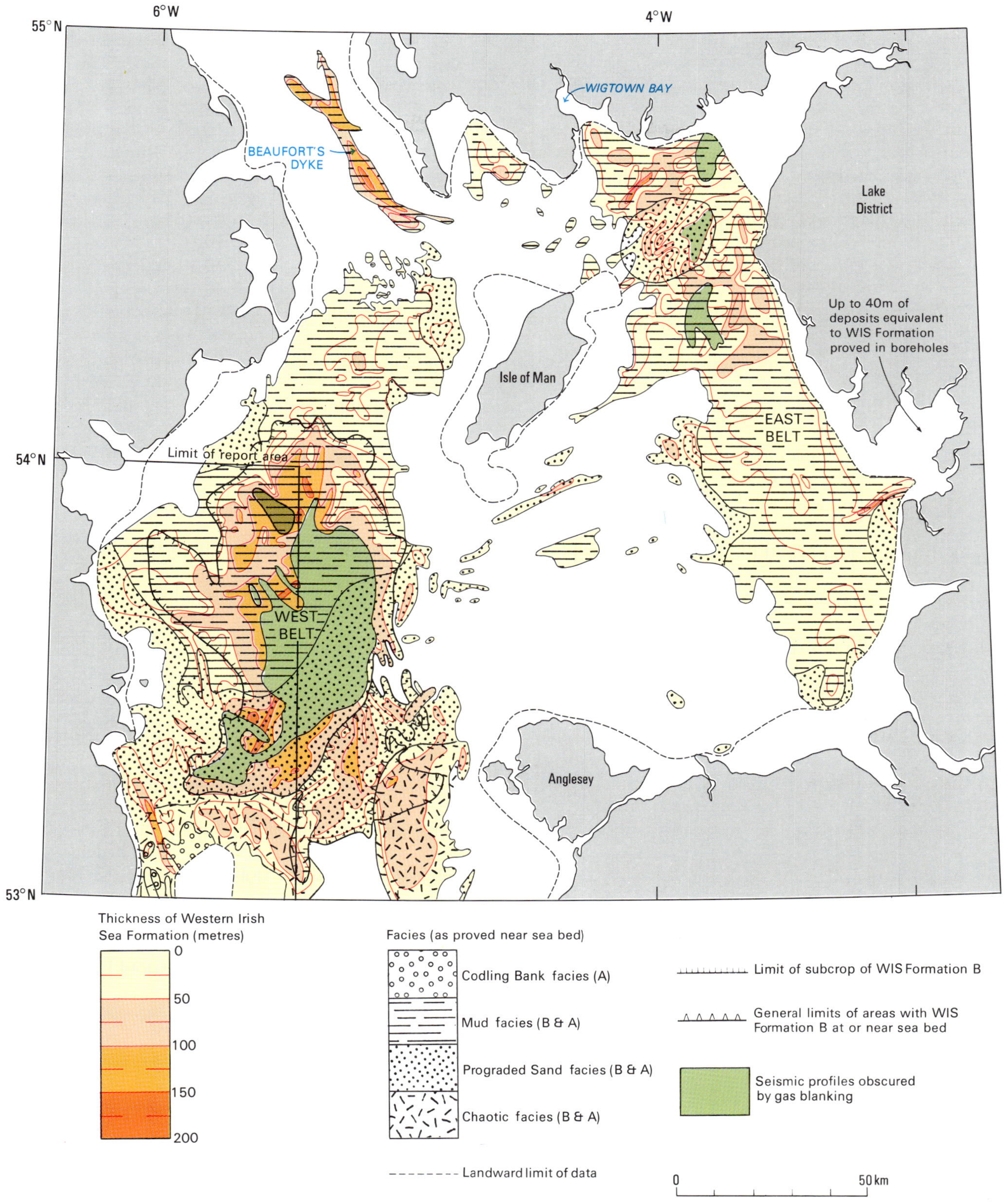

Figure 72 Distribution, thickness and facies of the Western Irish Sea Formation.

Many sea-bed cores prove these deposits to be black to greenish grey, shelly, sulphide- or glauconite-rich silts passing towards the margins of the belts into sandier deposits before interdigitating with the prograded facies. In the west belt, a number of cores show that from 1 to 5 m below sea bed, the foraminifera and dinoflagellate cysts may show a change downwards from temperate to cold, boreal conditions of deposition (R Harland and D M Gregory, written communications, 1984–1986). Borehole 89/15 (Figure 71) proved temperate, presumed Holocene, deposits to a depth of 38 m (R Harland, written communication, 1990).

Only borehole 89/15, in the west belt, has proved the mud facies in Western Irish Sea Formation B. Below an unconformity at 39 m beneath sea bed, clays with cold-water dinoflagellate cysts (R Harland, written communication 1990) overlie sands of the prograded facies (Figure 71). Shells of *Elphidium excavatum* from 50 m depth in the cold-water facies gave amino-acid ratios of 0.056 to 0.068, indicating an early Weichselian age for the uppermost 10 to 20 m of Western Irish Sea Formation B sediments in this borehole (Knudsen and Sejrup, 1988).

Many boreholes in the east belt have proved the mud facies of the upper member of the Western Irish Sea Formation A. These boreholes generally penetrated cold-water deposits, with some evidence of minor ameliorative episodes that appear to be comparable in duration with ones described in the Firth of Clyde by Deegan et al. (1973). A 46 m thickness of mud facies was proved in BGS borehole 70/12 in Wigtown Bay (Figures 71 and 72).

The mud facies is interpreted to comprise distal, glaciomarine deposits passing upwards into distal, marine deposits. Detailed micropalaeontological analyses of sea-bed cores show that the cold-water deposits were laid down in waters deeper than those of the present, whereas the temperate-water deposits were laid down in shallower waters (R Harland and D M Gregory, written communications, 1984–1986). Scattered pebbles and a very few cobbles are thought to be dropstones from floating ice.

Codling Bank facies

This facies occurs only in the upper member of Western Irish Sea Formation A, and displays an almost opaque acoustic signature. It is restricted to Irish waters west of the report area, in the region of the Codling Bank off County Wicklow, where it apparently forms the dissected remnants of a previously continuous mantle. The facies forms upstanding features including the shoals comprising the Codling Bank. The Codling Bank facies is up to some 15 m thick, and overlies deposits of both the prograded and chaotic facies of the lower member, and the chaotic facies of the upper member, of the Western Irish Sea Formation A.

Warren and Keary (1988) recorded that dredging operations on the Codling Bank have obtained large quantities of ballast, almost entirely in the cobble to boulder sizes. The Codling Bank facies may compare with the lateglacial, braidplain deposits described onshore in Leinster by Eyles and McCabe (1989), or with the 'capping diamicton' they attributed to subaqueous ice-rafting and sediment gravity flows. Alternatively, an analogue might be with the south Icelandic sandur of very coarse-grained deposits, which Maizels (1989) identified as dominantly formed by jökulhlaup floods.

SURFACE SANDS FORMATION

Deposits of this formation are less than 2 m thick, or absent, over the greater part of the area, where they include mobile sea-bed sediments (discussed below). However, they are locally thick in the nearshore and intertidal areas, and up to 40 m thick in sandbanks (Knight, 1977; Pantin, 1978). Thicker deposits of this formation, up to 100 m, occur as the youngest infill of some intra-Western Irish Sea Formation major incisions (Figures 68 and 69). The base of the Surface Sands Formation is everywhere an unconformity; this erosion surface has been proved in many sea-bed cores. Where the prograded facies of the Western Irish Sea Formation appears to pass up without break into present-day sand waves (Hession, 1988), it is inferred that there is a surface of reworking at their bases.

The Surface Sands Formation is divided into three members, all of which dominantly comprise sand. Their uppermost parts are the products of present-day marine processes at the sea bed, but each member includes sediments deposited in conditions very different from those of the present, including some subaerial deposits. In intertidal areas, muds make up an important, though subordinate, part of the Surface Sands Formation.

Sea Bed Depression member

This member comprises the partial or complete fill of hollows cut into deposits of the Western Irish Sea Formation A in some intra-Western Irish Sea Formation major incisions (Figure 68). Wingfield (1990) postulated that these hollows were exhumed as large kettle holes, some of which have an incomplete fill of Holocene sediment, and thus still form enclosed bathymetric deeps. Elsewhere, many filled kettle holes are capped by deposits of other members of this formation, as in the Solway Firth and the east Irish Sea. In borehole 70/08 in Tremadog Bay to the south (Figure 71; Tappin et al., 1994), the Sea Bed Depression member was proved as 27 m of sandy silt with shell debris; this section divulged a rich, temperate-marine, Holocene microbiota. Two cores of the upper part of the member from the Lambay Deep recovered shelly, sulphide-rich, muddy sands with dinoflagellate cysts indicative of continuous deposition over the last 4 ka (R Harland, written communication, 1986).

The SL1 and SL2 members

Following Pantin's (1977; 1978) work in the east Irish Sea, the extensive but thinner parts of the formation are divided into lower (SL2) and upper (SL1) members. SL stands for sediment layer.

The SL2 member is diachronous and of varied lithology; it comprises the deposits formed across a surface of erosion during the early Holocene marine transgression after 10.2 ka BP. This member includes sands, silts and clays in Morecambe Bay (Units ii/1 and ii/2 of Knight, 1977), and peaty-silt in borehole 70/07 off Llandudno (Figure 71), which Tooley (written communication, 1971) considered to represent reed swamps adjacent to open water before 9.2 ka BP. Intertidal muds below the Kish Bank sands of the SL1 member (Whittington, 1977; Harris, 1980) belong to the SL2 member (Figure 69), as do deposits in the east belt where shallow-marine and beach sediments have been recovered above Western Irish Sea Formation A in many sea-bed cores (Pantin, 1977; 1978). Moribund tidal sand ridges (see Figure 77) are also considered to be part of this member.

The SL1 member disconformably overlies an erosion surface across the SL2 member, or rests upon the Sea Bed Depression member or older strata. It represents the present-day mobile sediments. These range from transitory, thin deposits, through tabular-stratified accumulations up to 5 m

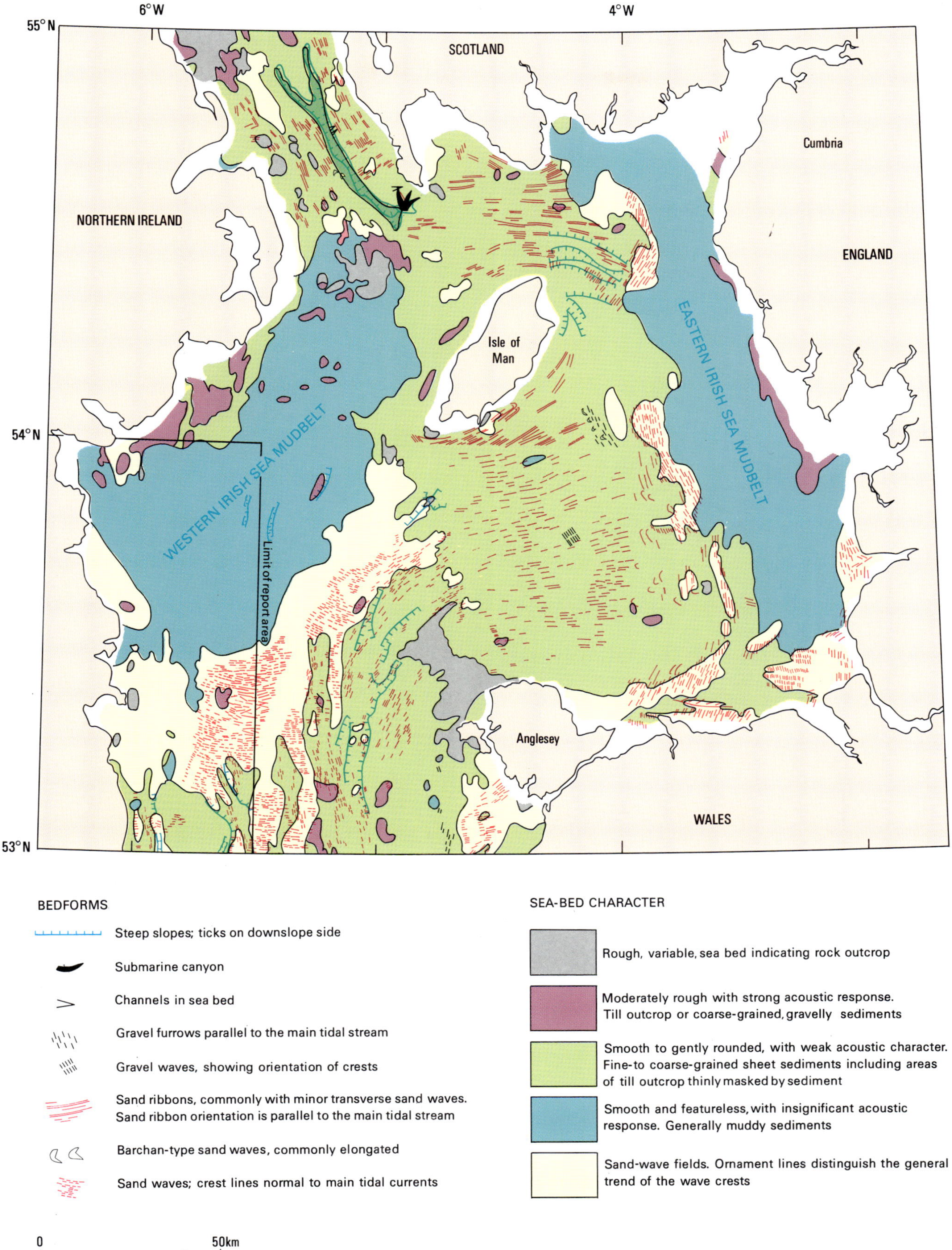

Figure 73 The character of the sea bed, and the distribution of active bedforms. See also Figure 77.

thick in the east Irish Sea (Pantin, 1977; 1978), to a variety of bedforms (Figure 73 and see Figure 77) ranging upwards in size to active tidal sand ridges and giant sand waves up to 40 m in height (BGS Sea Bed Sediments sheets). In the nearer-shore areas and inshore, the sediments pass landwards into intertidal sandy muds and saltmarsh organic clays (Naylor, 1964; Knight, 1977).

THE PRESENT SEA FLOOR AND ITS EVOLUTION

The sea floor in the Irish Sea is very much flatter than the surrounding land, even when compared with the lower-lying land areas. The extreme depth variation offshore is 315 m, which is modest compared to heights on the land that exceed 1000 m. The report area has probably had a highly fluctuating sea level during the later Pleistocene, and during the lowest stand of the Weichselian almost the whole report area may have been land. Eyles and McCabe (1989) have recognised late Weichselian and Holocene events in the Irish Sea associated with changes of sea level that were produced by the interactions, due to the growth and wastage of the ice-cap centres in Scotland, of glacioeustasy, glacioisostasy, forebulge development, and rebound. The bathymetric features of the Irish Sea (Figure 66) divide into five zones:

1. Estuaries and coastal embayments that have water depths between high water mark and about 10 m. These are valleys drowned by the Holocene sea-level rise, since when they have been areas of sedimentation.

2. Inner-shelf platforms that vary in width from 5 km in the North Channel to greater than 100 km in the east Irish Sea. These generally have gentle gradients of 1:100 to 1:2000, with steeper slopes only over sandbanks, near open coasts, and where sea-floor rock outcrop produces a rugged topography. Water depths on the inner shelves normally range from 10 to 60 m. These platforms were bevelled by repeated coastal wave attack during eustatic fluctuations of sea level between −25 m and −75 m OD during the early and mid-Weichselian (Figure 74), and probably during previous glacial stages also. Most pre-existing Pleistocene deposits were removed from the platforms by coastal and surf-zone processes during the repeated transgressions and regressions of such intervals, although they are preserved in incision infills below the base level.

3. The Western Trough, which is the broad trough of subdued slopes of 1:50 or less, runs sinuously from the North Channel to St George's Channel. It forms a zone up to 80 km wide, and varies from 60 m to over 140 m in depth. The disposition of the pre-Quaternary rock units does not indicate that the Western Trough, with its thick Pleistocene deposits, was controlled by the underlying geology. It formed during or before mid-Pleistocene times, and has since acted as an area of net deposition with intermittent erosion. The erosive episodes took place during the major falls of sea level produced by glacial maxima, and during the subsequent sea-level rises (Figure 74).

4. Enclosed deeps that form areas less than 5 km wide, generally up to 30 km long, and from 10 to 165 m deeper than the surrounding sea floor. All such deeps appear to display similar dimensions, have smooth sides and floors, are rather irregularly shaped in plan, and have side gradients less than 1:10 (Wingfield, 1989). They formed as kettle holes during the early Holocene, since when they have been partly infilled.

5. Rocky prominences that are generally areally restricted zones of rough and rugged topography caused by outcrop of rock at headlands, islets, and shoals; these generally occur in the coastal zone, but also farther offshore. In the Western Trough, rocky prominences occur at depths of 75 m or more, forming steep-sided (1:20 to 1:10) knolls above generally flat sea floors of 80 to 130 m depth. Those in deeper water formed during times of lower sea level, whereas those at the present-day coast are being continually modified.

Each bathymetric zone has differing controls on sedimentary processes due to water depths, gradients, and distances from sediment supply. Deeper waters reduce both wave- and current-induced bottom stresses, and gradients modify the effects of both tidal currents and waves on the sea bed. The deepest waters, in enclosed bathymetric deeps, have weak bottom currents largely incapable of lifting sediment up the low gradient and out of the deep. Thus, such deeps are areas of net sedimentation. The effect of waves towards both open and embayed coasts is strongly controlled by the gradient of the sea bed. In areas such as Liverpool Bay, the larger-amplitude waves dissipate their force well offshore against the gently shoaling gradients, whereas steeper gradients allow the transfer of wave energies close to shore, as around Anglesey and on the North Channel coasts.

Additionally, the overall bathymetric form of the area exerts control on sediment movement and deposition. Sediments derived from coastal erosion and rivers are carried alongshore to collect in embayments, or to pass offshore into general marine circulation. The length of marine sediment transport paths vary with grain size; the coarsest components (pebbles and larger clasts) are barely moved or remain as a lag, whereas the finest components (silt and clay) may remain in suspension to be carried along the transport path, perhaps out of the report area. The middle size fractions remain in circulation, largely in transient bedforms, although a comparatively minor proportion is abstracted into sediment sinks or estuaries.

SEA-BED SEDIMENTS

The sea bed can be divided into areas of net erosion and net deposition. In the former, sediments of Pleistocene age or older may form the sea bed, and there are relict bedforms, many of which formed at times of lower sea level. In areas of net deposition, the accumulation of sediments reaches its maximum in sand waves up to 40 m in height.

In the report area, BGS has taken some 1600 sea-bed samples, and other organisations have collected another 1200. These include grab or dredge samples, short cores, and many vibrocore stations cored down to 6 m depth. The BGS Sea Bed Sediments sheets of the report area (Figure 75 and see inside back cover) are based on particle-size analyses of the grab and shallow-core samples collected by BGS, but take account of sonar data and other information. Whereas grab samples take a scoop of sediment at sea bed to less than 0.15 m depth (which is usually homogenised when subsampled for analysis), concurrent shallow cores can recover more than one sediment layer intact.

Pantin and Evans (1984) suggested that the sediments exposed at the sea bed belong to three units. The oldest, layer C, is relict Quaternary sediment or solid rock. Above it lies layer B, a gravelly lag deposit that in part represents the SL2 member of the Surface Sands Formation. The uppermost layer, layer A, comprises mobile sediments that form the tops of both the Sea Bed Depression and SL1 members of the Surface Sands Formation. Only layer A, and to a limited extent

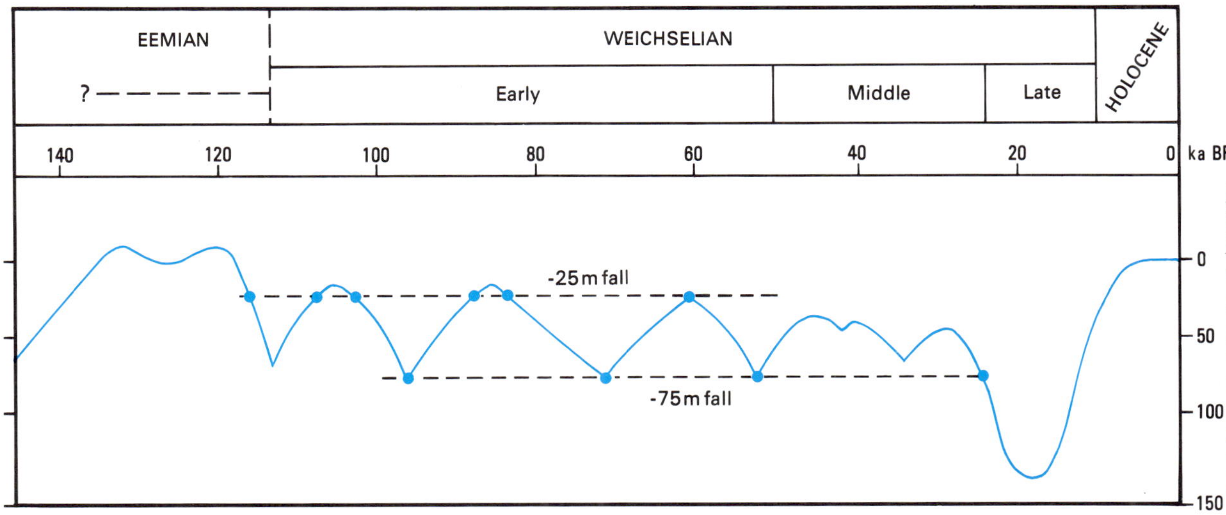

Figure 74 Worldwide glacioeustatic curve from Eemian times to the present. After Coleman and Roberts (1988).

layer B, are mobile and actively involved in the present hydraulic regime.

Layer B

Where layer A is absent or patchy, older Quaternary deposits or solid rock are mantled only by a discontinuous pebbly coquina or shelly gravel of layer B. Where layer A forms the upper, mobile part of the SL1 member, a relict layer B generally underlies it upon the basal disconformity. This layer B lag deposit is generally from 0.1 to 0.2 m thick, and comprises sandy, shelly, poorly sorted gravel, and coarse-grained sand. Across gravelly substrates, the base of layer B may form an armoured pavement of cobbles, whereas across fine-grained substrates, layer B may be absent or be represented by a shell lag with sand grains. Where gravels at sea bed are several metres thick, the base of layer B can be considered to form the active depth of reworking by the winnowing of fines.

Layer B represents both winnowed, immobile sediment, and mobile sediment that undergoes only infrequent reworking, largely by wave action during storms. A few cores indicate that layer B can interdigitate with layer A, for pebbles and shells from layer B may be incorporated within layer A.

Layer A

Layer A is commonly less than 0.3 m thick across the regions of largely gravelly sediment of layer B that make up the floors of the St George's and North channels, and occur to the north and south of the Isle of Man (Figure 75). However, it thickens locally as sand or gravel bedforms. Where there is patchy exposure of layers B or C, a mixture of two, or even three, sediment layers have been sampled.

Thin gravelly deposits pass into thicker sands towards both areas of widespread muddy deposits (Figure 75), a lateral transition effected by the patchy appearance of sands over the gravels, and then by the coalescing of these sand bodies into continuous sheets. The sands are proved in some cores to overlie thin gravels similar to layer B. The sands vary from 0.5 to 40 m thick, with the largest thicknesses in giant sand waves and tidal sand ridges (Wingfield, 1987; James and Wingfield, 1987).

The areas of extensive muddy sediments to the west and east of the Isle of Man (Figure 75) are respectively termed the Western Irish Sea Mudbelt (Belderson, 1964) and the Eastern Irish Sea Mudbelt (Vannin Sound of Pantin, 1977; 1978). In both mudbelts, as well as in the inner parts of estuaries, the sands pass progressively into muddy sands and sandy muds (Figures 73 and 75). Mud (less than 10 per cent sand) is restricted in the Eastern Irish Sea Mudbelt, but is extensivly distributed in the Western Irish Sea Mudbelt. In estuaries, fine-grained, mobile sediments occur as river-channel muds, and as intertidal and salting mudflat deposits (Harris, 1974; Knight, 1977; Jeffery et al., 1978; BGS Lake District and Isle of Man Sea Bed Sediments and Quaternary Geology sheets). These muddy sediments are greater than 10 m thick, and only their uppermost parts should be considered as mobile.

Nuclear power plants in Anglesey and Cumbria discharge their cooling waters into the Irish Sea, leading to concentrations of artificial radionuclides in the Eastern Irish Sea Mudbelt and in the mud and silt deposits of the estuaries around the east Irish Sea (Pentreath et al., 1986; Kershaw et al., 1988; Jones et al., 1986). No similar concentrations are found in the Western Irish Sea Mudbelt (Pentreath et al., 1986). Levels of concentration of waste are low; for the most part, even spot concentrations are only analagous to naturally occurring heavy-mineral or radionuclide concentrations.

Oceanography

The principal controls on the present hydraulic regime are bathymetry (as described above), climate, and tidal currents. Their interactions create the dynamic pattern of erosion and deposition. Other minor factors include the increasing effects of man-made construction, increased amounts of dredging, and the dumping of sewage and industrial wastes (Kirby et al., 1983).

Climate

The present climate is cool-temperate in a zone of westerly winds with frequent gales and storms. Winds of gale force and higher are recorded on 35 to 45 days a year (Hydrographic Department, 1960). No part of the report area is greater than 55 km offshore, and fetches greater than 100 km are rare, except from the south-west. Storms raise waves up to 6 m in height, and the 50-year maximum wave height is

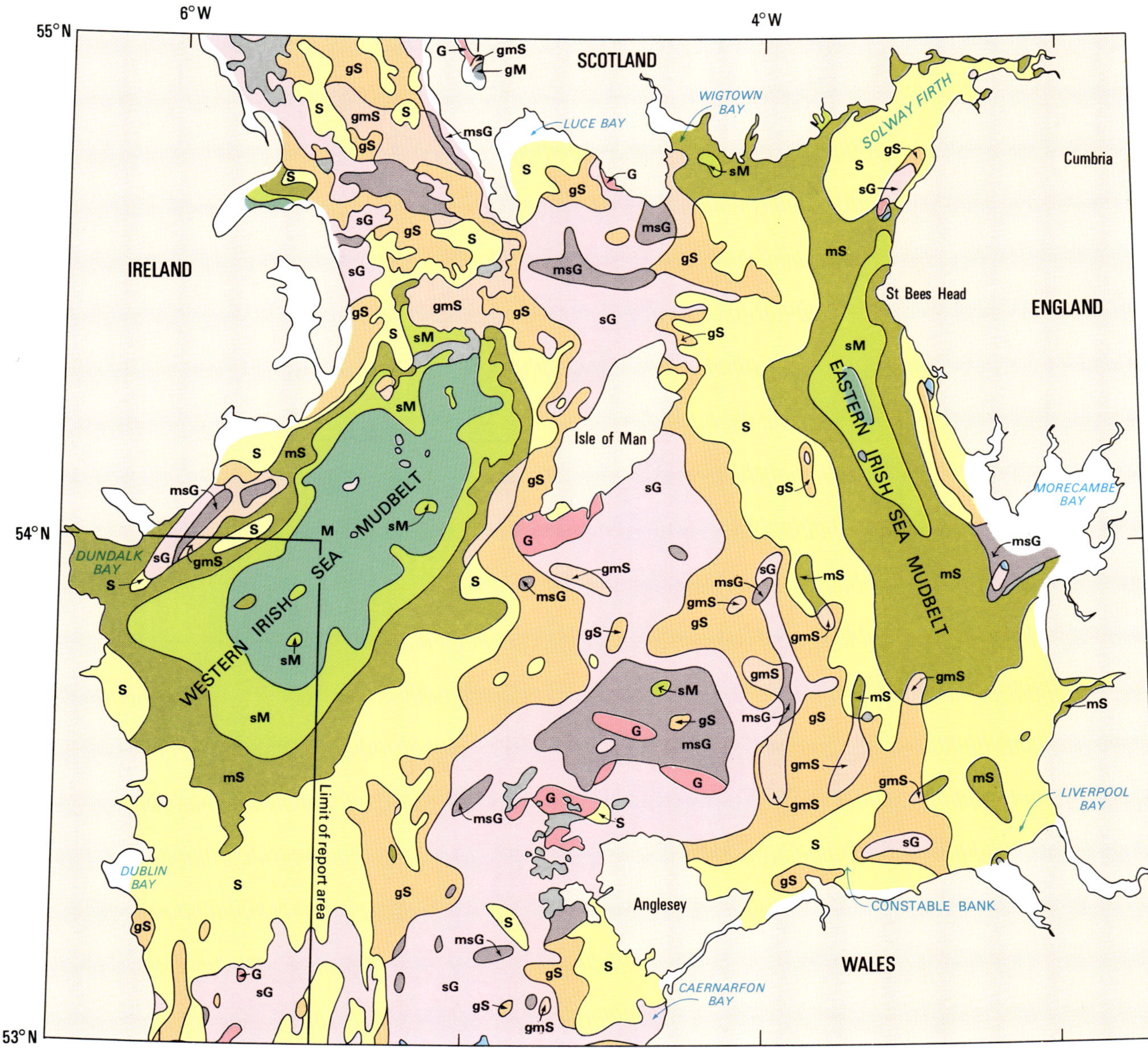

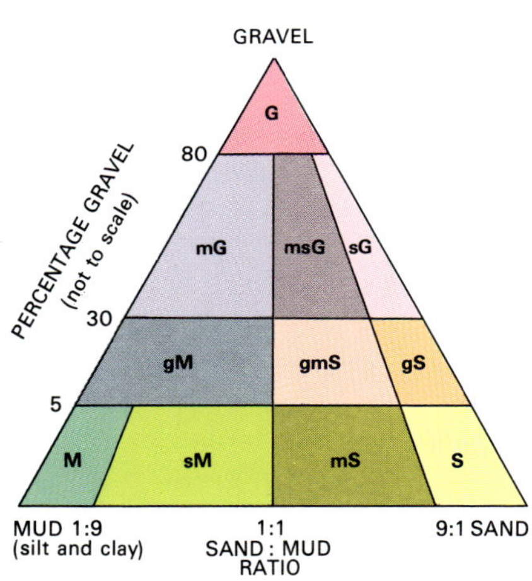

M	Mud
sM	Sandy mud
gM	Gravelly mud
S	Sand
mS	Muddy sand
gmS	Gravelly muddy sand
gS	Gravelly sand
G	Gravel
mG	Muddy gravel
msG	Muddy sandy gravel
sG	Sandy gravel

The sediment classification used on the map is modified after Folk (1954)

Outcrop of till (Quaternary) with local and/or intermittent cover of sediment

Outcrop of rock (pre-Quaternary) with local and/or patchy cover of sediment

Figure 75 Sea-bed sediment distribution in the Irish Sea.

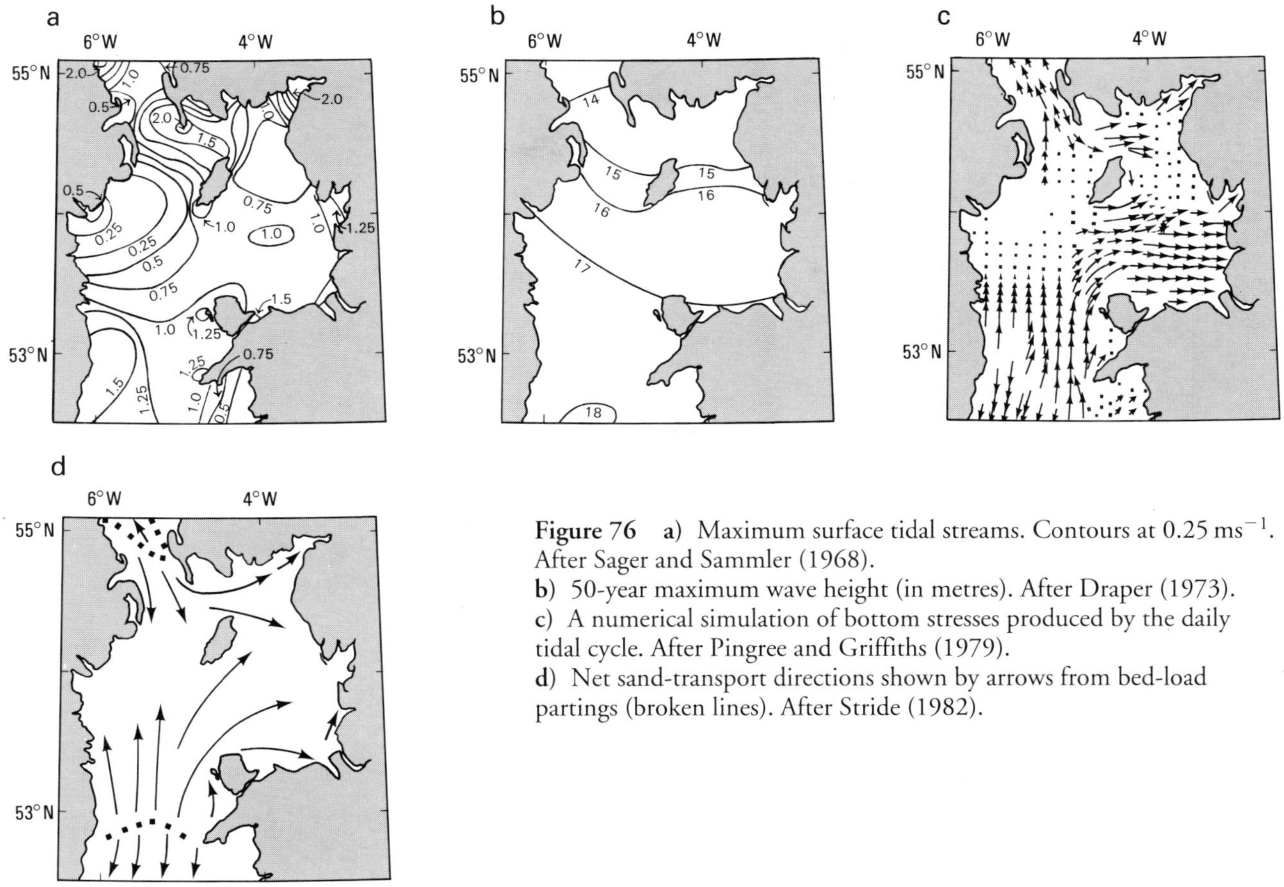

Figure 76 a) Maximum surface tidal streams. Contours at 0.25 ms^{-1}. After Sager and Sammler (1968).
b) 50-year maximum wave height (in metres). After Draper (1973).
c) A numerical simulation of bottom stresses produced by the daily tidal cycle. After Pingree and Griffiths (1979).
d) Net sand-transport directions shown by arrows from bed-load partings (broken lines). After Stride (1982).

assessed to be a factor of 2 to 3 times greater (Figure 76b). Gravel waves in water 50 m deep, with orientations unrelated to the tidal currents (BGS Isle of Man Sea Bed Sediments and Quaternary Geology sheet), may indicate that gravel can be mobilised at such depths by these storm waves.

The predominant effect of waves is confined to a limited envelope of depths between just above high water and a few metres below low water. This coastal wave attack acts to produce a surface of marine erosion by the wearing down of headlands and islands. The debris from this coastal erosion is carried into embayments protected from wave attack, and disposed to form bars and spits so as to smooth out the irregularities of the coast. That this process takes thousands of years to attain equilibrium is shown by the presence in the report area of numerous offshore islets, prominent headlands and open bays. Yet the present wave regime has operated under similar climatic conditions with a sea level little changed over the last 4 to 6 ka (Tooley, 1985).

It is only within the envelope of wave attack that effective erosion of lithified deposits occurs in the marine environment. Less consolidated or unconsolidated deposits are generally affected by wave motion to modest (c.15 m) depths; the predominant effect of wave action is to produce cyclic loading, leading to the consolidation of the clay fraction of the sediment.

Tidal currents

Tidal ranges in the Irish Sea vary from 3 m in the North and St George's channels, to over 9 m in Liverpool Bay. This macrotidal regime (Reading, 1978) produces wide areas of intertidal ground in the major embayments of the east Irish Sea. Offshore, the principal tidal effects result from the horizontal motions of the water mass set up by the tidal streams (Figure 76a). The tidal streams produce bottom stresses (Figure 76c) which create bed-load parting zones in the St George's and North channels. Net sand-transport paths (Figure 76d) diverge from these zones to converge in the slack tidal areas of fine-grained sediment, and form the Western and Eastern Irish Sea mudbelts. Comparison of Figure 75 with Figure 76c suggests that the distribution of the sea-bed sediments, essentially showing the mobile sediment distribution (layer A), matches well with the bottom stresses induced by tidal currents.

Zones of net erosion comprise: shoals fronting open coasts where wave action is most effective; bed-load parting zones of maximum bottom stress; and the areas of strong tidal streams north and south of the Isle of Man. These zones of net erosion act as sediment sources by provision of winnowed material, which is mostly sand sized. Little sediment, even mud, is derived from river inputs (Kirby et al., 1983). Sediment movements are by the suspension of silt and clay, and by the bottom traction of sand and fine-grained gravel.

BEDFORMS

Active bedforms

Active bedforms (Figure 73) are dominantly formed from sand, although some gravelly and muddy bedforms also occur. Stride (1982), describing the variation of bedforms along sediment transport paths under the influence of tidal currents on continental shelves, identified both 'low' and 'abundant' sediment-supply patterns.

In the 'low sediment-supply' pattern, there is a passage from furrows and waves in gravel, through a zone with isolated, sporadic sand ribbons, into a zone with horned, barchan-type, large sand waves. Farther down-path there are extensive sand patches with small sand waves (the megaripples of BGS Sea Bed Sediment sheets). Such bedform trains occur in the North Channel, north and south of the

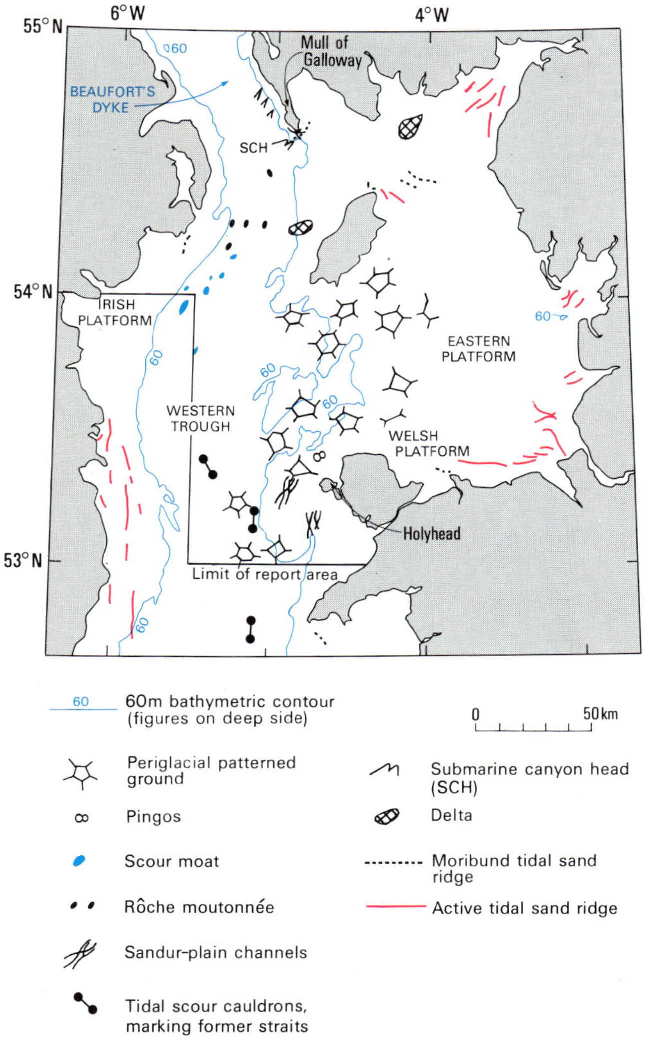

Figure 77 Distribution of relict bedforms and active tidal-sand ridges.

Isle of Man, and in north-east St George's Channel and Caernarfon Bay (Figure 73).

The 'abundant sediment-supply' pattern passes from a gravelly sea floor with sand ribbons and elongate sand patches, into a continuous carpet of sand. The sand carpet is shaped into both transverse sand waves and sandbanks (tidal sand ridges) subparallel to the current. Farther down-path, these pass into continuous sand-wave fields. Bedform trains of this sequence occur in north-west St George's Channel and southern Liverpool Bay. Active tidal sand ridges and giant sand waves are prominent features of the St George's Channel (Figure 77). Sandbanks, as finger-like splays paralleling tidal channels, form drying banks in the great estuaries of the east Irish Sea.

The offshore muddy areas (Figures 73 and 75) are generally featureless on sidescan sonar records; there are however artifacts such as trawl and anchor scars, and obstacle or wreck marks (Stride, 1982). Only a few fluid-escape structures, such as pockmarks up to 1.5 m deep or domes to less than 0.5 m high, have been noted in the mudbelts (Hession and Whittington, 1987; Hovland and Judd, 1988). Plumes of gas rising to the sea bed (Floodgate et al., 1986; Jones et al., 1986) from the very extensive, gasified, fine-grained, Western Irish Sea Formation sediments (Figure 72) apparently give rise to little sea-bed expression.

Relict bedforms

Unlike the bedforms described above which are attributed to active sedimentary processes, there is a suite of relict bedforms (Figure 77) considered to have formed subaerially or in water shallower than at present. It is also of interest that marine features from times of relatively higher sea levels have been recognised on the adjacent land (Eyles and McCabe, 1989).

Glacially influenced bedforms include roche moutonnés in the southern North Channel; these commonly have a crag-and-tail form, with the tails of till pointing south, which implies that they were formed by southward-moving ice. Wingfield (1987) described extensive development of polygonal textures on sidescan-sonar records across sea-bed outcrops north and west of Anglesey; these are interpreted to be periglacial, ice-wedge polygons up to 80 m in diameter. 15 km north-west of Holyhead, a 75 m-diameter circular structure on the sea bed has been interpreted as a drowned pingo scar of subaerial tundra origin (Wingfield, 1987). A second, similar, but less well-resolved feature occurs nearby.

Anastomosing nets of channels, individually up to 200 m wide and up to a few kilometres long, are observed on sidescan-sonar records; these occur in St George's Channel, and on the Eastern Platform both north of the Isle of Man and in the North Channel. Two of the channels south-west of the Mull of Galloway (Figure 77) debouch into canyons descending steeply into Beaufort's Dyke; the channels may have been formed by braided rivers on subaerial sandur.

Many of the rocky prominences of the southern North Channel and northern Manx Depression (Figures 66 and 77) are separated from the surrounding flat and muddy sea floor by moats up to 500 m wide and as much as 40 m deeper than the surrounding sea bed. The moats are interpreted to be scour pits produced around the upstanding obstacle by powerful tidal currents when sea level may have been 60 m or more below the present level. In the Western Trough, shallow-marine, tidal-scour cauldrons (Mogi, 1979) occur in water depths greater than 80 m; these are inferred to have formed on either side of narrow, high-tidal-volume straits during the early Holocene (Wingfield, 1993; 1994, in press).

Delta lobes of sands of the prograded facies of the Western Irish Sea Formation A are observed on seismic profiles and proved in boreholes west of the Isle of Man and south-east of Wigtown Bay (Figure 77). If the delta topsets approximated with contemporary sea levels, they record levels respectively 75 and 30 m below present.

Some tidal sand ridges north of the Isle of Man are thought to be moribund (Figure 77). They have least water depths of 7 m to more than 12 m over their crests, and were formed during sea level lower than at present (Kenyon et al., 1981).

11 Economic geology

Oil and gas are important resources in the Irish Sea; the Morecambe Gasfield was discovered in the mid-1970s, and other new finds have been made in the late 1980s and early 1990s. Historically, haematite, halite and coal have been worked beneath the Irish Sea coastal waters off north-west England and Wales. Today, geological resources are exploited both where attached to onshore workings and in entirely off-shore areas. In recent years, proposals to utilise virgin resources using advanced technology include in-situ coal gasification (McArdle and Keary, 1986) and the exploitation of geothermal energy (Downing and Gray, 1986). The economic geology of the onshore perimeter is treated in full in BGS memoirs and summarised in the BGS Regional Geology guides (Edwards and Trotter, 1954; Smith and George, 1961; Wilson, 1972; Greig, 1971; Taylor et al., 1971), as well as in Moseley (1978), Holland (1981a), Craig (1991) and Duff and Smith (1992).

ENERGY SOURCES

Oil and gas

Hydrocarbons are the prime economic resource of the Irish Sea. To date, offshore drilling has been confined to the East Irish Sea Basin (Figures 1, 78 and 79) although drilling has also taken place in the Kish Bank Basin to the west of the report area. Until the 1990s, only minor quantities of oil and condensate were encountered, although important gas discoveries were made. In addition to the North and South Morecambe gasfields (Figure 78) and the recently discovered Hamilton and Hamilton North gasfields of block 110/13, smaller gas finds include those in wells 110/2b-9, 110/3b-4, 110/7-3, 110/7a-4, 110/8a-4 and 113/26-1, none of which have been exploited. Following the award of blocks in the 11th UK round of licensing in 1989, several oil discoveries have been made in the southern part of the East Irish Sea Basin, which can now be regarded as a mixed, oil-and-gas province.

Oil, gas and condensate are widely distributed in the East Irish Sea Basin within sediments ranging from the Dinantian to the Upper Triassic Mercia Mudstone Group, although most discoveries have been made in the Lower Triassic Ormskirk Sandstone Formation. The prime oil-prone source rocks are Dinantian and early Namurian shales, but bituminous Dinantian limestones, and Westphalian oil shales, cannel coals and possibly marine bands have potential, and may contribute lesser amounts. Gas-prone source rocks largely comprise Westphalian coals and argillaceous beds, with a secondary contribution from Namurian mudstones (Johnson, 1981; Lawrence et al., 1987), and minor potential in the north from coals of the Yoredale facies.

An east–west oil fairway can probably be discerned in the southern part of the East Irish Sea Basin, extending westwards from the Formby Oilfield through the Lennox Oilfield (discovery well 110/15-6) to the Douglas Oilfield (discovery well 110/13-2) (Figure 78). The fairway may be sharply delineated to the north by the Hamilton and Hamilton North gasfields (discovery wells 110/13-1 and 110/13-5 respectively). The oil source rocks in this southern area are believed to be the Holywell Shales, Bowland Shales and equivalent Namurian to latest Dinantian deposits close to, or along, the axis of the Craven–Dublin Trough (Ramsbottom, 1969b; Collinson, 1988).

Hydrocarbon exploration plays in the East Irish Sea Basin may be grouped into three types, each requiring access to Carboniferous source rocks. Firstly, where the Ormskirk Sandstone Formation has a top and lateral seal of Mercia Mudstone Group. Secondly, where the Collyhurst Sandstone, and/or the overlying basal Upper Permian carbonates, are sealed by Manchester Marls/Upper Permian evaporites or juxtaposed against a lateral seal of Mercia Mudstone Group at major faults. Thirdly, where Carboniferous sandstones and carbonates retain primary porosity, or have acquired secondary porosity, and are sealed intraformationally by mudstones or at fault contacts.

Morecambe Gasfield

The Morecambe Gasfield (Colter, 1978; Bushell, 1986) is volumetrically the second largest gas find on the UK Continental Shelf, with 12.1 per cent of proven UK gas reserves (Department of Trade and Industry, 1992). It is a large, complex, faulted pericline trending north-north-west in the central part of the East Irish Sea Basin (Figure 79). It is divided into the North and South fields by a narrow, deep, east-north-easterly trending postdepositional graben filled with Mercia Mudstone Group, and along which limited sinistral movement may have taken place (Ebbern, 1981; Stuart and Cowan, 1991). The Western Boundary Fault marks the western margin of the field along its entire length. The South Morecambe structure bifurcates into eastern and western lobes at the Tynwald Fault Complex, which, like the dividing graben, is another postdepositional graben.

The field was discovered by Hydrocarbons Great Britain Limited in 1974 at well 110/2-1, although an earlier well, 110/8-2 drilled in 1969, also lies within the area of closure (Figure 79). The reservoir lies at a particularly shallow average subsea depth of 900 m. The 36-inch pipeline which transports the gas directly to Barrow-in-Furness (Figure 78) delivered the first gas in 1985. South Morecambe provides 1220 million cubic feet/day, but production will increase when the North Morecambe and two satellite fields come on stream (Bushell, 1986; Stuart and Cowan, 1991).

The chief reservoir is the 250 m-thick Ormskirk Sandstone Formation, although in the crestal parts of the South Morecambe Gasfield the top 200 m of the St Bees Sandstone Formation also lies above the gas–water contact. The top seal to the reservoir is provided by the basal Mercia Mudstone Group. In the Tynwald Fault Complex and the narrow dividing graben, an excellent lateral seal is afforded by younger mudstones and halites, particularly the Preesall Halite.

Geochemical studies demonstrate that the gas has been derived mainly from Westphalian coal and shale, with a minor, and later, contribution from Dinantian limestone (Stuart and Cowan, 1991). Barnard and Cooper (1983) have pointed out that up to 50 per cent of gas derived from the Westphalian may originate from the interbedded shales. At 7

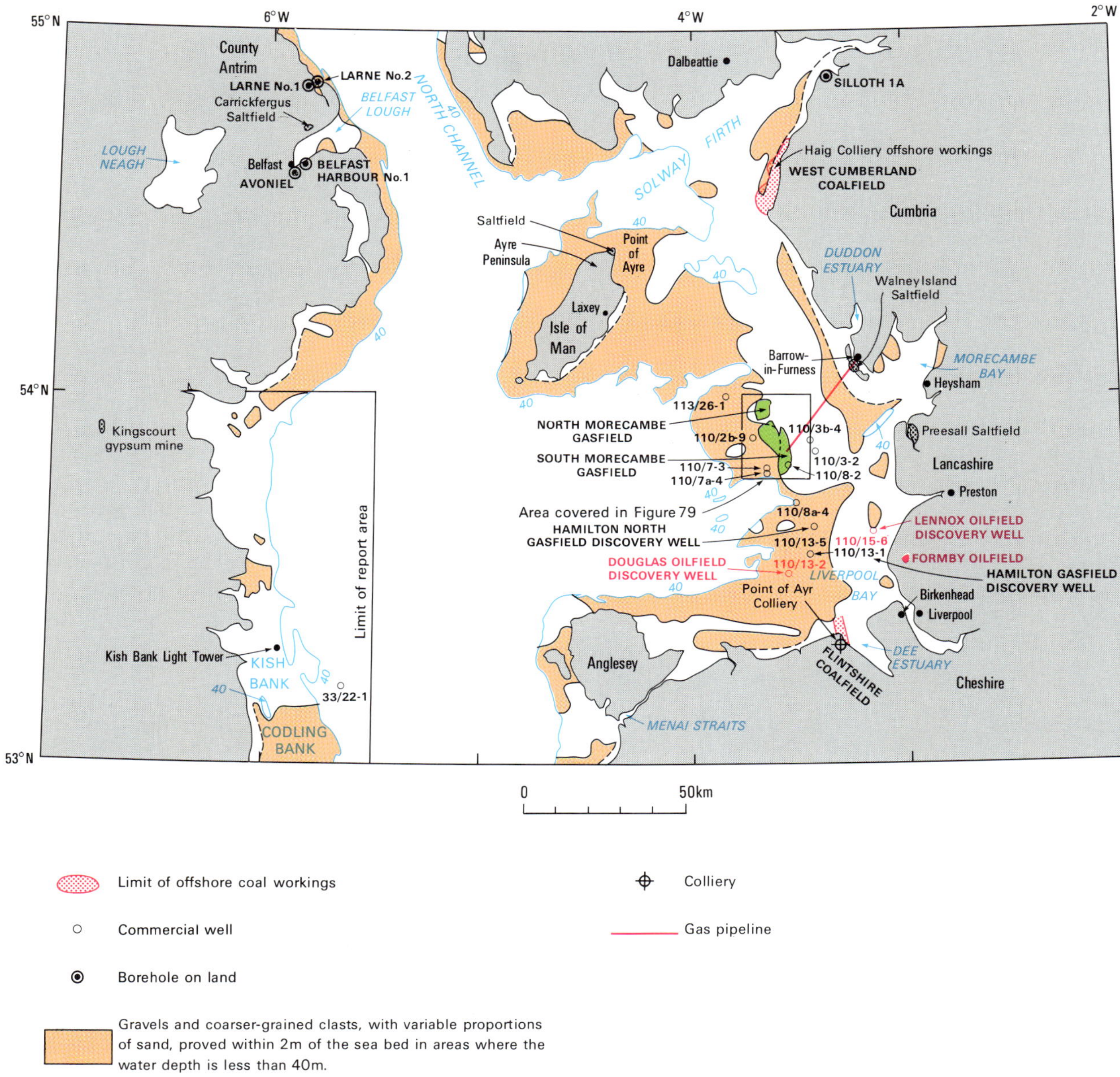

Figure 78 Selected economic resources of the Irish Sea and its coastal perimeter.

to 8 per cent, the nitrogen content is higher than the 3.6 per cent average for the southern North Sea (Cornford, 1990), which suggests either long migration routes or very mature source rocks. The CO_2 content is also high in North Morecambe.

For reservoir engineering purposes, the reservoir has been divided into a lower layer with poor permeability, and an upper layer with enhanced permeability (Ebbern, 1981; Bushell, 1986). The diagenetic history of the reservoir sandstones has been elucidated by a detailed study of the matrix and authigenic cements (Colter and Ebbern, 1978; Burley, 1984). Acidic porewaters, which were crucial in dissolving carbonates and feldspars to create secondary porosity, entered an early 'Morecambe structure' just before the emplacement of liquid hydrocarbons during the Early Jurassic (Stuart and Cowan, 1991). Beneath this 'palaeo' oil–water contact, authigenic platy illite was deposited during the late Early Jurassic (Stuart and Cowan, 1991), causing a drastic permeability reduction. Uplift, tilting and breaching of this early structure is thought to have occurred at the end of the Jurassic, and only after renewed burial and increased heat flow during the Cretaceous and early Tertiary was the structure refilled by the present gas.

Coastal occurrences

There are numerous brief references to minor hydrocarbon occurrences in the coastal perimeter of the Irish Sea. They include surface seepages, and underground oil and gas shows, in beds ranging from Dinantian limestones to Westphalian coal measures of the Lancashire Coalfield, parts of Flintshire, and areas bordering the Ramsey–Whitehaven Ridge (Strahan, 1920; Cope, 1939; Kent, 1954; Pinfold, 1958; Falcon and Kent, 1960; Rose and Dunham, 1977; Selley, 1992).

The products of surface oil seepages at Formby were being used as fuel before 1637. The small, shallow, now defunct, Formby Oilfield (Figure 78) can be regarded as Britain's first offshore oilfield since before Downholland Moss had been fully drained, the early production boreholes in the 1940s were drilled by rigs floated to the sites on rafts (Huxley, 1983). The multi-storey oilfield comprised an arrested surface seepage in Flandrian peat and silt, and a subdrift oil accumulation in the Tarporley Siltstone Formation, jointly termed

Figure 79 Simplified structure of the Morecambe Gasfield, showing the depth to the top of the Ormskirk Sandstone Formation. Modified from Bushell (1986). See Figure 78 for area covered by diagram.

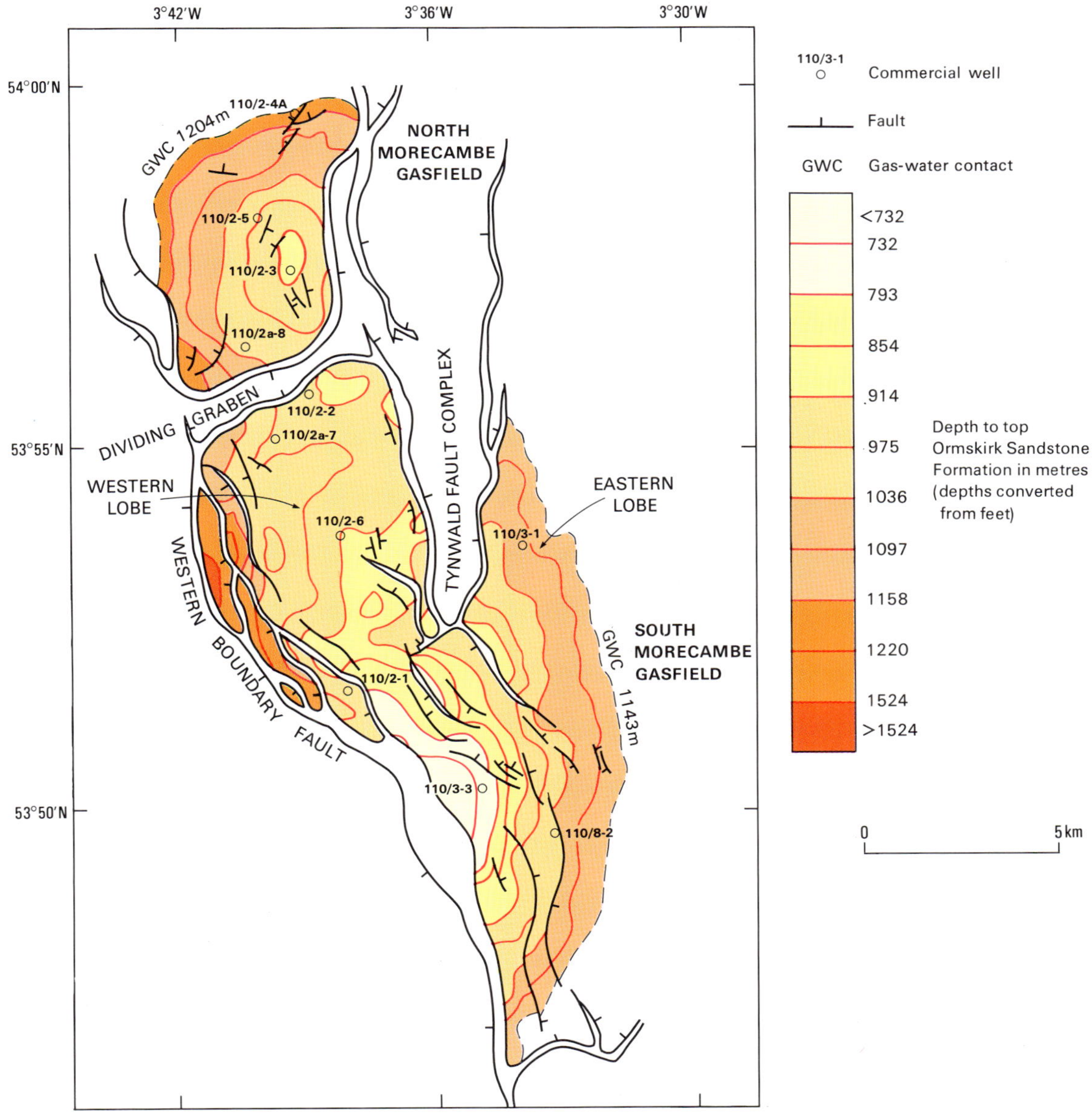

the Downholland oilpool. The producing beds in the lower reservoir were the soft, porous sandstones of the uppermost Ormskirk Sandstone Formation (Wray and Cope, 1948; Falcon and Kent 1960) contained in a faulted monocline with plunge culmination on the upthrow side of the Ince–Blundell Fault (Lees and Taitt, 1946; Warman et al., 1956). The light crude oil (37° API) was similar in composition to the Carboniferous crude from the East Midlands oilfields, and an origin from the marine Bowland Shales has been suggested (Colter, 1978). Minimal bacterial degradation of this light oil demonstrates post-Flandrian migration from a deep source.

Since 1957, there has been exploration and drilling for hydrocarbons in south-east County Antrim and in the Lough Neagh Basin. Although no commercial hydrocarbons have been found, exploration continues. Targets are in the Permo-Triassic sequences, and source rocks are likely to be underlying Namurian and/or Westphalian coals (Illing and Griffith, 1986).

Coal

Westphalian coal has been proved in several offshore areas, including extensions of worked coalfields (Figure 78). Point of Ayr Colliery in Flintshire (Lane, 1987) includes extensive reserves to the north-west of present workings under the Dee Estuary, although constraints of offshore underground mining preclude extraction of more than 3 or 4 out of the 10 thickest seams. In the West Cumberland Coalfield, numerous north-westerly trending faults gave rise to difficult working conditions in faces which formerly extended offshore for over 7 km in the Haig Colliery (Mitchell et al., 1978).

Productive measures are present in the East Irish Sea, West Irish Sea, and Solway Firth basins (Smith, 1985), although they are economically less attractive because of the overlying thick zone of pre-Permian secondary reddening. A concealed coalfield has been postulated beneath Belfast Lough (Griffith and Wilson, 1982), and coal measures are probably located over and surrounding much of the Ogham Platform east of the Isle of Man (e.g. well 113/26-1; Figure 33), as well as in the core of the Quadrant 109 Syncline (Smith, 1985; Jackson et al., 1987).

Cannel coals are widespread in Langsettian and Duckmantian rocks, and were formerly exploited for the distillation of paraffin oil (Strahan, 1920) and as a local supplement in the manufacture of town gas in the coalfields, but little is known about their offshore distribution.

In the Kish Bank Basin, two zones of significant coal-resource potential have been identified (McArdle and Keary, 1986) by extrapolation from well 33/22-1, which proved 26 seams of probable Duckmantian to Westphalian D age (Jenner, 1981). McArdle and Keary (1986) estimated that 215 million tonnes of coal could be mined in the Irish sector of the Irish Sea.

Tidal barrages

Tidal barrages have been considered for electricity generation across both Morecambe Bay and the Solway Firth (Knight, 1977). The Menai Straits also might also be a suitable location, and there is much interest in a 2 km-wide Mersey Barrage between Liverpool and Birkenhead (Figure 78).

Geothermal energy

There are two types of geothermal resource in the UK; firstly granitic batholiths with high heat production derived from the decay of radioactive elements and known as hot dry rocks (HDR), and secondly hot groundwaters in aquifers within deep sedimentary basins (low enthalpy reserves) (Downing and Gray, 1986). There are no known batholiths within the sedimentary basins of the Irish Sea, and thus there is no HDR potential.

Bottom-hole temperatures (largely uncorrected) and readings at intermediate levels taken from drill-stem tests in the East Irish Sea Basin suggest temperature gradients of 30°F/1000 feet within the Mercia Mudstone Group, and 15°F/1000 feet in wells drilled to the Carboniferous. Although temperatures within the Sherwood Sandstone Group in low-enthalpy basins surrounding the report area might be expected to be relatively high beneath the thick, low-thermal-conductivity Mercia Mudstone Group, investigations of geothermal resources to date have not proved encouraging (Downing et al., 1987). In the Cheshire Basin, the temperatures are not sufficiently elevated, even at depth, and in the West Lancashire and Carlisle basins the Mercia Mudstone Group is not sufficiently thick to raise the temperature in the Sherwood Sandstone Group aquifer (Downing and Gray, 1986). In Northern Ireland, the permeability of the Permo-Triassic rocks proved in the Larne No. 2 borehole (Figure 78) was too low to encourage economic development (Bennett, 1983). However, geothermal energy could supplement the local energy supply in coastal urban areas such as Barrow-in-Furness, Preston or Belfast.

The regional tilt of individual and linked half-grabens in the East Irish Sea Basin may determine the piezometric gradient and groundwater flow pattern, producing a generally outward and upward flow towards the basin margins, resulting in increased heat flow in these areas. Rising ferriferous hypersaline waters from the centre of the East Irish Sea Basin in post-Early Triassic times may account for the haematite deposits in the west and south of Cumbria (Rose and Dunham, 1977).

MINERALS

Many economic materials have been worked in the coastal perimeter (e.g. Ford, 1987; Warren et al., 1984); those most closely related to the offshore geology are discussed below.

Halite

On the coastal perimeter of the Irish Sea, Permian halites have not been exploited, but Triassic salt has been worked at Carrickfergus, Preesall, Walney Island, and the Ayre Peninsula on the Isle of Man (Figure 78). At Carrickfergus (Griffith and Wilson, 1982), the Larne Halite has three individual halites totalling at least 40 m in thickness; their average purity of 91 to 93 per cent compares with a minimum 95 per cent purity for the Northwich Halite in Cheshire (Earp and Taylor, 1986). The Larne Halite thickens to 331 m at Larne No. 1 borehole. Extraction at Carrickfergus was by pillar-and-stall mining; at Preesall, a refined method of controlled brine pumping is used to exploit the Preesall Salt, which has a maximum thickness of 185 m (Wilson and Evans, 1990). The disused saltworks on Walney Island were supplied by brine pumped from the Preesall Salt at Biggar (Rose and Dunham, 1977). At Point of Ayre, thin salt beds and a natural brine run (Lamplugh in Sherlock, 1921) are interpreted as correlatives of the Rossall Halite (upper leaf), and are expected to thicken northwards into the Solway Firth Basin.

The Irish Sea itself may contain the largest resource of Triassic halite within the British Isles. The Ballyboley/Rossall, Carnduff/Mythop and Larne/Preesall halites of the Mercia Mudstone Group are found throughout the North Channel and East Irish Sea basins (Jackson and Mulholland, 1993), and probably occur widely in the Solway Firth and Kish Bank basins; the Preesall Halite is 590 m thick in well 110/3-2. Upper Permian halites have been proved in the East Irish Sea (220 m in well 110/8-2) and North Channel basins.

Anhydrite and gypsum

Limited deposits of anhydrite occur below the zone of present-day meteoric water penetration, as a widespread facies within both Upper Permian evaporites (Chapter 6) and the basal Mercia Mudstone Group in the East Irish Sea Basin. Upper Permian anhydrite occurs in the nearshore zone and coastal fringe of south-west Cumbria (Rose and Dunham, 1977; Jackson et al., 1987), and is recorded in the Avoniel (Manning et al., 1970) and Belfast Harbour boreholes (Smith, 1986), from where it probably extends northwards under Belfast Lough to south-east County Antrim (Figure 78).

Gypsum is developed where anhydrite has been hydrated near the surface by meteoric waters, and also at depth as a subordinate constituent within blocky mudstones of the St Bees, Eden and Stanwix shales, and the Mercia Mudstone Group (Gregory, 1920; Sherlock and Hollingworth, 1938). Thinly bedded secondary and replacement gypsum after anhydrite occurs in the Dinantian Ballycultra Formation near Belfast (Smith, 1986), and gypsum and associated anhydrite are obtained from the Upper Permian beds at Kingscourt, County Cavan (Figure 78; Sevastopulo, 1981d). Disseminated secondary and fibrous gypsum (Holliday, 1970) is a persistent component in wet rockhead areas overlying halite in the Mercia Mudstone Group, both onshore (Dunham and Rose, 1949) and offshore.

Haematite and associated hydrocarbon and radioactive mineral occurrences

Large deposits of haematite metasomatically replacive of carbonate have been worked from Dinantian limestones in south and west Cumbria (Rose and Dunham, 1977; Trotter et al., 1937). Rose and Dunham (1977) have suggested that the deposits were formed from hypersaline brines driven through the Sherwood Sandstone Group from the centre of the East Irish Sea Basin, possibly stimulated by high heat flow and intrusion of the Fleetwood Dyke Group during the Tertiary. Undiscovered resources may exist offshore in the Duddon Estuary and Morecambe Bay.

Uranium, or pitchblende, mineralisation occurs in veins associated with minor, north-westerly trending tear faults and shatter belts south-east of Dalbeattie. The pitchblende is associated with hydrocarbons and haematite, and has been U-Pb dated as 185 ± 20 Ma — Early Jurassic (Miller and Taylor, 1966; Gallagher et al., 1971). Similar occurrences of uraniferous hydrocarbons dated as Early Jurassic are known from the Manx Slates at Laxey (Davidson and Bowie, 1951) and from the Dinantian limestones of north-east Wales (Eakin, 1989). Vanadiferous hydrocarbon nodules at the perimeter of the Irish Sea include examples from Namurian sandstone at Heysham (Harrison, 1970), and from the Mercia Mudstone Group at Larne (Parnell and Eakin, 1987). An Early Jurassic phase of oil accumulation in the Morecambe Gasfield predating deposition of diagenetic platy illite at about 180 Ma (Stuart and Cowan, 1991), and the post-Early Triassic age recognised for the Cumbrian haematites, possibly suggests a contemporary genesis with uranium mineralisation near the time of maximum burial of the basin.

AGGREGATES

Sand, gravel and coarser aggregates occur widely in the report area (Figure 75), where they show great variation in thickness and size grade. In the 1980s, some 15 per cent of UK aggregates was extracted offshore, of which a significant proportion was removed from the Liverpool Bay area (Ardus and Harrison, 1990). Areas off the Isle of Man have also been exploited, as has the Codling Bank in the Irish sector (Figure 78). It is likely that the periphery of the Irish Sea will become increasingly important for the mining of aggregates as onshore supplies become depleted.

The principal limitations controlling the profitability of sea-bed aggregates are distance from port and market, and depth of water. A 7 to 25 m depth range is preferred, although aggregate in water up to some 40 m deep can be exploited (Ardus and Harrison, 1990; Figure 78). The gravel reserves north and north-west of Anglesey are probably too remote to be economic at present, as are many of the sands and gravels off the Isle of Man. However, where particular aggregate qualities are required, longer transport paths become acceptable. For this reason the very coarse ballast up to 2 m in diameter on the Codling Bank has been shipped across the Irish Sea. Other limitations to exploitation result from too high a percentage of fines, excessive thickness of unusable overburden, high shell:lithic clast ratios, and environmental constraints. Although each limitation is significant, the offshore reserves are so extensive that sands or coarser material in the area may include economic aggregate in an environment suitable for extraction.

ENGINEERING AND WASTE DISPOSAL

Until the second half of the present century, the principal economic activity in the Irish Sea was shipping. Engineering works to assist maritime activity require knowledge of the coastal and intertidal geology; such works includes the construction of ports, offshore terminals, artificial islands and lighthouses. Most lighthouses are built on solid rock, but in the 1960s the Kish Bank Light tower (Figure 78) was constructed on a tidal sand ridge after an extensive study of its stability.

Cables and pipelines have been laid across the area; the first cables laid across the Irish Sea were telephone cables spooled onto the sea bed. Modern surveys have revealed that cable breaks occur primarily in zones of rough sea bed, where unsupported cable lengths fray in strong tidal currents. Modern hydrocarbon pipelines and cable lines for both fibre-optic telephone links and electricity power cables are trenched into the sea bed, requiring a detailed knowledge of the sea-bed geology.

Categories of terrain considered suitable for the underground storage of low- and intermediate-level nuclear waste include seaward-dipping sediments of Permo-Triassic age on the west coast of Britain (Chapman, 1986). Serious consideration has been given to underground storage within halite both in Germany and the USA, but the recognition of diapiric structures in the Mercia Mudstone Group of the East Irish Sea Basin militates against this method of disposal.

References

Most of the references listed below are held in the Library of the British Geological Survey at Keyworth, Nottingham. Copies of the references can be purchased subject to the current copyright legislation.

References to BGS offshore maps are not given here, but an index map showing their distribution in the report area is presented inside the back cover.

AGGER, H E, and CARPENTER, E W. 1964. A crustal study in the vicinity of the Eskdalemuir seismological array station. *Geophysical Journal of the Royal Astronomical Society*, Vol. 9, 69–83.

AITKENHEAD, N, BRIDGE, D Mc C, RILEY, N J, and KIMBELL, S F. 1992. Geology of the country around Garstang. *Memoir of the British Geological Survey*, Sheet 67 (England and Wales).

— CHISHOLM, J I, and STEVENSON, I P. 1985. Geology of the country around Buxton, Leek and Bakewell. *Memoir of the Geological Survey*, Sheet 111 (England and Wales).

AL-SHAIKH, Z D. 1969. Geophysical results from Caernarvon and Tremadoc Bays. *Nature, London*, Vol. 224, 897–899.

— 1970. The geological structure of part of the Central Irish Sea. *Geophysical Journal of the Royal Astronomical Society*, Vol. 20, 233–237.

ALLEN, J R L. 1965. The sedimentation and palaeogeography of the Old Red Sandstone of Anglesey, North Wales. *Proceedings of the Yorkshire Geological Society*, Vol. 35, 139–185.

— and CROWLEY, S F. 1983. Lower Old Red Sandstone fluvial dispersal systems in the British Isles. *Transactions of the Royal Society of Edinburgh: Earth Sciences*, Vol. 74, 61–68.

ANDERTON, R, BRIDGES, P H, LEEDER, M R, and SELLWOOD, B W. 1979. *A dynamic stratigraphy of the British Isles.* (London: Allen and Unwin.)

— GIBBONS, W, and NICHOLSON, P G. 1992. Precambrian. 5–12 *in* Atlas of palaeogeography and lithofacies. COPE, J C W, INGHAM, J K, and RAWSON, P F (editors). *Memoir of the Geological Society of London*, No. 13.

ARDUS, D A, and HARRISON, D J. 1990. The assessment of aggregate resources from the UK Continental Shelf. 113–128 in *Ocean Resources, Vol. 1. Assessment.* ARDUS, D A, and CHAMP, M A (editors). (Dordrecht: Kluver.)

ARTER, G, and FAGIN, S W. 1993. The Fleetwood Dyke and the Tynwald fault zone, Block 113/27, East Irish Sea Basin. 835–843 in *Petroleum Geology of Northwest Europe: Proceedings of the 4th Conference.* PARKER, J R (editor). (London: The Geological Society of London.)

ARTHURTON, R S. 1980. Rhythmic sedimentary sequences in the Triassic Keuper Marl (Mercia Mudstone Group) of Cheshire, northwest England. *Geological Journal*, Vol. 15, 43–58.

— BURGESS, I C, and HOLLIDAY, D W. 1978. Permian and Triassic. 189–206 *in* The geology of the Lake District. MOSELEY, F (editor). *Yorkshire Geological Society Occasional Publication*, No. 3.

— and HEMINGWAY, J E. 1972. The St Bees Evaporites — a carbonate–evaporite formation of Upper Permian age in West Cumberland, England. *Proceedings of the Yorkshire Geological Society*, Vol. 38, 565–592.

— JOHNSON, E W, and MUNDY, D J C. 1988. Geology of the country around Settle. *Memoir of the Geological Survey of Great Britain*, Sheet 60 (England and Wales).

— and WADGE, A J. 1981. Geology of the country around Penrith. *Memoir of the Geological Survey of Great Britain*, Sheet 24 (England and Wales).

AUDLEY-CHARLES, M G. 1970a. Stratigraphical correlation of the Triassic rocks of the British Isles. *Quarterly Journal of the Geological Society of London*, Vol. 126, 19–47.

— 1970b. Triassic palaeogeography of the British Isles. *Quarterly Journal of the Geological Society of London*, Vol. 126, 49–90.

BACON, M, and MCQUILLIN, R. 1972. Refraction seismic surveys in the north Irish Sea. *Journal of the Geological Society of London*, Vol. 128, 613–621.

BAMFORD, D, NUNN, K, PRODEHL, C, and JACOB, B. 1978. LISPB-IV. Crustal structure of Northern Britain. *Geophysical Journal of the Royal Astronomical Society*, Vol. 54, 43–60.

BANKS, R J. 1986. The interpretation of the Northumberland Trough geomagnetic variation anomaly using two-dimensional current models. *Geophysical Journal of the Royal Astronomical Society*, Vol. 87, 595–616.

BARNARD, P C, and COOPER, B S. 1983. A review of geochemical data related to the northwest European Gas Province. 19–33 *in* Petroleum geochemistry and exploration of Europe. BROOKS, J (editor). *Special Publication of the Geological Society of London*, No. 12.

BARNES, R P. 1989. Geology of the Whithorn district. *Memoir of the British Geological Survey*, Sheet 2 (Scotland).

— ANDERSON, T B, and MCCURRY, J A. 1987. Along-strike variation in the stratigraphical and structural profile of the Southern Uplands Central Belt in Galloway and Down. *Journal of the Geological Society of London*, Vol. 144, 807–816.

— LINTERN, B C, and STONE, P. 1989. Timing and regional implications of deformation in the Southern Uplands of Scotland. *Journal of the Geological Society of London*, Vol. 146, 905–908.

— ROCK, N M S, and GASKARTH, J W. 1986. The Caledonian dyke swarms in Southern Scotland: new field and petrological data for the Wigtown Peninsula, Galloway. *Geological Journal*, Vol. 21, 101–125.

BARR, K W, COLTER, V S, and YOUNG, R. 1981. The geology of the Cardigan Bay–St. George's Channel Basin. 432–443 in *Petroleum Geology of the Continental Shelf of North-West Europe.* ILLING, L V, and HOBSON, G D (editors). (London: Heyden and Son.)

BAZLEY, R A B. 1975. The Tertiary igneous and Permo-Triassic rocks of the Ballyalton Borehole, County Down. *Bulletin of the Geological Survey of Great Britain*, No. 50, 71–92.

BEAMISH, D. 1986. Deep crustal geoelectric structure beneath the Northumberland Basin. *Geophysical Journal of the Royal Astronomical Society*, Vol. 84, 619–640.

— and SMYTHE, D K. 1986. Geophysical images of the deep crust: the Iapetus suture. *Journal of the Geological Society of London*, Vol. 143, 489–497.

BELDERSON, R H. 1964. Holocene sedimentation in the western half of the Irish Sea. *Marine Geology*, Vol. 2, 147–163.

BENNETT, J R P. 1983. The sedimentary basins in Northern Ireland. *Investigation of the geothermal potential of the UK.* (London: Institute of Geological Sciences.)

BENTON, M J, WARRINGTON, G, NEWELL, A J, and SPENCER, P S. 1994. A review of British Mid Triassic tetrapod faunas. In *In the shadow of the dinosaurs: Early Mesozoic Tetrapods.* FRASER, N C, and SUES, H-D (editors). (Cambridge: Cambridge University Press.)

BESLY, B M. 1988. Palaeogeographic implications of late Westphalian to early Permian red-beds, Central England. 200–221 in *Sedimentation in a synorogenic basin complex: the Upper*

Carboniferous of Northwest Europe. Besly, B M, and Kelling, G (editors). (London: Blackie.)

Bluck, B J. 1985. The Scottish paratectonic Caledonides. *Scottish Journal of Geology,* Vol. 21, 437–464.

Blundell, D J, and Parks, R. 1969. A study of the crustal structure beneath the Irish Sea. *Geophysical Journal of the Royal Astronomical Society,* Vol. 17, 45–62.

— and Raynaud, B. 1986. Modelling lower crust reflections observed on BIRPS profiles. 287–295 *in* Reflection seismology: a global perspective. Barazangi, M, and Brown, L (editors). *Geodynamics Series, American Geophysical Union,* Vol. 14.

Bott, M H P. 1964. Gravity measurements in the north-eastern part of the Irish Sea. *Quarterly Journal of the Geological Society of London,* Vol. 120, 369–396.

— 1965. The deep structure of the Northern Irish Sea — a problem of crustal dynamics. 179–204 in *Submarine geology and geophysics: Proceedings of the 17th Symposium of the Colston Research Society, Bristol.* Whittard, W F, and Bradshaw, R (editors).

— 1967. Geophysical investigations of the northern Pennine basement rocks. *Proceedings of the Yorkshire Geological Society,* Vol. 36, 139–168.

— 1968. The geological structure of the Irish Sea Basin. 93–115 in *Geology of shelf seas.* Donovan, D T (editor). (Edinburgh: Oliver and Boyd.)

— 1978. Deep Structure. 25–40 *in* The geology of the Lake District. Moseley, F (editor). *Yorkshire Geological Society Occasional Publication,* No. 3.

— and Young, D G G. 1971. Gravity measurements in the north Irish Sea. *Quarterly Journal of the Geological Society of London,* Vol. 126, 413–434.

Bowen, D Q, Rose, J, McCabe, A M, and Sutherland, D G. 1986. Correlation of Quaternary glaciations in England, Ireland, Scotland and Wales. 229–340 *in* Quaternary glaciations in the Northern Hemisphere. Sibrava, V, Bowen, D Q, and Richmond, G M (editors). *Quaternary Science Reviews,* Vol. 5.

Boyd-Dawkins, W. 1894. On the geology of the Isle of Man. Part 1, the Permian, Carboniferous and Triassic rocks and the New Saltfield of the North. *Transactions of the Manchester Geological Society,* Vol. 22, 590–606.

— 1902. The Red Sandstone–Rocks of Peel (Isle of Man). *Quarterly Journal of the Geological Society of London,* Vol. 58, 633–646.

Branney, M J. 1988. The subaerial setting of the Ordovician Borrowdale Volcanic Group, English Lake District. *Journal of the Geological Society of London,* Vol. 145, 887–890.

Brewer, J A, and five others. 1983. BIRPS deep seismic reflection studies of the British Caledonides. *Nature, London,* Vol. 305, 206–210.

Brookfield, M E. 1978. Revision of the stratigraphy of Permian and supposed Permian rocks of Southern Scotland. *Geologische Rundschau,* Vol. 67, 110–149.

— 1980. Permian intermontane basin sedimentation in southern Scotland. *Sedimentary Geology,* Vol. 27, 167–194.

— 1989. Rotliegendes desert sedimentation in Scotland. *Zeitschrift Für Geologische Wissenschaften,* Vol. 17, 205–241.

Browne, B C, and Cooper, R I B. 1950. The British submarine gravity surveys of 1938 and 1946. *Philosophical Transactions of the Royal Society of London,* Vol. 242A, 243–310.

Brück, P M, and eight others. 1979. South-east Ireland: Lower Palaeozoic stratigraphy and depositional history. 533–544 *in* The Caledonides of the British Isles — reviewed. Harris, A L, Holland, C H, and Leake, B E (editors). *Special Publication of the Geological Society of London,* No. 8.

Burgess, I C. 1965. The Permo-Triassic rocks around Kirkby Stephen, Westmoreland. *Proceedings of the Yorkshire Geological Society,* Vol. 35, 91–101.

Burley, S D. 1984. Patterns of diagenesis in the Sherwood Sandstone Group (Triassic), United Kingdom. *Clay Minerals,* Vol. 19, 403–440.

Bushell, T P. 1986. Reservoir geology of the Morecambe Field. 189–208 *in* Habitat of Palaeozoic gas in NW Europe. Brooks, J, Goff, J C, and Van Hoorn, B (editors). *Special Publication of the Geological Society of London,* No. 23.

Butterworth, M A, and Butcher, C E. 1983. Basal Dinantian miospores from the Cockermouth area, west Cumbria. *Journal of Micropalaeontology,* Vol. 2, 13–16.

Calver, M A. 1968. Distribution of Westphalian marine faunas in northern England and adjoining areas. *Proceedings of the Yorkshire Geological Society,* Vol. 37, 1–72.

— and Smith, E G. 1974. The Westphalian of North Wales. 169–183 in *The Upper Palaeozoic and post-Palaeozoic rocks of Wales.* Owen, T R (editor). (Cardiff: University of Wales Press.)

Cameron, I B, and Stephenson, D. 1985. *British regional geology: the Midland Valley of Scotland.* (London: HMSO for British Geological Survey.)

Cameron, T D J, and six others. 1992. *United Kingdom offshore regional report: the geology of the southern North Sea.* (London: HMSO for British Geological Survey.)

Campbell, S D G, Howells, M F, Smith, M, and Reedman, A J. 1988. A Caradoc failed-rift within the Ordovician Marginal Basin of Wales. *Geological Magazine,* Vol. 125, 257–266.

Capewell, J G. 1955. The post-Silurian pre-marine Carboniferous sedimentary rocks of the eastern side of the English Lake District. *Quarterly Journal of the Geological Society of London,* Vol. 111, 23–46.

Caston, G F. 1975. Igneous dykes and associated scour hollows of the North Channel, Irish Sea. *Marine Geology,* Vol. 18, M77–M85.

— 1976. The floor of the North Channel, Irish Sea: a side-scan sonar survey. *Report of the Institute of Geological Sciences,* No. 76/7.

Chadwick, R A. 1985. Seismic reflection investigations into the stratigraphy and structural evolution of the Worcester Basin. *Journal of the Geological Society of London,* Vol. 142, 187–202.

— and Holliday, D W. 1991. Deep crustal structure and Carboniferous basin development within the Iapetus convergence zone, northern England. *Journal of the Geological Society of London,* Vol. 148, 41–53.

Chapman, N. 1986. Nuclear waste disposal. *British Geologist,* Vol. 12, 128–133.

Charlesworth, J K. 1937. The olivine dolerite sill of The Maidens, Co. Antrim. *Irish Naturalists' Journal,* Vol. 6, 265–273.

— 1963. *Historical geology of Ireland.* (Edinburgh and London: Oliver and Boyd.)

Charsley, T J. 1984. Early Carboniferous rocks of the Swinden No. 1 borehole, west of Skipton, Yorkshire. *Report of the British Geological Survey,* Vol. 16, No. 1, 5–12.

Clark, R A, and Stuart, G W. 1981. Upper mantle structure of the British Isles from Rayleigh wave dispersion. *Geophysical Journal of the Royal Astronomical Society,* Vol. 67, 59–75.

Clarke, A M, Chambers, R E, Allonby, R H, and Magraw, D. 1961. A marine geophysical survey of the undersea coalfields of Northumberland, Cumberland and Durham. *Transactions of the Institution of Mining Engineers,* Vol. 121, 197–215.

Clayton, G, Sevastopulo, G D, and Sleeman, A G. 1986. Carboniferous (Dinantian and Silesian) and Permo-Triassic rocks in south County Wexford, Ireland. *Geological Journal,* Vol. 21, 355–374.

Clegg, J A, Almond, M, and Stubbs, P H S. 1954. The remanent magnetism of some sedimentary rocks in Britain. *Philosophical Magazine,* Vol. 45, 583–598.

Clemmensen, L B. 1978. Alternating aeolian, sabkha and shallow-lake deposits from the Middle Triassic Gipsdalen Formation,

Scoresby Land, East Greenland. *Palaeogeography, Palaeoclimatology, Palaeoecology*, Vol. 24, 111–135.

Coleman, J M, and Roberts, H H. 1988. Sedimentary development of the Louisiana continental shelf related to sea level changes. *Geo-Marine Letters*, Vol. 8, 63–119.

Collar, F A. 1974. A geophysical interpretation of the structure of the Vale of Clwyd, North Wales. *Geological Journal*, Vol. 9, 65–76.

Collinson, J D. 1986. Alluvial sediments. 20–62 in *Sedimentary environments and facies* (2nd edition). Reading, H G (editor). (Oxford: Blackwell Scientific Publications.)

— 1988. Controls on Namurian sedimentation in the Central Province basins of northern England. 85–101 in *Sedimentation in a synorogenic basin complex: the Upper Carboniferous of Northwest Europe.* Besley, B M, and Kelling, G (editors). (London: Blackie.)

Colter, V S. 1978. Exploration for gas in the Irish Sea. 503–516 *in* Key-notes of the MEGS-II: Amsterdam, Netherlands. Loon, A J van (editor). *Geologie en Mijnbouw*, Vol. 57, 503–516.

— and Barr, K W. 1975. Recent developments in the geology of the Irish Sea and Cheshire Basins. 61–75 in *Petroleum and the Continental Shelf of North-West Europe.* Woodland, A W (editor). (London: Applied Sciences Publishers.)

— and Ebbern, J. 1978. The petrography and reservoir properties of some Triassic sandstones of the Northern Irish Sea Basin. *Journal of the Geological Society of London*, Vol. 135, 57–62.

— — 1979. SEM studies of Triassic reservoir sandstones from the Morecambe Field, Irish Sea, UK. *Scanning Electron Microscopy*, Vol. 1, 531–538.

Cooper, A H, and Molyneux, S G. 1990. The age and correlation of Skiddaw Group (early Ordovician) sediments in the Cross Fell inlier (northern England). *Geological Magazine*, Vol. 127, 147–157.

Cope, F W. 1939. Oil occurrences in south-west Lancashire. *Bulletin of the Geological Survey of Great Britain,* No. 2, 18–25.

Cope, J C W. 1984. The Mesozoic history of Wales. *Proceedings of the Geologists' Association,* Vol. 95, 373–385.

— Getty, T A, Howarth, M K, Morton, N, and Torrens, H S. 1980. A correlation of Jurassic rocks in the British Isles. Part One: Introduction and Lower Jurassic. *Special Report of the Geological Society of London*, No. 14.

Cornford, C. 1990. Source rocks and hydrocarbons of the North Sea. 294–361 in *Introduction to the petroleum geology of the North Sea* (3rd edition). Glennie, K W (editor). (Oxford: Blackwell Scientific Publications.)

Cornwell, J D. 1972. A gravity survey of the Isle of Man. *Proceedings of the Yorkshire Geological Society*, Vol. 39, 93–106.

Craig, G Y. 1956. The Lower Carboniferous outlier of Kirkbean, Kirkcudbrightshire. *Transactions of the Geological Society of Glasgow*, Vol. 22, 113–132.

— (editor). 1991. *Geology of Scotland* (3rd edition). (London: The Geological Society of London.)

Croker, P F, and Shannon, P M. 1987. The evolution and hydrocarbon prospectivity of the Porcupine Basin, offshore Ireland. 633–642 in *Petroleum geology of North West Europe.* Brooks, J, and Glennie, K W (editors). (London: Graham and Trotman.)

Crowley, S F. 1985. Lithostratigraphy of the Peel Sandstone, Isle of Man. *Mercian Geologist*, Vol. 10, 73–76.

Curry, D, and six others. 1978. A correlation of Tertiary rocks in the British Isles. *Special Report of the Geological Society of London*, No. 12.

Davidson, C F, and Bowie, S H U. 1951. On thucholite and related hydrocarbon-uraninite complexes, with a note on the origin of the Witwatersrand gold ores. *Bulletin of the Geological Survey of Great Britain*, No. 3, 1–18.

Davidson, K A S, Sola, M, Powell, D W, and Hall, J. 1984. Geophysical model for the Midland Valley of Scotland. *Transactions of the Royal Society of Edinburgh: Earth Sciences*, Vol. 75, 175–181.

Davies, J R, Riley, N J, and Wilson, D. 1989. The distribution of Chadian and earliest Arundian strata in North Wales: implications for Dinantian (Carboniferous) lithostratigraphy and palaeogeography. *Geological Journal,* Vol. 24, 31–47.

Deegan, C E. 1973. Tectonic control of sedimentation at the margin of a Carboniferous depositional basin in Kirkcudbrightshire. *Scottish Journal of Geology*, Vol. 9, 1–28.

— 1978. Interpretation of the shallow seismic data on sheet 55°N/6°W, with comments on the IGS seaborne magnetic results. 49–64 *in* The solid geology of the Clyde Sheet (55°N/6°W). McLean, A C, and Deegan, C E (editors). *Report of the Institute of Geological Sciences,* No. 78/9.

— Kirby, R, Rae, I, and Floyd, R. 1973. The superficial deposits of the Firth of Clyde and its sea lochs. *Report of the Institute of Geological Sciences*, No. 73/9.

Delair, J B. 1991. An unusual reptilian trackway from Locharbriggs, Dumfriesshire. *Transactions of the Dumfriesshire and Galloway Natural History and Antiquarian Society*, Vol. 31, 1–10.

Department of Trade and Industry. 1992. *Development of the oil and gas resources of the United Kingdom.* (London: HMSO.)

Dewey, J F, and Windley, B F. 1988. Palaeocene–Oligocene tectonics of NW Europe. 25–34 *in* Early Tertiary volcanism and the opening of the NE Atlantic. Morton, A C, and Parson, L M (editors). *Special Publication of the Geological Society of London,* No. 39.

Dickson, J A D, Ford, T D, and Swift, A. 1987. The stratigraphy of the Carboniferous rocks around Castletown, Isle of Man. *Proceedings of the Yorkshire Geological Society*, Vol. 46, 203–230.

Dixon, E E L, Maden, J, Trotter, F M, Hollingworth, S E, and Tonks, L H. 1926. The geology of the Carlisle, Longtown and Silloth district. *Memoir of the Geological Survey of Great Britain.*

Dobson, M R, Evans, W E, and Whittington, R J. 1973. The geology of the south Irish Sea. *Report of the Institute of Geological Sciences,* No. 73/11.

— and Whittington, R J. 1979. The geology of the Kish Bank Basin. *Journal of the Geological Society of London*, Vol. 136, 243–249.

— — 1987. The geology of Cardigan Bay. *Proceedings of the Geologists' Association*, Vol. 98, 331–353.

Downing, R A, and five others. 1982. Geothermal aspects of the Larne No. 2 borehole. *Investigations of the geothermal potential of the UK.* (London: Institute of Geological Sciences.)

— Edmunds, W M, and Gale, I M. 1987. Regional groundwater flow in sedimentary basins in the UK. 105–125 *in* Fluid flow in sedimentary basins and aquifers. Goff, J C, and Williams, B P J (editors). *Special Publication of the Geological Society of London,* No. 34.

— and Gray, D A (editors). 1986. *Geothermal energy — the potential in the United Kingdom.* (London: HMSO for British Geological Survey.)

Draper, L. 1973. Extreme wave conditions in British and adjacent waters. *Proceedings, 13th Coastal Engineering Conference, Vancouver, Canada.* 157–165.

Duff, P McL D, and Smith, A J (editors). 1992. *Geology of England and Wales.* (London: The Geological Society of London.)

Dunham, K C, and Rose, W C C. 1949. Permo-Triassic geology of south Cumberland and Furness. *Proceedings of the Geologists' Association*, Vol. 60, 11–40.

Eakin, P A. 1989. Isotopic and petrographic studies of uraniferous hydrocarbons from around the Irish Sea Basin. *Journal of the Geological Society of London*, Vol. 146, 663–673.

Earp, J R, Magraw, D, Poole, E G, Land, D H, and Whiteman, A J. 1961. Geology of the country around Clitheroe and Nelson. *Memoir of the Geological Survey of Great Britain*, Sheet 68 (England and Wales).

— and Taylor, B J. 1986. Geology of the country around Chester and Winsford. *Memoir of the British Geological Survey,* Sheet 109 (England and Wales).

Eastwood, T. 1930. The geology of the Maryport district. *Memoir of the Geological Survey of Great Britain,* Sheet 22 (England and Wales).

— Dixon, E E L, Hollingworth, S E, and Smith, B. 1931. The geology of the Whitehaven and Workington district. *Memoir of the Geological Survey of Great Britain,* Sheet 28 (England and Wales).

— Hollingworth, S E, Rose, W C C, and Trotter, F M. 1968. Geology of the country around Cockermouth and Caldbeck. *Memoir of the Geological Survey of Great Britain,* Sheet 23 (England and Wales).

Ebbern, J. 1981. The geology of the Morecambe Gas Field. 485–493 in *Petroleum geology of the Continental Shelf of North-West Europe.* Illing, L V, and Hobson, G D (editors). (London: Heyden and Son.)

Eden, R A, Deegan, C E, Rhys, G H, Wright, J E, and Dobson, M R. 1973. Geological investigations with a manned submersible in the Irish Sea and off western Scotland 1971. *Report of the Institute of Geological Sciences,* No. 73/2.

Edwards, W, and Trotter, F M. 1954. *British regional geology: the Pennines and adjacent areas* (3rd edition). (London: HMSO for the Institute of Geological Sciences.)

Elliott, T. 1986. Deltas. 113–154 in *Sedimentary environments and facies* (2nd edition). Reading, H G (editor). (Oxford: Blackwell Scientific Publications.)

Evans, A L, Fitch, F J, and Miller, J A. 1973. Potassium-argon age determinations on some British Tertiary igneous rocks. *Journal of the Geological Society of London,* Vol. 129, 419–443.

Evans, D, Hallsworth, C, Jolley, D W, and Morton, A C. 1991. Late Oligocene terrestrial sediments from a small basin in the Little Minch. *Scottish Journal of Geology,* Vol. 27, 33–40.

Evans, D J, Rees, J G, and Holloway, S. 1993. The Permian to Jurassic stratigraphy and structural evolution of the central Cheshire Basin. *Journal of the Geological Society of London,* Vol. 150, 857–870.

Evans, W B. 1970. The Triassic salt deposits of north-western England. *Quarterly Journal of the Geological Society of London,* Vol. 126, 103–123.

— Wilson, A A, Taylor, B J, and Price, D. 1968. Geology of the country around Macclesfield, Congleton, Crewe and Middlewich. *Memoir of the Geological Survey of Great Britain,* Sheet 110 (England and Wales).

Eyles, N C, and McCabe, A M. 1989. The late Devensian (<22 000 BP) Irish Sea basin: the sedimentary record of a collapsed ice sheet margin. *Quaternary Science Reviews,* Vol. 8, 307–351.

Falcon, N L, and Kent, P E. 1960. Geological results of petroleum exploration in Britain, 1945–1957. *Memoir of the Geological Society of London,* No. 2.

Faller, A M, and Briden, J C. 1978. Palaeomagnetism of Lake District rocks. 17–24 *in* The geology of the Lake District. Moseley, F (editor). *Yorkshire Geological Society Occasional Publication,* No. 3.

Fewtrell, M D, and Smith, D G. 1980. Revision of the Dinantian stratigraphy of the Craven Basin, N. England. *Geological Magazine,* Vol. 117, 37–49.

Fisher, M J, and Mudge, D C. 1990. Triassic. 191–218 in *Introduction to the petroleum geology of the North Sea* (3rd edition). Glennie, K W (editor). (Oxford: Blackwell Scientific Publications.)

Fletcher, B N, and Ransome, C R. 1978. Offshore Cumbria. 242–249 *in* The geology of the Lake District. Moseley, F (editor). *Yorkshire Geological Society Occasional Publication,* No. 3.

Fletcher, T P. 1977. Lithostratigraphy of the Chalk (Ulster White Limestone Formation) in Northern Ireland. *Report of the Institute of Geological Sciences,* No. 77/24.

Floodgate, G D, Jones, G W, and Yarrington, M R. 1986. Chemical and micro-biological aspects of acoustically turbid sediments. *GERBAM, Deuxième Colloque International de Bactériologie Marine, CNRS Brest 1–5 October 1984, IFREMER, Actes de Colloques,* Vol. 3, 35–42.

Folk, R L. 1954. The distinction between grain size and mineral composition in sedimentary rock nomenclature. *Journal of Geology,* Vol. 62, 344–359.

Ford, T D. 1971. Slump structures in the Peel Sandstone Series, Isle of Man. *Isle of Man Natural History and Antiquarian Journal,* Vol. 7, 440–448.

— 1987. Geological excursion guide 4: the Isle of Man. *Geology Today,* Vol. 3, 64–69.

Fowler, A, and Robbie, J A. 1961. Geology of the country around Dungannon. *Memoir of the Geological Survey of Northern Ireland.*

Francis, E H. 1982. Magma and sediment — 1. Emplacement mechanism of late Carboniferous tholeiite sills in Northern Britain. *Journal of the Geological Society of London,* Vol. 139, 1–20.

Freshney, E C, and Taylor, R T. 1980. The Variscides of southwest Britain. 49–57 *in* United Kingdom, introduction to general geology and guides to excursions 002, 055, 093 and 151. Owen, T R (editor). *Proceedings 26th International Congress, Paris, 1980.*

Frey, R W, and Pemberton, S G. 1987. The *Psilonichnus ichnocoenose,* and its relationship to adjacent marine and non-marine ichnocoenoses along the Georgia coast. *Bulletin of Canadian Petroleum Geology,* Vol. 35, 333–357.

Fuller, J G C M. 1958. The petrology of the Carboniferous rocks near Stranraer, Wigtownshire. *Proceedings of the Geologists' Association,* Vol. 69, 166–174.

Fyfe, J A, Long, D, and Evans, D. 1993. *United Kingdom offshore regional report: the geology of the Malin–Hebrides sea area.* (London: HMSO for British Geological Survey.)

Gale, I N, Carruthers, R M, and Evans, R B. 1984. *The Carlisle Basin and adjacent areas. Investigation of the geothermal potential of the UK.* (Keyworth: British Geological Survey.)

— and Downing, R A. 1988. *Heat flow and regional groundwater flow in the United Kingdom. Investigation of the geothermal potential of the UK.* (Keyworth: British Geological Survey.)

Gallagher, M J, Michie, U McL, Smith, R T, and Haynes, L. 1971. New evidence of uranium and other mineralization in Scotland. *Transactions of the Institute of Mining and Metallurgy,* Vol. 80B, 150–173.

Gardiner, P R R, and Visscher, H. 1971. Permian-Triassic transition sequence at Kingscourt, Ireland. *Nature, London,* Vol. 229, 209–210.

Garrard, R A. 1977. The sediments of the south Irish Sea and Nymphe Bank area of the Celtic Sea. 69–92 *in* The Quaternary history of the Irish Sea. Kidson, C, and Tooley, M J (editors). *Special Issue of the Geological Journal,* No. 7.

Gawthorpe, R L, Gutteridge, P, and Leeder, M R. 1989. Late Devonian and Dinantian basin evolution in northern England and North Wales. 1–23 *in* The role of tectonics in Devonian and Carboniferous sedimentation in the British Isles. Arthurton, R S, Gutteridge, P, and Nolan, S C (editors). *Yorkshire Geological Society Occasional Publication,* No. 6.

Geiger, M E, and Hopping, C A. 1968. Triassic stratigraphy of the southern North Sea Basin. *Philosophical Transactions of the Royal Society of London,* Vol. 254B, 1–36.

Geological Survey of Great Britain. 1965 and 1972. Aeromagnetic map of Great Britain, 1:625 000, Sheets 1 (1965) and 2 (1972). (Southampton: Ordnance Survey.)

George, T N. 1958. Lower Carboniferous palaeogeography of the British Isles. *Proceedings of the Yorkshire Geological Society,* Vol. 31, 227–318.

— 1974. Lower Carboniferous rocks in Wales. 85–115 in *The Upper Palaeozoic and post-Palaeozoic rocks of Wales.* OWEN, T R (editor). (Cardiff: University of Wales Press.)

GIBBONS, W. 1983. Stratigraphy, subduction and strike-slip faulting in the Mona Complex of North Wales — a review. *Proceedings of the Geologists' Association,* Vol. 94, 147–163.

— 1989. Suspect terrane definition in Anglesey, North Wales. 59–66 *in* Terranes in the Circum-Atlantic orogen. DALLMEYER, R D (editor). *Geological Society of America Special Paper,* No. 230.

GIBSON, D, MCCORMICK, A G, MEIGHAN, I G, and HALLIDAY, A N. 1987. The British Tertiary Igneous Province: young Rb-Sr ages for the Mourne Mountain Granites. *Scottish Journal of Geology,* Vol. 23, 221–225.

GIESE, P, JODICKE, H, PRODEHL, C, and WEBER, K. 1983. The crustal structure of the Hercynian Mountain System — a model for crustal thickening by stacking. 405–426 in *Intracontinental fold belts.* MARTIN, H, and EDER, F W (editors). (Berlin and Heidelberg: Springer-Verlag.)

GLENNIE, K W, and EVANS, G. 1976. A reconnaissance of the Recent sediments of the Ranns of Kutch, India. *Sedimentology,* Vol. 23, 625–647.

GREENLY, E. 1919. The geology of Anglesey. *Memoir of the Geological Survey of Great Britain.*

GREGORY, J W. 1915. A deep bore at Seascale in Cumberland. *Geological Magazine,* Vol. 62, 146–149.

— 1920. The red rocks of a deep bore at the north end of the Isle of Man. *Transactions of the Institution of Mining Engineers,* Vol. 59, 156–168.

GREIG, D C. 1971. *British regional geology: the south of Scotland* (3rd edition). (London: HMSO for the Institute of Geological Sciences.)

GRIFFITH, A E. 1983. The search for petroleum in Northern Ireland. 213–222 *in* Petroleum geochemistry and exploration of Europe. BROOKS, J (editor). *Special Publication of the Geological Society of London,* No. 12.

— and WILSON, H E. 1982. Geology of the country around Carrickfergus and Bangor. *Memoir of the Geological Survey of Northern Ireland,* Sheet 29 (Northern Ireland).

GRIFFITHS, D H, KING, R F, and WILSON, C D V. 1961. Geophysical investigations in Tremadoc Bay, North Wales. *Quarterly Journal of the Geological Society of London,* Vol. 117, 171–191.

GUION, P D, and FIELDING, C R. 1988. Westphalian A and B sedimentation in the Pennine Basin, UK. 153–177 in *Sedimentation in a synorogenic basin complex: the Upper Carboniferous of Northwest Europe.* BESLY, B M, and KELLING, G (editors). (London: Blackie.)

HALL, J. 1986a. Nature of the lower continental crust: evidence from BIRPS work on the Caledonides. 223–231 *in* Reflection seismology: a global perspective. BARAZANGI, M, and BROWN, L (editors). *Geodynamics Series, American Geophysical Union,* Vol. 14.

— 1986b. The physical properties of layered rocks in deep continental crust. 51–62 *in* The nature of the lower continental crust. DAWSON, J B, CARSWELL, D A, HALL, J, and WEDEPOHL, K H (editors). *Special Publication of the Geological Society of London,* No. 24.

— BREWER, J A, MATTHEWS, D H, and WARNER, M R. 1984. Crustal structure across the Caledonides from the 'WINCH' seismic reflection profile: influences on the evolution of the Midland Valley of Scotland. *Transactions of the Royal Society of Edinburgh: Earth Sciences,* Vol. 75, 97–109.

HAQ, B U, HARDENBOL, J, and VAIL, P R. 1987. The chronology of fluctuating sea-level since the Triassic. *Science, New York,* Vol. 235, 1136–1167.

HARDMAN, M. 1992. The East Irish Sea Basin. ?Direct evidence for Permo-Carboniferous rifting. *Abstracts for conference on Permo-Triassic rifting,* Geological Society of London.

HARRIS, C R. 1974. The evolution of the north Bull Island, Dublin Bay. *Scientific Proceedings of the Royal Dublin Society,* Vol. A5, 237–252.

— 1980. Recent sediment distribution in Dublin Bay and its approaches. *Journal of Earth Sciences, Royal Dublin Society,* Vol. 3, 41–52.

HARRISON, R K. 1970. Hydrocarbon-bearing nodules from Heysham, Lancashire. *Geological Journal,* Vol. 7, 101–110.

HERBERT-SMITH, M. 1979. The age of the Tertiary deposits of the Llanbedr (Mochras Farm) Borehole as determined from palynological studies. *Report of the Institute of Geological Sciences,* No. 78/24, 15–29.

HESSION, M A I. 1988. Quaternary geology of the South Irish Sea. Unpublished PhD thesis, University of Wales.

— and WHITTINGTON, R J. 1987. Aspects of the Quaternary sediments of the Anglesey sheet. *Proceedings of the Geologists' Association,* Vol. 98, 398–400.

HOLLAND, C H (editor). 1981a. *A geology of Ireland.* (Edinburgh: Scottish Academic Press.)

— 1981b. Cambrian and Ordovician of the paratectonic Caledonides. 41–64 in *A geology of Ireland.* HOLLAND, C H (editor). (Edinburgh: Scottish Academic Press.)

— 1981c. Silurian. 65–81 in *A geology of Ireland.* HOLLAND, C H (editor). (Edinburgh: Scottish Academic Press.)

— 1981d. Devonian. 121–146 in *A geology of Ireland.* HOLLAND, C H (editor). (Edinburgh: Scottish Academic Press.)

HOLLIDAY, D W. 1970. The petrology of secondary gypsum rocks: a review. *Journal of Sedimentary Petrology,* Vol. 40, 734–744.

HOLLOWAY, S, and CHADWICK, R A. 1986. The Sticklepath–Lustleigh fault zone: Tertiary sinistral reactivation of a Variscan dextral strike-slip fault. *Journal of the Geological Society of London,* Vol. 143, 447–452.

HOLMES, R. 1977. Quaternary deposits of the central North Sea, 5. The Quaternary geology of the UK sector of the North Sea between 56° and 58°N. *Report of the Institute of Geological Sciences,* No. 77/14.

HORBURY, A D. 1989. The relative roles of tectonism and eustacy in the deposition of the Urswick Limestone in south Cumbria and north Lancashire. 153–169 *in* The role of tectonics in Devonian and Carboniferous sedimentation in the British Isles. ARTHURTON, R S, GUTTERIDGE, P, and NOLAN, S C (editors). *Yorkshire Geological Society Occasional Publication,* No. 6.

HOVLAND, M, and JUDD, A G. 1988. *Seabed pockmarks and seepages.* (London: Graham and Trotman.)

HOWELLS, M F, REEDMAN, A J, and CAMPBELL, S D G. 1991. *Ordovician (Caradoc) marginal basin volcanism in Snowdonia (north-west Wales).* (London: HMSO for the British Geological Survey.)

HSÜ, K J. 1972. Origin of saline giants: a critical review after the discovery of the Mediterranean Evaporite. *Earth Science Reviews,* Vol. 8, 371–396.

HUDDART, D, TOOLEY, M J, and CARTER, P A. 1977. The coasts of north-west England. 119–154 *in* The Quaternary history of the Irish Sea. KIDSON, C, and TOOLEY, M J (editors). *Special Issue of the Geological Journal,* No. 7.

HUXLEY, J. 1983. Britain's onshore oil industry. *Globe Industry Report,* No. 1.

HYDROGRAPHIC DEPARTMENT. 1960. *West coast of England pilot* (10th edition). (Taunton: Hydrographer of the Navy.)

ILLING, L V, and GRIFFITH, A E. 1986. Gas prospects in the "Midland Valley" of Northern Ireland. 73–84 *in* Habitat of Palaeozoic gas in NW Europe. BROOKS, J, GOFF, J C, and VAN HOORN, B (editors). *Special Publication of the Geological Society of London,* No. 23.

INSTITUTE OF GEOLOGICAL SCIENCES. 1975. IGS boreholes 1974. *Report of the Institute of Geological Sciences,* No. 75/7.

IRELAND, R J, POLLARD, J E, STEEL, R J, and THOMPSON, D B. 1978. Intertidal sediments and trace fossils from the Waterstones (Scythian–Anisian?) at Daresbury, Cheshire. *Proceedings of the Yorkshire Geological Society*, Vol. 41, 399–436.

IVIMEY-COOK, H C. 1971. Stratigraphical palaeontology of the Lower Jurassic of the Llanbedr (Mochras Farm) Borehole. 87–92 *in* The Llanbedr (Mochras Farm) Borehole. WOODLAND, A W (editor). *Report of the Institute of Geological Sciences*, No. 71/18.

— 1975. The stratigraphy of the Rhaetic and Lower Jurassic in east Antrim. *Bulletin of the Geological Survey of Great Britain*, No. 50, 51–69.

— WARRINGTON, G, HOLLOWAY, S, WORLEY, N, and YOUNG, B. In press. Rocks of Late Triassic and Early Jurassic age in the Carlisle Basin, Cumbria (north-west England). *Proceedings of the Yorkshire Geological Society.*

JACKSON, D I. 1994. *Discussion of* Geophysical log signatures in the Eden Shales (Permo-Triassic) of Cumbria and their regional significance. *Proceedings of the Yorkshire Geological Society*, Vol. 50.

— and MULHOLLAND, P. 1993. Tectonic and stratigraphic aspects of the East Irish Sea Basin and adjacent areas: contrasts in their post-Carboniferous structural styles. 791–808 in *Petroleum Geology of Northwest Europe: Proceedings of the 4th Conference.* PARKER, J R (editor). (London: The Geological Society of London.)

— — JONES, S M, and WARRINGTON, G. 1987. The geological framework of the East Irish Sea Basin. 191–203 in *Petroleum geology of North West Europe.* BROOKS, J, and GLENNIE, K W (editors). (London: Graham and Trotman.)

JACKSON, J S. 1965. The Upper Carboniferous (Namurian and Westphalian) of Kingscourt, Ireland. *Scientific Proceedings of the Royal Dublin Society*, Vol. 2, 131–152.

JACOB, A W B, KAMINSKI, W, MURPHY, T, PHILLIPS, W E A, and PRODEHL, C. 1985. A crustal model for a northeast–southwest profile through Ireland. *Tectonophysics*, Vol. 113, 75–103.

JAMES, J W C, and WINGFIELD, R T R. 1987. Aspects of sea bed sediments in the southern Irish Sea. *Proceedings of the Geologists' Association*, Vol. 98, 404–406.

JASIN, B. 1976. Quaternary microfauna in 5 BGS boreholes in Cardigan Bay. Unpublished MSc thesis, University of Wales.

JEFFERY, D W, PATKIN, P H, and WEST, A B. 1978. Intertidal environment of north Dublin Bay. *Estuarine and Coastal Marine Science*, Vol. 7, 163–171.

JENNER, J K. 1981. The structure and stratigraphy of the Kish Bank Basin. 426–431 in *Petroleum geology of the Continental Shelf of North-West Europe.* ILLING, L V, and HOBSON, G D (editors). (London: Heyden and Son.)

JENYON, M K. 1985. Basin-edge diapirism and updip salt flow in the Zechstein of Southern North Sea. *Bulletin of the American Association of Petroleum Geologists*, Vol. 69, 53–64.

— and CRESSWELL, P M. 1987. The southern Zechstein salt basin of the British North Sea, as observed in regional seismic traverses. 277–292 in *Petroleum geology of North West Europe.* BROOKS, J, and GLENNIE, K W (editors). (London: Graham and Trotman.)

JOHNSON, E W. 1981. A tunnel section through a prograding Namurian (Arnsbergian E_{2a}) delta, in the western Bowland Fells, north Lancashire. *Geological Journal*, Vol. 16, 93–110.

JOHNSON, G A L, and TARLING, D H. 1985. Continental convergence and closing seas during the Carboniferous. *Dixième Congrès International de Stratigraphie et de Géologie du Carbonifère, Madrid, 1983*, Vol. 4, 163–168.

JONES, G B, FLOODGATE, G D, and BENNELL, J D. 1986. Chemical and microbiological aspects of acoustically turbid sediments: preliminary investigations. *Marine Geotechnology*, Vol. 6, 315–332.

JONES, R C B, TONKS, L H, and WRIGHT, W B. 1938. Wigan district. *Memoir of the Geological Survey of Great Britain*, Sheet 84 (England and Wales).

KELLING, G, and WELSH, W. 1970. The Loch Ryan Fault. *Scottish Journal of Geology*, Vol. 6, 266–271.

KENT, P E. 1948. A deep borehole at Formby, Lancashire. *Geological Magazine*, Vol. 85, 253–264.

— 1954. Oil occurrences in Coal Measures of England. *Bulletin of the American Association of Petroleum Geologists*, Vol. 38, 1699–1713.

KENYON, N H, BELDERSON, R H, STRIDE, A H, and JOHNSON, M A. 1981. Offshore tidal sandbanks as indicators of net sand transport and as potential deposits. *Special Publication of the International Association of Sedimentologists*, Vol. 55, 257–268.

KERSHAW, P J, SWIFT, D J, and DENOON, D C. 1988. Evidence of recent sedimentation in the eastern Irish Sea. *Marine Geology*, Vol. 85, 1–14.

KIRBY, R, PARKER, W R, PENTREATH, R J, and LOVETT, M B. 1983. Sedimentation studies relevant to low-level radioactive effluent dispersal in the Irish Sea. Part III. An evaluation of possible mechanisms for the incorporation of radionuclides into marine sediments. *Report of the Institute of Oceanographic Sciences*, No. 178.

KIRTON, S R, and DONATO, J A. 1985. Some buried Tertiary dykes of Britain and surrounding waters deduced by magnetic modelling and seismic reflection methods. *Journal of the Geological Society of London*, Vol. 142, 1047–1057.

KLEMPERER, S L. 1989. Seismic reflection evidence for the location of the Iapetus suture west of Ireland. *Journal of the Geological Society of London*, Vol. 146, 409–412.

— and HOBBS, R. 1992. *The BIRPS Atlas, deep seismic reflection profiles around the British Isles.* (Cambridge: Cambridge University Press.)

KNELLER, B C. 1991. A foreland basin on the southern margin of Iapetus. *Journal of the Geological Society of London*, Vol. 148, 207–210.

KNIGHT, D J. 1977. Morecambe Bay feasibility study — sub-surface investigations. *Quarterly Journal of Engineering Geology*, Vol. 10, 303–320.

KNUDSEN, K L, and SEJRUP, H P. 1988. Amino-acid geochronology of selected interglacial sites in the North Sea area. *Boreas*, Vol. 17, 347–354.

KOKELAAR, B P, HOWELLS, M F, BEVINS, R E, ROACH, R A, and DUNKLEY, P N. 1984. The Ordovician marginal basin of Wales. 245–269 *in* Marginal basin geology. KOKELAAR, B P, and HOWELLS, M F (editors). *Special Publication of the Geological Society of London*, No. 16.

LAMPLUGH, G W. 1903. The geology of the Isle of Man. *Memoir of the Geological Survey of the United Kingdom.*

LANE, D E. 1987. Point of Ayr Colliery. *Mining Magazine*, Vol. 158, 226–237.

LAWRENCE, S R, COSTER, P W, and IRELAND, R J. 1987. Structural development and petroleum potential of the northern flanks of the Bowland Basin (Carboniferous), north-west England. 225–233 in *Petroleum geology of North West Europe.* BROOKS, J, and GLENNIE, K W (editors). (London: Graham and Trotman.)

LEE, A G. 1988. Carboniferous basin configuration of central and northern England modelled using gravity data. 69–84 in *Sedimentation in a synorogenic basin complex: the Upper Carboniferous of Northwest Europe.* BESLY, B M, and KELLING, G (editors). (London: Blackie.)

LEE, M K. 1986. A new gravity survey of the Lake District and three-dimensional model of the granite batholith. *Journal of the Geological Society of London*, Vol. 143, 425–435.

— 1989. Structure of the granitic batholith and Lower Palaeozoic succession in the Lake District from modelling and image processing of potential field data. *British Geological Survey Technical Report*, WK/89/1.

LEEDER, M R. 1973. Sedimentology and palaeogeography of the Upper Old Red Sandstone in the Scottish Border Basin. *Scottish Journal of Geology*, Vol. 9 117–144.

— 1982. Upper Palaeozoic basins of the British Isles: Caledonide inheritance versus Hercynian plate margin processes. *Journal of the Geological Society of London*, Vol. 139, 479–491.

— and Hardman, M. 1990. Carboniferous geology of the Southern North Sea Basin and controls on hydrocarbon prospectivity. 87–105 *in* Tectonic events responsible for Britain's oil and gas reserves. Hardman, R F P, and Brooks J (editors). *Special Publication of the Geological Society of London*, No. 55.

Lees, A. 1964. The structure and origin of the Waulsortian (Lower Carboniferous) 'reefs' of west-central Eire. *Philosophical Transactions of the Royal Society of London*, Vol. 247B, 483–531.

— and Miller, J. 1985. Facies variations in Waulsortian buildups, Part 2; Mid-Dinantian buildups from Europe and North America. *Geological Journal*, Vol. 20, 159–180.

Lees, G M, and Taitt, A H. 1946. The geological results of the search for oilfields in Great Britain. *Quarterly Journal of the Geological Society of London*, Vol. 101, 255–317.

Leggett, J K, McKerrow, W S, and Eales, M H. 1979. The Southern Uplands of Scotland: A Lower Palaeozoic accretionary prism. *Journal of the Geological Society of London*, Vol. 136, 755–770.

— — and Soper, N J. 1983. A model for the crustal evolution of southern Scotland. *Tectonics*, Vol. 2, 187–210.

Lewis, C L E, Green, P F, Carter, A, and Hurford, A J. 1992. Elevated K/T palaeotemperatures throughout Northwest England: three kilometres of Tertiary erosion? *Earth and Planetary Science Letters*, Vol. 112, 131–145.

Long, D, and Stoker, M S. 1986. Channels in the North Sea: the nature of a hazard. 339–351 in *Advances in underwater technology, ocean sciences and offshore engineering*, Vol. 6. (London: Graham and Trotman.)

MacDonald, R, and Walker, B H. 1985. Geochemistry and tectonic significance of the Lower Carboniferous Cockermouth lavas, Cumbria. *Proceedings of the Yorkshire Geological Society*, Vol. 45, 141–146.

Maizels, J. 1989. Sedimentology, palaeoflow dynamics and flood history of jökulhlaup deposits: palaeohydrology of Holocene sediment sequences in southern Iceland sandur deposits. *Journal of Sedimentary Petrology*, Vol. 59, 204–223.

Manning, P I, Robbie, J A, and Wilson, H E. 1970. Geology of Belfast and the Lagan Valley. *Memoir of the Geological Survey of Northern Ireland*, Sheet 36 (Northern Ireland).

— and Wilson, H E. 1975. The stratigraphy of the Larne Borehole, County Antrim. *Bulletin of the Geological Survey of Great Britain*, No. 50, 1–50.

Mansfield, J, and Kennett, P. 1963. A gravity survey of the Stranraer sedimentary basin. *Proceedings of the Yorkshire Geological Society*, Vol. 34, 139–151.

Marshall, J E A. 1991. Palynology of the Stonehaven Group, Scotland: evidence for a Mid Silurian age and its geological implications. *Geological Magazine*, Vol. 128, 283–286.

Matthews, D H. 1986. Seismic reflections from the lower crust around Britain. 11–21 *in* The nature of the lower continental crust. Dawson, J B, Carswell, D A, Hall, J, and Wedepohl, K H (editors). *Special Publication of the Geological Society of London*, No. 24.

— and Cheadle, M J. 1986. Deep reflections from the Caledonides and Variscides west of Britian and comparison with the Himalayas. 5–19 *in* Reflection seismology: a global perspective. Barazangi, M, and Brown, L (editors). *Geodynamics Series, American Geophysical Union*, Vol. 14.

Max, M D, Ryan, P D, and Inamdar, D D. 1983. A magnetic deep structural geology interpretation of Ireland. *Tectonics*, Vol. 2, 431–451.

— Barber, A J, and Martinez, J. 1990. Terrane assemblages of the Leinster Massif, SE Ireland, during the Lower Palaeozoic. *Journal of the Geological Society of London*, Vol. 147, 1035–1050.

McArdle, P, and Keary, R. 1986. Offshore coal in the Kish Bank Basin: its potential for commercial exploitation. *Geological Survey of Ireland Report Series*, No. 86/3.

McKerrow, W S. 1988. The development of the Iapetus Ocean from the Arenig to the Wenlock. 405–412 *in* The Caledonian–Appalachian orogen. Harris, A L, and Fettes, D J (editors). *Special Publication of the Geological Society of London*, No. 38.

— and Soper, N J. 1989. The Iapetus suture in the British Isles. *Geological Magazine*, Vol. 126, 1–8.

McLean, A C. 1978. Evolution of fault-controlled ensialic basins in northwestern Britain. 325–346 *in* Crustal evolution in northwestern Britain and adjacent regions. Bowes, D R, and Leake, B E (editors). *Special Issue of the Geological Journal*, No. 10.

— and Deegan, C E (editors). 1978. The solid geology of the Clyde sheet (55°N/6°W). *Report of the Institute of Geological Sciences*, No. 78/9.

McQuillin, R, Wright, J E, Owens, B, and Lister, T R. 1969. Recent geological investigations in the Irish Sea. *Nature, London*, Vol. 222, 365–366.

Meighan, I G, McCormick, A G, Gibson, D, Gamble, J A, and Graham, I J. 1988. Rb-Sr isotopic determinations and the timing of Tertiary central complex magmatism in NE Ireland. 349–360 *in* Early Tertiary volcanism and the opening of the NE Atlantic. Morton, A C, and Parson, L M (editors). *Special Publication of the Geological Society of London*, No. 39.

Meissner, R, Matthews, D, and Wever, Th. 1986. The 'Moho' in and around Great Britain. *Annales Geophysicae*, Vol. 4B, 659–663.

Miller, J, and Grayson, R F. 1982. The regional context of Waulsortian facies in northern England. 17–33 in *Symposium on the environmental setting and distribution of the Waulsortian facies*. Bolton, K, Lane, H R, and LeMone, D V (editors). (El Paso: El Paso Geological Society and University of Texas.)

Miller, J M, and Taylor, K. 1966. Uranium mineralization near Dalbeattie, Kirkcudbrightshire. *Bulletin of the Geological Survey of Great Britain*, No. 25, 1–18.

Millward, D, Moseley, F, and Soper, N J. 1978. The Eycott and Borrowdale volcanic rocks. 99–120 *in* The geology of the Lake District. Moseley, F (editor). *Yorkshire Geological Society Occasional Publication*, No. 3.

— and Molyneux, S G. 1992. Field and biostratigraphic evidence for an unconformity at the base of the Eycott Volcanic Group in the English Lake District. *Geological Magazine*, Vol. 129, 77–92.

Mitchell, M, Taylor, B J, and Ramsbottom, W H C. 1978. Carboniferous. 168–188 *in* The geology of the Lake District. Moseley, F (editor). *Yorkshire Geological Society Occasional Publication*, No. 3.

Mogi, A. 1979. *An atlas of the sea floor around Japan: aspects of submarine geomorphology.* (Tokyo: University of Tokyo Press.)

Molyneux, S G. 1979. New evidence for the age of the Manx Group, Isle of Man. 415–421 *in* The Caledonides of the British Isles — reviewed. Harris, A L, Holland, C H, and Leake, B E (editors). *Special Publication of the Geological Society of London*, No. 8.

— 1988. Micropalaeontological evidence for the age of the Borrowdale Volcanic Group. *Geological Magazine*, Vol. 125, 541–542.

Morton, A C, and Parson, L M (editors). 1988. Early Tertiary volcanism and the opening of the NE Atlantic. *Special Publication of the Geological Society of London*, No. 39.

Moseley, F. 1972. A tectonic history of northwest England. *Journal of the Geological Society of London*, Vol. 128, 561–598.

— (editor). 1978. The geology of the Lake District. *Yorkshire Geological Society Occasional Publication*, No. 3.

Murphy, F C. 1987. Late Caledonian granitoids and timing of deformation in the Iapetus suture zone of eastern Ireland. *Geological Magazine*, Vol. 124, 135–142.

MUSSETT, A E, DAGLEY, P, and SKELHORN, R R. 1988. Time and duration of activity in the British Tertiary Igneous Province. 337–348 *in* Early Tertiary volcanism and the opening of the NE Atlantic. MORTON, A C, and PARSON, L M (editors). *Special Publication of the Geological Society of London*, No. 39.

MYKURA, W. 1960. The replacement of coal by limestone and the reddening of Coal Measures in the Ayrshire Coalfield. *Bulletin of the Geological Survey of Great Britain*, No. 16, 69–109.

NAYLOR, D. 1964. Pleistocene and post-Pleistocene sediments in Dublin Bay. *Scientific Proceedings of the Royal Dublin Society*, Vol. A2, 175–188.

— HAUGHEY, N, CLAYTON, G, and GRAHAM, J R. 1993. The Kish Bank Basin, offshore Ireland. 845–855 in *Petroleum Geology of Northwest Europe: Proceedings of the 4th Conference.* PARKER, J R (editor). (London: The Geological Society of London.)

— and SHANNON, P. 1982. *The geology of offshore Ireland and west Britain.* (London: Graham and Trotman.)

ORD, D M, CLEMMEY, H, and LEEDER, M R. 1988. Interaction between faulting and sedimentation during Dinantian extension of the Solway basin, SW Scotland. *Journal of the Geological Society of London*, Vol. 145, 249–259.

OWEN, T R (editor). 1974. *The Upper Palaeozoic and post-Palaeozoic rocks of Wales.* (Cardiff: University of Wales Press.)

PANTIN, H M. 1977. Quaternary sediments from the northern Irish Sea. 27–54 *in* The Quaternary history of the Irish Sea. KIDSON, C, and TOOLEY, M J (editors). *Special Issue of the Geological Journal*, No. 7.

— 1978. Quaternary sediments from the north-east Irish Sea, Isle of Man to Cumbria. *Bulletin of the Geological Survey of Great Britain*, No. 64.

— and EVANS, C D R. 1984. The Quaternary history of the central and south-western Celtic Sea. *Marine Geology*, Vol. 57, 259–293.

PARKIN, M, and CROSBY, A. 1982. Geological results of boreholes drilled on the southern United Kingdom shelf by the Institute of Geological Sciences 1969–1981. *Institute of Geological Sciences Marine Report*, No. 82/2.

PARNELL, J, and EAKIN, P A. 1987. The replacement of sandstones by uraniferous hydrocarbons: significance for petroleum migration. *Mineralogical Magazine*, Vol. 51, 505–515.

PATERSON, I B, and HALL, I H S. 1986. Lithostratigraphy of the late Devonian and early Carboniferous rocks in the Midland Valley of Scotland. *Report of the British Geological Survey*, Vol. 18, No. 3.

PATTISON, J. 1970. A review of the marine fossils from the Upper Permian rocks of Northern Ireland and North-West England. *Bulletin of the Geological Survey of Great Britain*, No. 32, 123–165.

— SMITH, D B, and WARRINGTON, G. 1973. A review of late Permian and early Triassic biostratigraphy in the British Isles. 220–260 *in* The Permian and Triassic systems and their mutual boundary. LOGAN, A, and HILLS, L V (editors). *Memoir of the Canadian Society of Petroleum Geologists*, No. 2.

PENN, I E, and EVANS, C D R. 1976. The Middle Jurassic (mainly Bathonian) of Cardigan Bay and its palaeogeographical significance. *Report of the Institute of Geological Sciences*, No. 76/6.

— and seven others. 1983. The Larne No. 2 Borehole: discovery of a new Permian volcanic centre. *Scottish Journal of Geology*, Vol. 19, 333–346.

PENTREATH, R J, WOODHEAD, D S, KERSHAW, P J, JEFFRIES, D F, and LOVETT, M B. 1986. The behaviour of plutonium and americium in the Irish Sea. 60–69 in *Rapport et procès-verbeaux des réunions, Commission permanent internationale pour l'exploration Scientifique de la mer, 1986.*

PHILLIPS, W E A. 1981. The pre-Caledonian basement. 7–16 in *A geology of Ireland.* HOLLAND, C H (editor). (Edinburgh: Scottish Academic Press.)

— STILLMAN, C J, and MURPHY, T. 1976. A Caledonian plate tectonic model. *Journal of the Geological Society of London*, Vol. 132, 579–609.

PICKEN, G S. 1988. The concealed coalfield at Canonbie: an interpretation based on boreholes and seismic surveys. *Scottish Journal of Geology*, Vol. 24, 61–71.

PINFOLD, E S. 1958. The search for oil in Lancashire. *Liverpool and Manchester Geological Journal*, Vol. 2, 106–123.

PINGREE, R D, and GRIFFITHS, D K. 1979. Sand transport paths around the British Isles resulting from M_2 and M_4 tidal interactions. *Journal of the Marine Biological Association of the United Kingdom*, Vol. 58, 211–226.

PIPER, J D A. 1978. Palaeomagnetism and paleogeography of the Southern Uplands block in Ordovician times. *Scottish Journal of Geology*, Vol. 14, 93–107.

POLLARD, J E. 1981. A comparison between the Triassic trace fossils of Cheshire and south Germany. *Palaeontology*, Vol. 24, 555–588.

— 1985. *Isopodichnus*, related arthropod trace fossils and notostracans from Triassic fluvial sediments. *Transactions of the Royal Society of Edinburgh*, Vol. 76, 273–285.

POOLE, E G, and WHITEMAN, A J. 1955. Variations in the thickness of the Collyhurst Sandstone in the Manchester area. *Bulletin of the Geological Survey of Great Britain*, No. 9, 33–41.

— — 1966. Geology of the country around Nantwich and Whitchurch. *Memoir of the Geological Survey of Great Britain*, Sheet 122 (England and Wales).

POWELL, D W. 1956. Gravity and magnetic anomalies in North Wales. *Quarterly Journal of the Geological Society of London*, Vol. 111, 375–397.

PRESTON, J. 1967. A Tertiary feeder dyke in County Fermanagh, Northern Ireland. *Scientific Proceedings of the Royal Society of Dublin, Series A*, Vol. 3, 1–16.

— 1981. Tertiary igneous activity. 213–223 in *A geology of Ireland.* HOLLAND, C H (editor). (Edinburgh: Scottish Academic Press.)

PRICE, D, WRIGHT, W B, JONES, R C B, TONKS, L H, and WHITEHEAD, T H. 1963. Geology of the country around Preston. *Memoir of the Geological Survey of Great Britain*, Sheet 75 (England and Wales).

RAMSBOTTOM, W H C. 1969a. The Namurian of Britain. *Compte Rendu Sixième Congrès International de stratigraphie et de Géologie du Carbonifère, Sheffield*, 1967, Vol. 1, 219–232.

— 1969b. Reef distribution in the British Lower Carboniferous. *Nature, London*, Vol. 222, 765–766.

— 1973. Transgressions and regressions in the Dinantian: A new synthesis of British Dinantian stratigraphy. *Proceedings of the Yorkshire Geological Society*, Vol. 39, 567–607.

— 1977. Major cycles of transgression and regression (mesothems) in the Namurian. *Proceedings of the Yorkshire Geological Society*, Vol. 41, 261–293.

— and six others. 1978. A correlation of Silesian rocks in the British Isles. *Special Report of the Geological Society of London*, No. 10.

READING, H G (editor). 1986. *Sedimentary environments and facies* (2nd edition). (Oxford: Blackwell Scientific Publications.)

REEDMAN, A J, LEVERIDGE, B E, and EVANS, R B. 1984. The Arfon Group ('Arvonian') of North Wales. *Proceedings of the Geologists' Association*, Vol. 95, 313–321.

RILEY, N J. 1990. Stratigraphy of the Worston Shale Group (Dinantian), Craven Basin north-west England. *Proceedings of the Yorkshire Geological Society*, Vol. 48, 163–187.

ROSE, W C C, and DUNHAM, K C. 1977. Geology and haematite deposits of South Cumbria. *Geological Survey of Great Britain Economic Memoir*, Sheets 58 and 48 part (England and Wales).

RUNDLE, C C. 1981. The significance of isotopic dates from the English Lake District for the Ordovician–Silurian time-scale. *Journal of the Geological Society of London*, Vol. 138, 569–572.

SAGER, G, and SAMMLER, R. 1968. *Atlas der gezeitenströme für die Nordsee, den Kanal, und die Irische See.* (Rostok:

Seehydrographischer Dienst der Deutschen Demokratischen Republik.) [In German.]

Scott, J, and Colter, V S. 1987. Geological aspects of current onshore Great Britain exploration plays. 95–107 in *Petroleum geology of North West Europe.* Brooks, J, and Glennie, K W (editors). (London: Graham and Trotman.)

Selden, P A, and White, D E. 1983. A new Silurian arthropod from Lesmahagow, Scotland. *Special Papers in Palaeontology,* No. 30, 43–49.

Selley, R C. 1992. Petroleum seepages and impregnations in Great Britain. *Marine and Petroleum Geology,* Vol. 9, 226–244.

Sevastopulo, G D. 1981a. Lower Carboniferous. 147–171 in *A geology of Ireland.* Holland, C H (editor). (Edinburgh: Scottish Academic Press.)

— 1981b. Upper Carboniferous. 173–187 in *A geology of Ireland.* Holland, C H (editor). (Edinburgh: Scottish Academic Press.)

— 1981c. Hercynian structures. 189–199 in *A geology of Ireland.* Holland, C H (editor). (Edinburgh: Scottish Academic Press.)

— 1981d. Economic geology. 273–301 in *A geology of Ireland.* Holland, C H (editor). (Edinburgh: Scottish Academic Press.)

Shaw, R W L. 1971. The faunal stratigraphy of the Kirkby Moor Flags of the type area near Kendal, Westmoreland. *Geological Journal,* Vol. 7, 359–380.

Sherlock, R L. 1921. Rock-salt and brine. *Memoir of the Geological Survey of Great Britain. Special Reports on the Mineral Resources of Great Britain,* Vol. 18.

— and Hollingworth, S E. 1938. Gypsum and anhydrite; celestine and strontianite (3rd edition). *Memoir of the Geological Survey of Great Britain. Special Reports on the Mineral Resources of Great Britain,* Vol. 3.

Simon, J B. 1984a. Provenance and depositional history of the Lower Old Red Sandstone of north-east Antrim. *Irish Journal of Earth Science,* Vol. 6, 1–13.

— 1984b. Sedimentation of a small complex alluvial fan of possible Upper Old Red Sandstone age, northeast County Antrim. *Irish Journal of Earth Science,* Vol. 6, 109–119.

— and Bluck, B J. 1982. Palaeodrainage of the southern margin of the Caledonian mountain chain in the northern British Isles. *Transactions of the Royal Society of Edinburgh,* Vol. 73, 11–15.

Simpson, A. 1963. The stratigraphy and tectonics of the Manx Slate Series, Isle of Man. *Quarterly Journal of the Geological Society of London,* Vol. 119, 367–400.

Smith, A G, Hurley, A M, and Briden, J C. 1981. *Phanerozoic palaeocontinental world maps.* (Cambridge: Cambridge University Press.)

Smith, B. 1924. On the West Cumberland Brockram and its associated rocks. *Geological Magazine,* Vol. 61, 289–308.

— 1927. On the Carboniferous Limestone Series of the northern part of the Isle of Man. *Summary of Progress of the Geological Survey of Great Britain* (for 1926), 108–119.

— 1931. Borings through the glacial drifts of the northern plain of the Isle of Man. *Summary of Progress of the Geological Survey of Great Britain,* (for 1930), No. 3, 14–23.

— and George, T N. 1961. *British regional geology: North Wales* (3rd edition). (London: HMSO.)

Smith, D B. 1972. The British Isles. 1–33 in *Rotliegend; essays on European Lower Permian.* Falke, H (editor). (Leiden: Brill.)

— Brunstrom, R G W, Manning, P I, Simpson, S, and Shotton, F W. 1974. A correlation of Permian rocks in the British Isles. *Special Report of the Geological Society of London,* No. 5.

— and Taylor, J C M. 1992. Permian. 87–96 *in* Atlas of palaeogeography and lithofacies. Cope, J C W, Ingham, J K, and Rawson, P F (editors). *Memoir of the Geological Society of London,* No. 13.

Smith, N J P (compiler). 1985. Pre-Permian geology of the United Kingdom (south). *British Geological Survey 150th Anniversary Publication.* (Keyworth: British Geological Survey.)

Smith, R A. 1986. Permo-Triassic and Dinantian rocks of the Belfast Harbour Borehole. *Report of the British Geological Survey,* Vol. 18, No. 6.

— Johnston, T P, and Legg, I C. 1991. Geology of the country around Newtownards. *Memoir of the Geological Survey of Northern Ireland,* Sheets 37 and part of 38 (Northern Ireland).

Snelling, N J (editor). 1985. The chronology of the geological record. *Memoir of the Geological Society of London,* No. 10.

— 1987. Measurements of geological time and the geological time scale. *Modern Geology,* Vol. 11, 365–374.

Somerville, I D, Mitchell, M, and Strank, A R E. 1986. An Arundian fauna from the Dyserth area, North Wales and its correlation within the British Isles. *Proceedings of the Yorkshire Geological Society,* Vol. 46, 57–75.

— Strank, A R E, and Welsh, A. 1989. Chadian floras and faunas from Dyserth: depositional environments and palaeogeographic setting of Viséan strata in northeast Wales. *Geological Journal,* Vol. 24, 49–66.

Soper, N J, and Hutton, D H W. 1984. Late Caledonian sinistral displacements in Britain: implications for a three-plate collision model. *Tectonics,* Vol. 3, 781–794.

— Webb, B C, and Woodcock, N H. 1987. Late Caledonian (Acadian) transpression in north-west England: timing, geometry and geotectonic significance. *Proceedings of the Yorkshire Geological Society,* Vol. 46, 175–192.

Stacey, A P. 1965. Marine magnetic survey off the coast of County Dublin. MSc thesis, University of Durham.

Steel, R J, and Thompson, D B. 1983. Structures and textures in Triassic braided stream conglomerates ("Bunter" Pebble Beds) in the Sherwood Sandstone Group, North Staffordshire, England. *Sedimentology,* Vol. 30, 341–367.

Stone, P. 1988. The Permian successions at Ballantrae and Loch Ryan, south-west Scotland. *Report of the British Geological Survey,* Vol. 19, No. 2, 13–18.

— Floyd, J D, Barnes, R P, and Lintern, B C. 1987. A sequential back-arc and foreland basin thrust duplex model for the Southern Uplands of Scotland. *Journal of the Geological Society of London,* Vol. 144, 753–764.

Strahan, A. 1901. On the passage of a seam of coal into a seam of dolomite. *Quarterly Journal of the Geological Society of London,* Vol. 57, 297–306.

— 1920. Mineral oil, Kimmeridge oil-shale, lignites, jets, cannel coals, natural gas. England and Wales. *Memoir of the Geological Survey of Great Britain, Special Report on the Mineral Resources of Great Britain,* Vol. 7.

Stride, A H (editor). 1982. *Offshore tidal sands — processes and deposits.* (London: Chapman and Hall.)

Stuart, I A, and Cowan, G. 1991. The South Morecambe Field, Blocks 110/2a, 110/3a, 110/8a, UK East Irish Sea. 527–541 *in* United Kingdom oil and gas fields, 25 years Commemorative Volume. Abbotts, I L (editor). *Memoir of the Geological Society of London,* No. 14.

Tappin, D R, Chadwick, R A, Jackson, A A, Wingfield, R T R, and Smith, N J P. 1994. *United Kingdom offshore regional report: the geology of Cardigan Bay and the Bristol Channel.* (London: HMSO for the British Geological Survey.)

Tate, M P, and Dobson, M R. 1989. Late Permian to early Mesozoic rifting and sedimentation, offshore NW Ireland. *Marine and Petroleum Geology,* Vol. 6, 49–59.

Taylor, B J. 1961. The stratigraphy of exploratory boreholes in the West Cumberland coalfield. *Bulletin of the Geological Survey of Great Britain,* No. 17, 1–74.

— Price, R H, and Trotter, F M. 1963. Geology of the country around Stockport and Knutsford. *Memoir of the Geological Survey of Great Britain,* Sheet 98 (England and Wales).

— and five others. 1971. *British regional geology: northern England.* (London: HMSO for Institute of Geological Sciences.)

Taylor, J C M. 1990. Late Permian–Zechstein. 153–190 in *Introduction to the petroleum geology of the North Sea* (3rd edition). Glennie, K W (editor). (Oxford: Blackwell Scientific Publications.)

Taylor, S R. 1983. A stable isotope study of the Mercia Mudstones (Keuper Marl) and associated sulphate horizons in the English Midlands. *Sedimentology,* Vol. 30, 11–31.

Thirlwall, M F. 1988. Geochronology of Late Caledonian magmatism in northern Britain. *Journal of the Geological Society of London,* Vol. 145, 951–967.

— and Fitton, J G. 1983. Sm-Nd garnet age for the Ordovician Borrowdale Volcanic Group, English Lake District. *Journal of the Geological Society of London,* Vol. 140, 511–518.

Thomas, G S P. 1985. The Quaternary of the northern Irish Sea basin. 143–158 in *The geomorphology of north-west England.* Johnson, R H (editor). (Manchester: Manchester University Press.)

Thompson, D B. 1969. Dome-shaped aeolian dunes in the Frodsham Member of the so-called "Keuper" Sandstone Formation (Scythian-?Anisian: Triassic) at Frodsham, Cheshire (England). *Sedimentary Geology,* Vol. 3, 263–289.

— 1970a. Sedimentation of the Triassic (Scythian) red pebbly sandstones in the Cheshire Basin and its margins. *Geological Journal,* Vol. 7, 183–216.

— 1970b. The stratigraphy of the so-called Keuper Sandstone Formation (Scythian–?Anisian) in the Permo-Triassic Cheshire Basin. *Quarterly Journal of the Geological Society of London,* Vol. 126, 151–181.

Thompson, P, Mussett, A E, and Dagley, P. 1987. Revised ^{40}Ar-^{39}Ar age for granites of the Mourne Mountains, Ireland. *Scottish Journal of Geology,* Vol. 23, 215–220.

Todd, S P, Murphy, F C, and Kennan, P S. 1991. On the trace of the Iapetus suture in Ireland and Britain. *Journal of the Geological Society of London,* Vol. 148, 869–880.

Tooley, M J. 1985. Sea-level changes and coastal morphology in north-west England. 94–121 in *The geomorphology of North-West England.* Johnson, R H (editor). (Manchester: Manchester University Press.)

Trotter, F M. 1939. Reddened Carboniferous beds in the Carlisle Basin and Edenside. *Geological Magazine,* Vol. 76, 408–416.

— 1953. Reddened beds of Carboniferous age in North-West England and their origin. *Proceedings of the Yorkshire Geological Society,* Vol. 29, 1–20.

— 1954. Reddened beds in the Coal Measures of South Lancashire. *Bulletin of the Geological Survey of Great Britain,* No. 5, 61–80.

— Hollingworth, S E, Eastwood, T, and Rose, W C C. 1937. The geology of the Gosforth District. *Memoir of the Geological Survey of Great Britain,* Sheet 37 (England and Wales).

Tubb, S R, Soulsby, A, and Lawrence, S R. 1986. Palaeozoic prospects on the northern flanks of the London–Brabant Massif. 55–72 *in* Habitat of Palaeozoic gas in NW Europe. Brooks, J, Goff, J C, and Van Hoorn, B (editors). *Special Publication of the Geological Society of London,* No. 23.

Tucker, R D, and Pharaoh, T C. 1991. U-Pb zircon ages for Late Precambrian igneous rocks in southern Britain. *Journal of the Geological Society of London,* Vol. 148, 435–443.

Turner, P. 1980. Continental red beds. *Developments in Sedimentology,* Vol. 29. (Amsterdam: Elsevier.)

Upton, B G J, Fitton, J G, and Macintyre, R M. 1987. The Glas Eilean lavas: evidence of a Lower Permian volcano-tectonic basin between Islay and Jura, Inner Hebrides. *Transactions of the Royal Society of Edinburgh: Earth Sciences,* Vol. 77, 289–293.

Visscher, H. 1971. The Permian and Triassic of the Kingscourt Outlier, Ireland. A palynological investigation related to regional stratigraphic problems in the Permian and Triassic of Western Europe. *Geological Survey of Ireland, Special Paper,* No. 1.

Wade, A. 1910. A deep boring at Heswall, Cheshire, and its bearing upon the underground geology of the Liverpool–Wirral areas. *Transactions of the Institution of Mining Engineers,* Vol. 39, 163–185.

Wadge, A J. 1978. Devonian. 164–167 *in* The geology of the Lake District. Moseley, F (editor). *Yorkshire Geological Society Occasional Publication,* No. 3.

— Gale, N H, Beckinsale, R D, and Rundle, C C. 1978. A Rb-Sr isochron for the Shap Granite. *Proceedings of the Yorkshire Geological Society,* Vol. 42, 297–305.

Wagner, R H. 1983. A lower Rotliegend flora from Ayrshire. *Scottish Journal of Geology,* Vol. 19, 135–155.

Walker, I M, and Cooper, W G. 1987. The structural and stratigraphic evolution of the northeast margin of the Sole Pit Basin. 263–275 in *Petroleum Geology of North West Europe.* Brooks, J, and Glennie, K W (editors). (London: Graham and Trotman.)

Walker, T R. 1967. Formation of red beds in modern and ancient deserts. *Bulletin of the Geological Society of America,* Vol. 78, 353–368.

Walsh, P T. 1966. Cretaceous outliers in south-west Ireland and their implications for Cretaceous palaeogeography. *Quarterly Journal of the Geological Society of London,* Vol. 122, 63–84.

— and Brown, E H. 1971. Solution subsidence outliers containing probable Tertiary sediment in north-east Wales. *Geological Journal,* Vol. 7, 299–320.

Warman, H R, Roberts, K H, Brunstrom, R G W, and Adcock, C M. 1956. Report on oil and gas in the United Kingdom. 317–357 *in* Symposium sobre yacimientos de petroleo y gas, Vol. 5. Guzman, E J (editor). XX *Congreso Geológico Internacional, México.*

Warren, P T, Price, D, Nutt, M J C, and Smith, E G. 1984. Geology of the country around Rhyl and Denbigh. *Memoir of the British Geological Survey,* Sheets 95 and 107 (England and Wales).

Warren, W P. 1985. Stratigraphy. 39–65 in *The Quaternary history of Ireland.* Edwards, K J, and Warren, W P (editors). (London: Academic Press.)

— and Keary, R. 1988. The sand and gravel resources of the Irish Sea Basin. 66–79 *in* The Irish Sea: a resource at risk. Sweeney, J C (editor). *Geographic Society of Ireland Special Publication,* No. 3.

Warrington, G. 1970a. The stratigraphy and palaeontology of the "Keuper" Series of the central Midlands of England. *Quarterly Journal of the Geological Society of London,* Vol. 126, 183–223.

— 1970b. The "Keuper" Series of the British Trias in the Northern Irish Sea and neighbouring areas. *Nature, London,* Vol. 226, 254–256.

— 1970c. Palynology of the Trias of the Langford Lodge borehole. 44–52 *in* Geology of Belfast and the Lagan Valley. Manning, P I, Robbie, J A, and Wilson, H E. *Memoir of the Geological Survey of Northern Ireland,* Sheet 36 (Northern Ireland).

— 1974. Les évaporites du Trias britannique. *Bulletin de la Société géologique de France (Septième Series),* Vol. 16, 708–723.

— and eight others. 1980. A correlation of Triassic rocks in the British Isles. *Special Report of the Geological Society of London,* No. 13.

— and Harland, R. 1975. Palynology of the Trias and Lower Lias of the Larne Borehole. 37–50 *in* The stratigraphy of the Larne Borehole, County Antrim. Manning, P I, and Wilson, H E. *Bulletin of the Geological Survey of Great Britain,* No. 50.

— and Ivimey-Cook, H C. 1992. Triassic. 97–106 *in* Atlas of palaeogeography and lithofacies. Cope, J C W, Ingham, J K, and Rawson, P F (editors). *Memoir of the Geological Society of London,* No. 13.

WEBB, B C. 1983. Caledonian structure in the Cambrian Slate Belt, Gwynedd, North Wales. *Report of the Institute of Geological Sciences,* No. 83/1.

— and COOPER, A H. 1988. Slump folds and gravity slide structures in a Lower Palaeozoic marginal basin sequence (the Skiddaw Group), NW England. *Journal of Structural Geology,* Vol. 10, 463–472.

WEDD, C B, and KING, W B R. 1924. The geology of the country around Flint, Hawarden and Caergwrle. *Memoir of the Geological Survey of Great Britain,* Sheet 108 (England and Wales).

WENZEL, F, SANDMEIER, K-J, and WÄLDE, W. 1987. Properties of the lower crust from modelling refraction and reflection data. *Journal of Geophysical Research,* Vol. 92, 11B 583–975.

WEST, R G. 1977. *Pleistocene geology and biology.* (London: Longman.)

WHITTAKER, A, HOLLIDAY, D W, and PENN, I E. 1985. Geophysical logs in British stratigraphy. *Special Report of the Geological Society of London,* No. 18.

WHITTINGTON, R J. 1977. A late-glacial drainage pattern in the Kish Bank area and post-glacial sediments in the Central Irish Sea. 55–58 *in* The Quaternary History of the Irish Sea. KIDSON, C, and TOOLEY, M J (editors). *Special Issue of the Geological Journal,* No. 7.

— 1980. Geophysical studies in the South Irish Sea. Unpublished PhD thesis, University of Wales.

WILKINSON, G C, BAZLEY, R A B, and BOULTER, M C. 1980. The geology and palynology of the Oligocene Lough Neagh Clays, Northern Ireland. *Journal of the Geological Society of London,* Vol. 137, 65–75.

WILKINSON, I P, and HALLIWELL, G P (compilers). 1979. Offshore micro-palaeontological biostratigraphy of southern and western Britain. *Report of the Institute of Geological Sciences,* No. 79/9.

WILLS, L J. 1956. *Concealed coalfields.* (London and Glasgow: Blackie.)

WILSON, A A. 1990. The Mercia Mudstone Group (Trias) of the East Irish Sea Basin. *Proceedings of the Yorkshire Geological Society,* Vol. 48, 1–22.

— and EVANS, W B. 1990. Geology of the country around Blackpool. *Memoir of the British Geological Survey,* Sheet 66 (England and Wales).

WILSON, C D V. 1959. Geophysical investigations in the Vale of Clwyd. *Liverpool and Manchester Geological Journal,* Vol. 2, 253–270.

WILSON, H E. 1972. *British regional geology: Northern Ireland.* (Belfast: HMSO for the Geological Survey of Northern Ireland.)

— 1981. Permian and Mesozoic. 201–212 in *A geology of Ireland.* HOLLAND, C H (editor). (Edinburgh: Scottish Academic Press.)

— and MANNING, P I. 1978. Geology of the Causeway Coast. *Memoir of the Geological Survey of Northern Ireland,* Sheet 7 (Northern Ireland).

WINGFIELD, R T R. 1987. Giant sand waves and relict periglacial features on the sea bed west of Anglesey. *Proceedings of the Geologists' Association,* Vol. 98, 400–404.

— 1989. Glacial incisions indicating Middle and Upper Pleistocene ice limits off Britain. *Terra Nova,* Vol. 1, 538–548.

— 1990. The origin of major incisions within the Pleistocene deposits of the North Sea. *Marine Geology,* Vol. 91, 31–52.

— In press. Modelling Holocene sea levels in the Irish and Celtic seas. *Proceedings of the International Coastal Congress, Kiel 1992.* (Kiel: Christian-Albrechts-Universitat Press.)

WOOD, A (editor). 1969. *The Pre-Cambrian and Lower Palaeozoic rocks of Wales.* (Cardiff: University of Wales Press.)

— and WOODLAND, A W. 1968. Borehole at Mochras, west of Llanbedr, Merionethshire. *Nature, London,* Vol. 219, 1352–1354.

WOODLAND, A W (editor). 1971. The Llanbedr (Mochras Farm) Borehole. *Report of the Institute of Geological Sciences,* No. 71/18.

WRAY, D A, and COPE, F W. 1948. Geology of Southport and Formby. *Memoir of the Geological Survey of Great Britain,* Sheets 74 and 83 (England and Wales).

WRIGHT, J E. 1975. The geology of the Irish Sea. 295–312 *in* Canada's continental margins and offshore petroleum exploration. YORATH, C J, PARKER, E R, and GLASS, D J (editors). *Memoir of the Canadian Society of Petroleum Geologists,* Vol. 4.

— HULL, J H, MCQUILLIN, R, and ARNOLD, S E. 1971. Irish Sea investigations 1969–70. *Report of the Institute of Geological Sciences,* No. 71/19.

YOUNG, D G G. 1965. Gravity and magnetic investigations of the deep structure of the north Irish Sea. Unpublished PhD thesis, University of Durham.

ZIEGLER, P A. 1990. *Geological atlas of Western and Central Europe 1990.* (Amsterdam: Elsevier for Shell Internationale Petroleum Maatschappij B V.)

Index

BRITISH GEOLOGICAL SURVEY

Keyworth, Nottingham NG12 5GG
0115 936 3100

Murchison House, West Mains Road, Edinburgh
EH9 3LA 0131-667 1000

London Information Office, Natural History Museum
Earth Galleries, Exhibition Road, London SW7 2DE
0171-589 4090

The full range of Survey publications is available through the Sales Desks at Keyworth and at Murchison House, Edinburgh, and in the BGS London Information Office in the Natural History Museum (Earth Galleries). The adjacent bookshop stocks the more popular books for sale over the counter. Most BGS books and reports can be bought from HMSO and through HMSO agents and retailers. Maps are listed in the BGS Map Catalogue, and can be bought together with books and reports through BGS-approved stockists and agents as well as direct from BGS.

The British Geological Survey carries out the geological survey of Great Britain and Northern Ireland (the latter as an agency service for the government of Northern Ireland), and of the surrounding continental shelf, as well as its basic research projects. It also undertakes programmes of British technical aid in geology in developing countries as arranged by the Overseas Development Administration.

The British Geological Survey is a component body of the Natural Environment Research Council.

HMSO

HMSO publications are available from:

HMSO Publications Centre
(Mail, fax and telephone orders only)
PO Box 276, London SW8 5DT
Telephone orders 0171-873 9090
General enquiries 0171-873 0011
Queuing system in operation for both numbers
Fax orders 0171-873 8200

HMSO Bookshops
49 High Holborn, London WC1V 6HB
(counter service only)
0171-873 0011 Fax 0171-831 1326
68–69 Bull Street, Birmingham B4 6AD
0121-236 9696 Fax 0121-236 9699
33 Wine Street, Bristol BS1 2BQ
0117-9264306 Fax 0117-9294515
9 Princess Street, Manchester M60 8AS
0161-834 7201 Fax 0161-833 0634
16 Arthur Street, Belfast BT1 4GD
01232-238451 Fax 01232-235401
71 Lothian Road, Edinburgh EH3 9AZ
0131-228 4181 Fax 0131-229 2734
HMSO Oriel Bookshop,
The Friary, Cardiff CF1 4AA
01222-395548 Fax 01222-384347

HMSO's Accredited Agents
(see Yellow Pages)

And through good booksellers

1: 250 000 MAP SERIES

All sheets have been given names (see index map opposite), but can also be referred to by the latitude and longitude of their south-western corner. Thus the Isle of Man sheet may also be referred to as 54°N 06°W. The list below shows the types of maps available for each sheet in the report area using the following key:

BA	Bouguer Gravity Anomaly
A	Aeromagnetic Anomaly
S	Solid Geology
Q	Quaternary Geology
SBS	Sea Bed Sediments
SBS & Q	Sea Bed Sediments and Quaternary

LIVERPOOL BAY (53°N 04°W)
BA, A, S, SBS & Q

ANGLESEY (53°N 06°W)
BA, A, S, Q, SBS

LAKE DISTRICT (54°N 04°W)
BA, A, S, SBS & Q

ISLE OF MAN (54°N 06°W)
BA, A, S, SBS & Q

EAST IRISH SEA
(SPECIAL SHEET EDITION)
S

NORTHERN IRELAND
(SPECIAL SHEET EDITION)
S, Q

LAND GEOLOGY MAPS

The index map shows sheet numbers of published maps at 1:63 360 and 1:50 000 scales. Maps in Scotland and Northern Ireland belong to separate series, and are numbered accordingly.

The diagrams in this report were produced by the Cartographic Production Group, BGS, Keyworth under the supervision of R J Parnaby. Cartography was by R J Demaine, D Daley, C Mawer and I J Wilkinson and reprography by C F Pope and S C Wilkinson

Book production was supervised by S M Murphy. The typesetting and page make-up was carried out by Ms J B A Evans.

OTHER BGS MAPS

Other smaller-scale maps produced by BGS that cover all or part of the region include:

Sea bed sediments around the United Kingdom (South Sheet). Scale 1:1 000 000

Quaternary sediments around the United Kingdom (South Sheet). Scale 1:1 000 000

Aeromagnetic map of Great Britain, south sheet. Scale 1:625 000

Bouguer anomaly map of the British Isles, south sheet. Scale 1:625 000

Geological map of the United Kingdom, South. Scale 1:625 000

Geology of the United Kingdom, Ireland and the adjacent Continental Shelf (South Sheet). Scale 1:1 000 000

Pre-Permian geology of the United Kingdom (South). Scale 1:1 000 000